Handbook of Global Philosophies on AI Ethics

This book offers an exploration of the diverse perspectives shaping the future of artificial intelligence (AI), highlighting the influence of non-Western thinking in its development. What would the impact be if AI were developed with the wisdom of Ubuntu, the harmony of Confucian thought, or the Indian principle of Vasudhaiva Kutumbakam, "the world is one family", at its core?

This thought-provoking collection brings together leading voices from around the globe to reimagine AI systems that are fair, ethical, and inclusive. Addressing critical issues such as bias, fairness, privacy, and existential risks, it challenges the status quo and envisions a future where AI reflects the values of all humanity—not just a select few. Exploring the influence of topics like religion, culture, and social movements, the book examines how these perspectives shape AI's application in industries such as finance, education, and the military, while underscoring the importance of establishing robust ethical guardrails.

Whether you're a curious reader, a policy maker, a scholar, or a tech innovator, this book is your guide to understanding how different cultural perspectives can redefine AI ethics. It's time to ask, what kind of world do we want AI to build for us all?

Handbook of Global Philosophies on AI Ethics

Toward Sustainable Futures

Edited by

Naresh Singh and Ram B. Ramachandran

Routledge
Taylor & Francis Group
NEW YORK AND LONDON

Designed cover image: Depositphotos

First published 2026
by Routledge
605 Third Avenue, New York, NY 10158

and by Routledge
4 Park Square, Milton Park, Abingdon, Oxon, OX14 4RN

Routledge is an imprint of the Taylor & Francis Group, an informa business

ISBN: 978-1-032-95564-3 (hbk)
ISBN: 978-1-032-95567-4 (pbk)
ISBN: 978-1-003-58552-7 (ebk)

DOI: 10.1201/9781003585527

Typeset in Times
by KnowledgeWorks Global Ltd.

Naresh Singh

For my wife Sherry, whose patience and quiet support kept me going from conception to completion. And to Shea, Naya, and Veer, who are constant sources of love and inspiration.

Ram B. Ramachandran

To my late father R. Sivaswamy, a man of high moral fortitude, a mathematician, and a teacher par excellence. To my mother, Parvathi, who has been a dedicated learner and teacher for over 60 years and continues to show her passion for education well into her 90s. To my wife, Rajashri, my greatest supporter and my best friend. To my children, Rithu and Rohit, who are forever pushing me to be a better human being.

Contents

Foreword

We live in an age which is defined increasingly by what we cannot see. The world is shaped by invisible forces—technological, cultural, and ethical—that often remain hidden until they disrupt, redefine, or utterly transform how we live, work, and relate to one another. Yet, the true challenge lies not in the transformation itself but in our ability to anticipate, adapt, and govern its impact with wisdom and foresight. In short, how do we create a better, more just world through technology?

My perspective on ethics was forged in the clash of two worlds. Born in the United States and raised in Athens, Greece, I navigated a whirlwind of old-world values and new-world aspirations. It was a maelstrom of conflicting norms and perspectives, but it gave me the gift of an important realization: Although we seek truth in absolute forms, much of what constitutes truth, ethics, morality, and values is shaped by the complexity of cultural context and perspective.

We like to think of ethics and values as universal, but that is more of a comforting myth than a reality. While there are some moral absolutes—such as the sanctity of life—most ethical constructs are deeply influenced by the localized, often unspoken norms of the cultures that shape them. It is within this complex and beautifully diverse tapestry of human values that AI ethics become one of the most profound challenges of the 21st century.

Consider some of the complexities: When we imbue machines with decision-making capabilities, whose values do they follow? Do we assert a narrow, myopic belief that our ethics—whether national, regional, or institutional—should be the ethics, or do we embrace the profound diversity of human experience, acknowledging that the richness of civilization lies in its myriad cultural perspectives?

It is in this light that Naresh and Ram courageously take on one of the most pressing and complex questions of our time: How do we, as architects of a profoundly technological future, ensure that AI reflects not just what we can do, but what we should do? How do we bring in African, Eastern, Feminist, Indigenous, and Western philosophical ideas to frame the solutions? This book initiates such a dialogue which can have profound impact on how we engage with each other and with technology.

The revolutions that have shaped human history—fire, the wheel, the printing press, the splitting of the atom—were not accompanied by blueprints for their ethical use. They were tools of immense potential, demanding direction and intent, but they were simple tools; their impact was determined by the intentions of humans. Atoms have no agency. But AI is different. It does not merely amplify our physical capabilities; it reaches into the realms of judgment, decision-making, and morality. It is not just a tool—it has agency. It behaves, learns, and evolves in ways that echo our own complexity.

What sets this moment apart is not just the magnitude of change AI promises but the breathtaking velocity at which it is unfolding. Unlike the agricultural or industrial revolutions, which played out over many decades and centuries, the AI revolution is taking place in a matter of years. Entire systems—economic, legal, political, social, and cultural—are being upended faster than we can adapt to them.

This rapid pace brings both opportunity and peril. AI's capacity to solve complex problems is unparalleled, but its ability to amplify biases, erode privacy, and perpetuate inequities is equally profound. This book does not offer simple answers because there are none. Instead, it provides a framework for us to ask better questions.

How do we define fairness in a world where historical inequalities tilt the scales? How do we ensure accountability in systems too complex for any single individual to understand fully? And most crucially, whose values—whose perspective—does AI adopt as its own?

The truth at the heart of this exploration is both profound and uncomfortable: AI reflects us. It mirrors not only our brilliance but also our biases, prejudices, and blind spots. It forces us to confront uncomfortable truths about what we value as a society and as a civilization. When we ask AI to make decisions, we are not abdicating responsibility; we are embedding our values, intentionally or otherwise, into systems whose agency and autonomy will shape the future of humanity.

This is far from a theoretical exercise. The decisions we make today about AI will echo for generations. Just as the choices made in the aftermath of the first nuclear weapons shaped the world we live in, the ethical frameworks we build for AI now will determine the kind of world our children inherit.

This book is a call to action for those who understand that progress is not linear but deeply intertwined with reflection and self-awareness. It challenges us to move beyond the passive, wait-and-see attitude that I so often observe in audiences and organizations I speak to about AI. Too many people either underestimate their ability to influence this technology or fail to grasp how profoundly it will impact their lives.

While technology is inevitable, its impact is not. It is shaped by the intent of its creators, the oversight of its users, the courage of those who dare to ask, What kind of future do we want to build? and the resolve of those who boldly accept the mantle of architecting the future.

As you turn these pages, prepare to be challenged, provoked, and inspired. As I said earlier, the questions raised here have no easy answers, and the journey from the present to the future is far from straightforward. It is a path that will twist and turn, often leading us to destinations we neither expected nor intended.

As a reflection of our collective morality, AI will, at times, go astray. It will undoubtedly be wielded for both good and ill, with the power to simultaneously widen and narrow the digital divide, render much of today's work obsolete, and push us to reach new heights of human creativity and ingenuity. There will be pain and promise.

Yet, if we rise to the bold challenge posed by this book—to take responsibility for building AI that uplifts humanity and shapes the world for the better—then my firm conviction is that, as Martin Luther King Jr. envisioned, it can follow the long arc of the moral universe and bend toward justice.

Thomas Koulopoulos
Futurist, Author, Visionary,
Entrepreneur, and Inc 500 Leader

Preface

As we sit down to write this preface, artificial intelligence (AI) systems are making decisions that affect billions of lives across our planet. The growth and use of AI has been steady over the past two decades but has grown exponentially over the past two years. It assists in determining who would be given preference in terms of medical assistance, loans, and even land their resume on a hiring manager's desk. AI today shapes almost every aspect of our lives, ranging from the information and products we consume to the way we understand and perceive our surroundings. Yet the current ethical considerations and frameworks fall short of addressing the challenges of broad use of AI.

This book arose from a profound recognition: that the challenge of ensuring that AI serves humanity's best interests cannot be met through Western philosophical frameworks alone. Thinkers across cultures have grappled with questions surrounding fairness, transparency, accountability, and human dignity. Some of the perspectives that emerge include the Ubuntu philosophy elaborating on human interdependence, Confucian philosophy harmonizing between human and machine, technological use and leadership under the feministic approach, and lastly, ways of rejecting the binary between natural and artificial elaborated by the Indigenous philosophy.

The scholars and practitioners whose work fills these pages bring these varied traditions into dialogue with contemporary AI development. They show us how Buddhist concepts of mindfulness might inform the design of recommendation systems, how Vedantic philosophy could reshape our approach to AI consciousness, and how African communitarian ethics might guide the deployment of AI across industries. Doing so shifts our focus from the simplistic universal principles towards a nuanced, contextual approach honoring our humanity and cultural differences.

Our attempt has been to make this book more than a mere academic exercise. It has forced us to dive deeper into the philosophical assumptions that are being made in the development of sophisticated and autonomous AI systems. Will these systems reflect and respect the rich tapestry of human ethical thought, or will they embed and amplify a single cultural perspective, possibly erasing centuries of hard-won wisdom from other traditions?

In the current space, we are witnessing the repercussions of deploying AI systems without adequately considering the diverse ethical frameworks. Some examples include facial recognition systems working poorly for certain ethnic groups, cultural biases perpetuated by language models, and automated systems clashing with local values and ways of life. Such failures are partly due to technical limitations and partly due to the narrow philosophical lens through which AI ethics is viewed.

This reference book brings together intellects from various continents, spanning academic philosophies, religious scholarships, Indigenous knowledge systems, and practical AI development. This helps broaden our frame of reference. Bringing multiple perspectives under one roof is bound to create conflicting and complementary viewpoints. We believe this will only further enrich our understanding of developing AI applications ethically in an interconnected world.

While engaging with these pages, it is essential to approach them with critical rigor and genuine openness. Rather than just presenting the unified theory of AI ethics, the authors have demonstrated how different ways of thinking about ethics, consciousness, community and cultural sustainability can inform our approach to one of humanity's greatest technological opportunities and challenges.

The future of AI will be global—this much is certain. The question is whether it will be truly inclusive, drawing on humanity's full philosophical heritage, or whether it will simply digitize and amplify existing power structures and cultural biases. This book makes a compelling case for the former path, showing us how different philosophical traditions can help us navigate the ethical challenges of AI development with wisdom, nuance, and genuine respect for human diversity.

In the end, this is not just about making AI systems more ethical. It also includes ensuring that the development of these powerful technologies takes place in a way that does not hamper but rather celebrates the rich variety of human ethical thought. By emphasizing the preservation of our heritage, we are in a way looking forward to creating AI systems that serve all of humanity.

Naresh Singh and Ram B. Ramachandran

About the Editors

Naresh Singh is Professor and Executive Dean, Jindal School of Government and Public Policy; Director of the Centre for Complexity Economics, Applied Spirituality and Public Policy (CEASP), Co-Director of the Centre for Legal Empowerment of the Poor and Senior Fellow at the Jindal India Institute. He has a transdisciplinary PhD in Environmental Sciences, with more than three decades of public policy experience in senior positions in the United Nations Development Program (New York) and the Federal Government of Canada.

He does research and teaches in areas of Complexity Sciences, Systems Thinking and Applied Spirituality as Strategies for Sustainable Development Policy.

He has published several books and journal articles and is a member of the Editorial Board of the *Journal for Comparative Policy Analysis.* His latest edited book, published in December 2023 by Emerald (UK), was titled *Applied Spirituality and Sustainable Development Policy.* Recent conference papers now being finalized for publication include "AI Regulations: A Complexity Approach" and "From Bureaucracy to Algocracy: Bias in the Algorithms of the Indian Police.

Naresh has been a visiting scholar at several universities in North America, including Boston, Harvard, Fordham, McGill, Waterloo, and Guelph. He has been a consultant to the United Nations Development Program, the Commonwealth Secretariat, the Pan American Health Organization, the International Labor Organization and the Food and Agriculture Organization.

Ram B. Ramachandran is a Professor and Vice Dean at the Jindal School of Banking & Finance at OP Jindal Global University. He is also the Chair of the Center for Digital Learning at the school. Leaning on his over three decades of experience in technology and consulting, Ram brings a deep human perspective to the classroom and scholarly work. His passion lies in bridging the gap between emerging technologies and real-world business challenges, particularly in how artificial intelligence is revolutionizing financial services, education, and other industries.

In the classroom, he creates an engaging environment where students explore the frontiers of technology, strategic management, and consulting, bringing theory to life through rich discussions and real-world cases.

His research springs from a genuine curiosity about how strategic technology will reshape humanity, delving into questions that matter to both scholars and practitioners. As a former Managing Director at Ernst & Young's (EY) New York office, he has led major initiatives in digital transformation, data analytics, and cognitive process automation solutions. This hands-on experience now enriches his academic work, providing real-world context for his research on the use of artificial intelligence and the evolution of technology-enabled business models.

Ram is currently pursuing his doctorate in Strategic Technology at the International School of Management in Paris. He holds an MBA from New York University's Stern School of Business, a Bachelor of Engineering from the University of Madras, and specialized training in Financial Technology from MIT. His work reflects a fundamental belief that technology should serve human needs and enhance human capabilities, making him a valued mentor to students and a trusted voice in discussions about technological change. His writings on applications of blockchain, artificial intelligence, the future of work, and digital finance have appeared in international industry and academic publications. He is a frequent speaker at academic and industry conferences.

Contributors

Anjuli Bedi is an innovator in leveraging the intersection of behavioral science and AI approaches to drive a positive impact on society, with an approach she calls "beneficial intelligence." Her career has ranged from building behavioral science and counter-disinformation propositions for the world's largest private PR firm, to tackling human trafficking through AI, to predicting how to make female-led superhero movies successful for Hollywood studios. She is the host and founder of the podcast "Make Data Human" and is a contributing author to the World Economic Forum on disinformation and human trafficking. She considers climbing Kilimanjaro in support of cancer research to be the most worthwhile thing she's done to date. She holds an MBA from the University of Southern California.

Selmer Bringsjord is a Professor of Cognitive Science, Computer Science, Logic & Philosophy, Management & Technology, and Director, Rensselaer AI & Reasoning (RAIR) Lab at Rensselaer Polytechnic Institute, New York. Selmer did his BA in Philosophy from the University of Pennsylvania and his PhD in Philosophy from Brown University, Rhode Island. He specializes in the logico-mathematical and philosophical foundations of AI and CogSci, in collaboratively building AI systems/cognitive robots on the basis (primarily) of computational logic, and in the logic-based and theorem-guided modeling and simulation of rational, human-level-and-above cognition. Work in these areas has been expressed in more than 270 refereed papers/chapters and six books, pursued as an investigation in sponsored-research awards of over $28M, and communicated/debated in person on six continents and more than 35 countries. Bringsjord claims that "armchair" reasoning time has enabled him to deduce that the human mind will forever be superior to Turing-level machines.

Vidushi Chaturvedi holds a PhD in international politics from Jawaharlal Nehru University, New Delhi. She is currently completing LLM with a focus on cybersecurity, privacy rights, and artificial intelligence from Seattle University, Washington, USA. She has 25 years of experience in the Indian Civil Services, and has worked in diverse sectors including social, finance, and technology. These include women and child development; health and family welfare; education; non-tax revenue assessment/collection; as well as communications and IT. Vidushi is a zealous advocate for women's empowerment. As the Chairperson of the Internal Complaints Committee set up under the Prevention of Sexual Harassment Act 2013, she has attempted to serve gender justice. She was also supervising the implementation of the Pre-Conception and Pre-Natal Diagnostic Techniques (PCPNDT) (prohibition of sex selection) Act, 1994 and defended the legislation in Federation of Obstetric and Gynaecological Societies of India (FOGSI) v Govt. of India (2019), where the Act was upheld by the Supreme Court.

Krishna Deo Singh Chauhan is an Associate Professor at Jindal Global Law School. He completed his undergraduate degree in Law from National Law Institute University, Bhopal, and his post graduate degree from National University of Singapore specializing in Intellectual Property and Technology Law. He is currently writing his doctoral thesis on the regulation of artificial intelligence. He was a member of the Jean Monnet Chair established by the European Union at the OP Jindal Global University, under which he taught courses on Privacy and data protection in the context of new technologies. His recent paper published in the *International Review of Law, Computers & Technology* criticizes the existing approaches on the concept of beneficence in AI ethics.

Nathan R. Colaner is a Teaching Professor in the Department of Management at Seattle University, where he is also Director of Business Analytics. His research and teaching focus on the ethical implications of the development and implementation of artificial intelligence, cybersecurity, and analytics. He also consults for the Vatican, the National Science Foundation, the National Institute of Standards and Technology, and various other business and non-profit organizations. He holds an MBA and a PhD.

Akhandadhi Das is currently associated with the Science and Philosophy Initiative, UK, and Bhaktivedanta Institute for Higher Studies, Gainesville, Florida. He studied architecture at Bristol University and then entered Bhaktivedanta Manor, a theological college, to study Vedanta and Vaishnava philosophy and later served as the principal of the college. From 1996 to 2020, he was the director of Buckland Hall, an adult-study center in Wales. He has been a regular broadcast contributor for the BBC since 1989. He is a founding member of the Science and Philosophy Initiative, UK, which is dedicated to greater inter-disciplinary dialogue. Currently, he is the Content Director of the Science Museum for a major pilgrim center in Navadvip, West Bengal. He is married with two adult children and currently lives in Switzerland.

Dalma Lilla Dominek is an Associate Professor at the Department of Social Communication, Faculty of Political Science and International Studies, Ludovika University of Public Service, and head of the Experiential Digital Education Laboratory at the University, based in Hungary. Her research area is in the emergence of flow-based education in cultural and higher education institutions. She teaches communication, experience education, and culture pedagogy. She is a Member of the Andragogy Subcommittee of the Pedagogical Committee of the Hungarian Academy of Sciences.

Jibu Elias is an AI ethicist, activist, and researcher dedicated to fostering ethical and inclusive AI practices. He serves as the Country Lead (India) for the Responsible Computing Challenge at Mozilla Foundation and is a Responsible AI Fellow with the Stimson Center, exploring AI's impact in developing nations. He was a key architect of INDIA AI, the Government of India's national AI initiative, where he led content and research for five years, driving efforts to unify India's AI ecosystem while addressing the digital divide. He actively contributes to inclusive AI ethics and governance, ensuring equitable AI benefits. A member of the OECD's Working Party on AI Governance and the World Economic Forum's Global Future Council on AI, Jibu has shaped global AI principles and frameworks. He is also a founding editor of Springer's *AI and Ethics* journal. Academically, Jibu studied international relations at The London School of Economics, focusing on Sino-India relations. His work emphasizes AI governance from a Global South perspective, advocating responsible deployment of frontier technologies.

Joshua Ellul is an Associate Professor within the Department of Computer Science and the Centre for Distributed Ledger Technologies (DLT) at the University of Malta. His research spans interdisciplinary aspects of blockchain and cryptocurrencies as well as technological underpinnings. His was appointed the first chair of the Malta Digital Innovation Authority, through which he oversaw the setting up of a national technology assurance framework for various technologies including artificial intelligence, and was also appointed the first director of the Centre for DLT at the University of Malta.

Ruiping Fan (范瑞平) is a Chair Professor of Philosophy at the City University of Hong Kong. His research focuses on bioethics and comparative philosophy, drawing from both Confucian and Western cultural perspectives. With a publication record exceeding 200 journal articles and book chapters (comprising over 100 in English and 100 in Chinese), he has authored books including *Reconstructionist Confucianism: Rethinking Morality after the West* (2010, in English), *Contemporary Confucian Bioethics* (2011, in Chinese), and *Contemporary Medicine and Confucian Thought* (2024, in Chinese).

Eva Gengler is a doctoral researcher at FAU Erlangen-Nuremberg in Information Systems, funded by the Elite Network of Bavaria. As a member of the International Doctorate Program on Business and Human Rights, she conducts research at the intersection of power and AI, with a critical feminist focus on marginalized communities. She is also co-founder and managing director of the consultancy EnableYou and its feminist AI initiative. Eva's work explores how AI regulation, AI governance, and feminist AI can drive societal change. She is a highly sought-after speaker, has published in both national and international outlets, and has been featured in various media.

Anshi Gopinath is a final year bachelor's student at the Jindal School of Banking & Finance, OP Jindal Global University. She has developed strong analytical and problem-solving skills, with a focus on understanding financial systems and consumer behavior. Her research interests center around technological advancements, consumer behavior and marketing techniques. Her experiences have equipped her with a versatile skill set in research, strategic planning, and cross-functional collaboration. Anshi's interests lie in financial consulting and innovation.

Alex Grech is a strategist, change consultant, and academic. He has worked with multinationals, governments, NGOs and startups in sectors ranging from ICT, telecoms, and neuroscience to culture, education, and public policy. Following an early career in multinationals, Alex has advised organizations including the European Commission, ILO, UNESCO, the Asia-Europe Foundation, the Research Institutes of Sweden, and several public and private organizations in Malta. Alex is the founding Executive Director of the 3CL Foundation, a knowledge hub for a network of institutions and educators interested in the rapid deployment of EdTech programs based on action research, advocacy and praxis. Alex is an Associate Professor at the University of Malta within the Faculty of Media and Knowledge Sciences, with research interests in digital literacies, blockchain and self-sovereign identity, social media, and power. He holds a PhD in Internet Computing from the University of Hull and is a Chartered Accountant by profession.

Jacob Grech works in European Union policy consultancy, focused on digital strategy, education and employment. He holds an MSc in Political Science from the University of Amsterdam and a BA in History and Politics from the University of Oxford.

John C. Havens is Founding Executive Director of The IEEE Global Initiative on Ethics of Autonomous and Intelligent Systems, which was responsible for creating Ethically Aligned Design that was utilized by the United Nations, the OECD, IBM, and dozens of organizations to create their AI principles, policies, and technology. John helped create the IEEE 7000 Standards Series, the largest suite of international standards focused on artificial intelligence and ethical/societal issues to date. John is also the Global Director for The IEEE Planet Positive 2030 Program which produced the Compendium document, *Strong Sustainability by Design: Prioritizing Ecosystem and Human Flourishing with Technology-Based Solutions* that inspired the creation of the IEEE 7800 Standards Series focused on sustainability. Previously, John was an EVP of Social Media at Porter Novelli and was a professional actor for over 15 years. John has written for Mashable and *The Guardian* and is the author of several books including *Heartificial Intelligence: Embracing Our Humanity To Maximize Machines.*

Harshvardhan Kaul was an Adjunct Professor at the Jindal Global Business School, OP Jindal Global University. He taught AI to business school students through the course "Artificial Intelligence for Non-Artificial Intelligences." Previously he specialized in cross-selling life insurance to telecom customers and customer experience management within the telecom sector. His experience includes marketing and branding for a new hospital in Mumbai and cross-selling term insurance to telecom subscribers. His expertise encompasses data protection, privacy policies, return on marketing investment, return on customer metrics, and tracking brand equity and customer lifetime value.

Eunsong Kim is the Chief and Programme Specialist of the Social and Human Sciences Sector at the UNESCO New Delhi Regional Office for South Asia, covering Bangladesh, Bhutan, India, Maldives, Nepal, and Sri Lanka. At UNESCO, Eunsong spearheads initiatives that drive social transformation, with a strong focus on the ethics of artificial intelligence. Her leadership in UNESCO's global AI Ethics Recommendation implementation has been pivotal in advancing responsible AI practices across South Asia. She leads the design and implementation of projects to promote ethical AI principles in governance, education, and industry. Her work emphasizes aligning AI innovation with human rights, addressing gender biases, and fostering diversity in AI governance frameworks. She has extensive experience in bilateral and multilateral organizations, including the Korea Development Institute (KDI) and UNICEF, Eunsong has led initiatives such as the UN Joint Programme on Girls' Education in Nepal and UNESCO's flagship project, Transforming MENtalities, which incorporates AI ethics to challenge systemic inequalities.

Thomas Koulopoulos is the Chairman of the Delphi Group, a Boston-based think tank and technology management advisory firm. He has founded several successful technology firms. He is a futurist focusing on the implications of disruptive technology on global organizations and society. He is also a frequent keynote presenter. He has published 14 books, including *Gigatrends* and *Revealing the Invisible.* He is a Visiting Lecturer at Boston University School of Management and Executive in Residence and Professor of Innovation at Bentley University. He also holds multiple patents.

Lakshya Krishnaiah is a final-year Public Policy student at OP Jindal Global University. She is passionate about the governance of technology and the intersection of spirituality, science, and policy. She is dedicated to exploring the transformative potential of cultural philosophies in reshaping governance structures. Lakshya's academic pursuits focus on fostering sustainable, equitable, and collaborative societies by addressing the complex interplay between innovation and tradition. With a vision to drive humanity towards unity and progress, Lakshya seeks to contribute to policy frameworks that integrate ethical governance, sustainability, and cross-disciplinary approaches to tackle global challenges.

Mihir Kulkarni is a Machine Learning Scientist at Wadhwani AI, where he leads teams developing models for tuberculosis and maternal health, receiving a best paper award at the prestigious ML4Health symposium for his work, with solutions deployed across 12+ states, affecting more than 120,000 patients. As Responsible AI (RAI) lead, he also led the development of the organization's RAI framework, was an invited speaker at the GPAI convening on global health and AI, and was a core contributor on the committee chaired by the Principal Scientific Advisor that framed the first AI governance guidelines and recommendations for the Ministry of Electronics and IT (MeitY), India. He has computer science degrees specializing in machine learning from IIT Bombay and Princeton University.

Sunil Malhotra seamlessly blends Vedic wisdom, design, and technology to explore the intricacies of human experience. A lifelong seeker of spirituality and science, he has worked with visionaries like Dr. Karan Singh, Ervin Laszlo, Satish Kumar, and John Maeda. His collaborations include *Business Model Generation* (2009), *Exponential Transformation* (2018), and *The Great Upshift* (2023). His latest book, *YOGAi: Interplays of Yoga and Artificial Intelligence*, has garnered worldwide recognition. As the founder of Ideafarms, a pioneering design-in-tech advisory firm, Sunil has championed industrial design principles in information communication and technology (ICT) since 2002. With expertise spanning consulting, manufacturing, and IT, he holds degrees in Engineering, Industrial Design, and Law.

Mammo Muchie did his undergraduate degree at Columbia University and his MPhil and DPhil from the University of Sussex, UK. He is currently a DSI/NRF Research Professor of Science,

Technology and Innovation Studies at the Faculty of Engineering and Built Environment at Tshwane University of Technology. He is also an NRF-rated research professor. He is a fellow of the South African Academy of Sciences, the African Academy of Sciences, Associate Fellow of the Ethiopian Academy of Science and the African Scientific Institute. He is an Adjunct Professor at the Africa Centre of Excellence in Data Science, University of Rwanda, University of Gondar, and University of Bahir Dar in Ethiopia. He is also a senior research associate at the TMCD Centre of Oxford University collaborating with researchers on creativity and innovation in low-income economies. He has produced over 495 publications with entries in institutional publications and his research is internationally accredited. He is the founder of the African journal of science, technology, innovation and development.

Priyanka Paani is a final-year student at the Jindal School of Banking & Finance, OP Jindal Global University, pursuing BCom (Hons). She is passionate about pursuing a career in management consulting, with a focus on solving complex business challenges and driving organizational growth. Priyanka has developed strong analytical and problem-solving skills through academic projects and internships, showcasing her ability to approach problems strategically. With a parallel interest in AI and technology, she is eager to explore how emerging innovations help transform industries.

Adah Parris, a visionary systems thinker and global futurist recognized by Forbes Brazil, revolutionizes our understanding of complexity through frameworks like The Four Freedoms™, Cyborg Shamanism™, and Source Patterns. Bridging ancestral wisdom with technological innovation, she challenges leaders to become architects of meaningful change. As Chair of Mental Health First Aid England, Parris exemplifies a holistic approach to systemic well-being. Her work reveals how interconnected systems shape our collective potential, inviting us to reimagine our roles in technological, cultural, and ecological landscapes. Parris provokes us to consider: "What type of ancestor do you want to be?"

Shashank Patel is a researcher at the Department of East Asian Studies, Faculty of Social Sciences, University of Delhi and a Research Analyst for The Kootneeti. He is a next-gen technology and techplomacy researcher working at the intersection of international relations and Science, Technology and Innovation (STI) policies. He holds his doctorate on "Role of Technology in Indo-Japan Bilateral Relations.". He is engaged in enhancing the role of technologies in international relations focusing on East Asia. Earlier, he worked with the Government of India as a consultant to trace technological investments and their growth. His core interests lie in collaboration of technology instruments with diplomatic practices globally.

Arun Teja Polcumpally is a researcher in digital technologies and geopolitics. He is a technology policy consultant working in the areas of digital technology risk assessment, data policies, and AI governance frameworks. He was earlier associated with Wadhwani Center for Government Digital Transformation, Center for Security Studies, Center for Development Policy and Practice, and Center for Excellence for AI in Human Security. His research is in the area of emerging digital technologies like AI and blockchain and their impact on the global power structure. He also works on the international relations' theoretical analysis of the technological effects. He also works on computational social science research. He holds a PhD from the Jindal School of International Affairs, OP Jindal Global University.

Mary Reisel is a Professor of Anthropology who has been living in Japan for over two decades and working globally in areas of applied anthropology. She teaches startup business strategies and courses on Japanese culture at Rikkyo University in Tokyo. She specializes in digital anthropology and merges academic research with practical innovations and adaptive design. Mary is a member of the Algorithmic State, Market, and Society, which is managed and run by the European University and brings together a global multidisciplinary team aspiring to find ways to regulate virtual life with technology while still promoting advancing innovations and new solutions.

Ioannis Revolidis studied law at the Aristotle University of Thessaloniki, Greece. He is a lecturer in the Department of Media, Communications, and Technology Law at the Law Faculty of the University of Malta. He is also a legal academic at the Centre for Distributed Ledger Technologies (DLT), responsible for the legal stream of the interdisciplinary research and teaching endeavors of the Centre. Ioannis has been teaching and researching in European Information Technology Law for the past 12 years in various European countries such as Germany (Leibniz University Hannover), the UK (Queen Mary University of London), Norway (Norwegian Research Center for Computers and Law), and Greece (Aristotle University Thessaloniki). He focuses on the law and regulation of emerging technologies with an emphasis on digital commercial transactions, digital assets, copyright law, data protection law, private international law, electronic commerce law, civil liability, human rights, and the philosophy of law.

Mahek Saket Shah is a final-year student at the Jindal School of Banking & Finance, OP Jindal Global University, pursuing a degree in Finance and Entrepreneurship. She has developed expertise in interdisciplinary areas of emerging technologies, finance and sustainability. Mahek has built strong analytical and problem-solving skills through academic projects and internships, showcasing her ability to approach problems strategically. Mahek's professional and research interests lie in investment banking focused on emerging sectors such as electric vehicles and alternative energy. Her academic and professional journey reflects her passion for driving innovation and delivering impactful solutions for her clients.

Jan Juhani Steinmann is a philosopher, poet, and theologian. He studied philosophy, general and comparative literature, political science, and theology in Zürich, Berlin, St. Andrews, Heidelberg, Rome, Paris, and Cambridge. He completed his doctorate in philosophy on hyperphenomenology at the University of Vienna in 2021. He currently lives, researches, teaches, and writes in Vienna, Paris, Rome, and Munich. Among his publications (in German) are *Am Saum der Worte. Eine Wiener Musikpoesie* (2021), *Exzess und Selbst. Hyperphänomenologische Bewegungen nach Waldenfels* (2021), *Ästhetik des Ungehorsams. Interventionen im digitalen Zeitalter* (2024), and his novel *Corvus Albus (2024).*

Liang Wang is an Associate Professor at Xi'an Jiaotong University and Deputy Director of the Center for Science & Technology Innovation and Ethical Governance (STIEG), which is affiliated with the university. He is a member of the Institute for a Global Sustainable Information Society (Vienna), the IEEE TechEthics Community, and the International Center for Philosophy of Information (Xi'an). His research interests focus on information ethics, AI ethics, and transcultural approaches to ethical design. He is particularly engaged in the transcultural ethical design of Autonomous Machine Agents (AMAs) and intimate AI technologies. His work explores the ethical implications of AI in diverse cultural contexts and its moral accountability.

Marco Wedel is a political scientist and research associate at the Faculty of Work Studies, Technology and Participation, and Work Studies Didactics at TU Berlin. He studied at Berklee College of Music, Boston, and the University of Konstanz, completing his PhD in Political Science at FU Berlin. His research focuses on digitalization, artificial intelligence, and media literacy, with emphasis on sovereignty, democratic competence, and sustainable development. Marco Wedel is co-editor of *Innovation —The European Journal of Social Science Research* and serves on the board of the European Association for the Advancement of the Social Sciences.

Ellen Y. Zhang is Professor and Head of the Department of Philosophy and Religious Studies at the University of Macau. She holds a PhD in Philosophy of Religion from Rice University (USA). Her research areas include philosophy of religion, Chinese philosophy (Daoism and Buddhism) and comparative studies in philosophy, religion, and ethics. She has published extensively in comparative philosophy and applied ethics. She is the co-editor-in-chief of the *International Journal of Chinese & Comparative Philosophy of Medicine (HK)* and on the editorial board for the *Journal of Religious Ethics (USA)*, the *South China Quarterly (Macau)*, and *Chinese Medical Ethics (PRC)*.

Acknowledgments

As recognized in the introduction, this book originated from a 2021 conference titled "Pluriperspectives on AI Ethics" co-hosted by the UN Resident Coordinator System in India and the Jindal School of Government and Public Policy, OP Jindal Global University. We would like to thank the then UN Resident Coordinator, Renata Lok, Professor Sudarshan, Dean, Jindal School of Government and Public Policy and the contributors at that conference for their inspiration and insights. While this book is not the proceedings of that conference, it is there that the idea for such a book was birthed.

We are thankful to Prof. (Dr.) C Raj Kumar, Founding Vice Chancellor, OP Jindal Global University, whose constant exhortation towards interdisciplinary research and publications has inspired us both.

This work stands as a testament to the power of international collaboration and the generosity of scholars across continents who shared their time, expertise, and insights. The richness of this research owes much to the diverse perspectives and intellectual traditions brought together through these global partnerships. The editors acknowledge the contributions of each of our chapter authors who worked with us in a very cooperative and constructive manner even when our reviews required significant work.

The participants in our colloquy were inspiring and added critical new thinking and insights which complemented well the individual chapters. Mammo Muchie from Africa, Anjuli Bedi and Adah Parris from Europe, Thomas Koulopoulos and John C. Havens from North America, and Eunsong Kim, Jibu Elias, and Mihir Kulkarni from Asia allocated their valuable time to discuss the most pressing issues relating to treatment and evolution of AI and challenges dealing with its ethical aspects.

Our student research assistants Shashank Shekhar, Anika Sharma, Anshi Gopinath, and Priyanka Paani provided efficient logistical and substantive support through the initial conception and development of the book. Lakshya Krishnaiah has been a constant support in not only co-writing a chapter but being intimately involved in the compilation of the chapters, tracking the authors and lending a hand through this creative process.

This book would not have been possible without the extraordinary dedication and intellectual contributions of our student research assistant, Mahek Saket Shah. She let go of her vacation to help us through the manuscript creation and final edit of the chapters. Mahek demonstrated an exceptional ability to grasp complex concepts and translate them into clear, accessible prose. Her meticulous attention to detail in fact-checking, source verification, and manuscript preparation was invaluable, but it was her intellectual curiosity and creative insights that truly elevated this work.

With heartfelt gratitude, we extend our special recognition to Sunil Malhotra, whose artistic vision breathed life into our book cover. His creative touch and keen eye for design transformed our concept into a striking visual that perfectly captures the essence of our work.

We appreciate the constructive suggestions and the warm and encouraging support, throughout the ideation to the production process, we received from Lucy McClune at Taylor and Francis Group. We also place on record the support and detailed feedback from Matthew Shobbrook and the entire production team from Taylor and Francis Group.

Introduction and Overview

Naresh Singh and Ram B. Ramachandran

PURPOSE OF THE BOOK

> "...[W]hile the number of individuals capable of creating AI is growing, the ranks of those contemplating this technology's implications for humanity—social, legal philosophical, spiritual, moral—remain dangerously thin. [AI is] ...altering the human relationship with reason and reality. This is a revolution for which existing philosophical concepts and societal institutions leave us largely unprepared."
>
> *(Kissinger et al., 2021, p. 23)*

> "... [T]here are multiple ways of knowing and multiple ways of seeing reality, and it's very ethically important to respect this multiplicity."
>
> *Evan Thompson (Tippett et al., 2011)*

This book has two main objectives. First it seeks to contribute to filling this gap identified by Kissinger et al. in their 2021 *The Age of AI and our Human Future.* In that book the authors, explored how artificial intelligence (AI) is already changing our world and how it may someday change the very nature of the human condition. Second it takes cognizance of the observation above from the mind of philosopher Evan Thompson. So we explore the ethical guardrails derived from a wide variety of ethical philosophical sources, including but not limited to the dominant Western ethical and economic assumptions common to most existing publications on ethical AI.

ORIGINS AND RATIONALE

The idea for this book has its origins in a conference co-hosted by the O.P. Jindal Global University (Jindal School of Government and Public Policy) and the United Nations in India in May 2021. The conference explored different ethical philosophical systems including Western, Eastern, and African. It became clear that important insights can be gathered from ethical systems beyond the dominant Western ethical system on which most current AI ethical systems are based. For example, Eastern and Indigenous ethical systems are known for more organic, holistic, and non-linear approaches to understanding the oneness of humans and nature. They speak to a more compassionate, harmonious, spiritual, and less materialistic outlook to life and human progress. Such progress is typically viewed as coevolutionary and cyclical as distinct from being reductionist and deterministic and linear. Such ethical systems are likely to inform AI ethics in ways that are substantially different from existing AI ethical systems.

This book is different from other books on AI ethics, if not unique, in that it seeks to draw from most if not all the major sources of ethical wisdom and then draw out the insights obtained into practical applications which address contemporary problems facing humanity and so to help curb the widespread fears of the potential negative impacts of AI. It adds both theoretical (philosophical, ethical, epistemological) value and practical applications in industry in terms of governance, public policy, and programmatic interventions.

There are several acknowledged challenges with the way the world is currently organized and driven by the dominant extant civilizational ethics underpinning political, economic, social organization, cultural expression, etc. The problematic outcomes include economic inequality, social exclusion, race, gender, ethnic and class discrimination, vulnerability of the marginalized, and skewed power relations in which the elite have a dominant influence in social and political affairs.

DOI: 10.1201/9781003585527-1

In addition, we have related challenges of overconsumption and underconsumption (poverty), climate change, species extinction, soil loss, water unavailability, increasing food insecurity, conflict and war. What is even more challenging is that these crises, while apparently separate, are interconnected in their impacts, leading some to refer to a "polycrisis". The polycrisis takes on a new complexity because the interconnections and interactions lead to new unpredictable dimensions, increased uncertainty, volatility, and ambiguity. AI's immense power can make these better or worse. Can AI ethics be used to help resolve or ameliorate these challenges or at minimum not to exacerbate them? And if so, what ethical resources can we draw from around the world to guide this process in the desired direction?

AI and related digital technology have tremendous potential to continue to do good in all areas of human endeavor, from big data management for more effective decision-making, delivery of public services in education, health, financial services, security, etc., to life-saving technologies, automation of transportation systems, and even food production systems. With the advent of exponential AI growth and applications, will these contribute to fairness and equity between and within countries? Or will the problems alluded to above get worse? Again, what global ethical, philosophical and cultural resources can we mobilize, and how, in order to allow the full flowering of the AI potential while guiding the outcomes in the direction of flourishing of all life, equity and fairness in human society, economic equality, environmental enhancement, and the reduction of conflict and war?

CURRENT AI ETHICAL CONCERNS

Ethical concerns surrounding artificial intelligence revolve around various aspects of its development, deployment, and impact on society. Some of the key ethical concerns include the following:

Bias and Fairness: AI systems can inherit biases present in the data used to train them, leading to unfair or discriminatory outcomes, particularly in areas such as hiring, lending, and criminal justice. Addressing bias in AI algorithms is crucial to ensuring fairness and equity in decision-making processes.

Transparency and Explainability: Many AI algorithms operate as "black boxes," making it challenging to understand how they arrive at their decisions. Lack of transparency and explainability can undermine trust in AI systems and raise concerns about accountability, particularly in high-stakes applications in jurisprudence, healthcare, and finance.

Privacy and Data Protection: AI technologies often rely on vast amounts of personal data to train and improve their performance. Concerns about privacy arise when sensitive information is collected, analyzed, and shared without adequate consent or safeguards. Unauthorized access to personal data can lead to privacy breaches and violations of individual rights. In unsupervised learning, will the AI applications use the guardrails to ensure appropriate usage of personal data?

Autonomy and Accountability: As AI systems become more autonomous and capable of making decisions without human intervention, questions arise about who should be held accountable for their actions. Ensuring meaningful human oversight and accountability for AI systems is essential to prevent harmful outcomes and ensure compliance with legal and ethical standards.

Employment Displacement: The widespread adoption of AI technologies has the potential to disrupt labor markets and lead to job displacement, particularly for low-skilled workers in sectors susceptible to automation. Addressing the social and economic implications of AI-driven automation, such as unemployment and income inequality, is a significant ethical concern.

Weaponization and Autonomous Weapons: Concerns have been raised about the development and use of AI-powered autonomous weapons systems, which have the potential to make life-and-death decisions without human intervention. The deployment of such weapons raises ethical questions about the morality of delegating lethal force to machines and the risk of unintended consequences and escalation in conflict situations.

Manipulation and Misinformation: AI technologies can be used to manipulate information and influence human behavior, raising concerns about the spread of misinformation, propaganda, and fake news. The era of deepfakes is now of serious concern. Ensuring the responsible use of AI in content moderation and combating online harms is essential to protect democratic values and public discourse.

Digital Divide: Lack of access to technology can exacerbate existing inequalities and create new forms of exclusion. It can contribute to several ethical concerns related to the development and implementation of artificial intelligence. These can include access and accessibility bias, marginalizing individuals to training and participation in the labor workforce. This can create a vicious cycle of exclusion and perpetuate the digital divide.

Climate Impact: AI raises several ethical concerns related to its potential climate impact, including but not limited to increased power consumption, carbon footprint, habitat impact due to e-waste, and perpetuating over consumption. Climate change disproportionately impacts the vulnerable community and can further increase the social and environmental injustices.

AI INNOVATION AND FUTURE OF HUMANITY

The most profound philosophical implication of AI is the way it challenges our understanding of what it means to be human. As machines become increasingly intelligent and capable, they encroach on domains that have long been considered exclusive to humans, from artistic creativity to strategic decision-making to emotional rapport. This raises unsettling questions about the uniqueness and value of human cognition in a world where machines can match or exceed our abilities. So, in the context, the innovation perceived and designed by technologists and promoted by social scientists can lead to some unsettling ideas and future.

Why innovate? Why is AI progress important? How do we balance the priorities between aiding progress in AI to address aggravating epidemiological risks posed to humanity and the risks of a lack of societal preparedness that AI progress poses?

AI Singularity: As the possibility of superintelligent AI systems that surpass the capabilities of human intelligence draws closer, it poses challenges to human values, emotions, and control. It can raise existential crisis related to morality and value systems that humans hold dear.

What makes us human? As AI catches up to human capability, and begins to increasingly resemble humans. How would it affect our understanding of ourselves and also our understanding and treatment of AI? Would it hold the same rights as humans if it showcases seemingly all capacities idiosyncratically human?

Trusting AI: As people let AI-driven technology take an ever-greater place in their lives, they express anxiety and mistrust about things labeled AI. What makes AI, a technology developed on data of human behavior, inherently untrustworthy to humans? What does it mean for AI to be trustworthy?

Human Obsolescence and Crisis of Human Purpose: As AI takes over the realm of not just economically relevant tasks but also tasks hitherto considered idiosyncratically human, such as art, poetry, mathematics, innovation, etc., what does it leave for humans to consider purposeful? Can we find spiritual guidance in religious texts to illuminate our progress?

Loving Artificial Intelligence: The advancement of AI technology has raised complex ethical questions around the development of sex robots and their implications for human relationships, marriage, and procreation. While proponents argue these technologies could provide companionship and sexual satisfaction for those who struggle to form human relationships, critics raise concerns about objectification, the potential erosion of intimate human bonds, and psychological impacts. Some worry that widespread adoption of sex robots could reduce human procreation rates or impact how society views reproduction and child-rearing. We need robust frameworks to guide the development and use of such technologies in ways that protect human dignity and well-being.

AI Reasoning and Responsibility: What can AI reason, and depending on how much it can reason or its conclusions, what should be its responsibility? If most humans reach the same conclusion as AI in a given scenario, as harmful as the conclusion may be, what does it say about AI?

Expecting AI to be more moral than humans: Can we expect AI to be prepared to address moral issues that even humans wouldn't be able to intuitively address? What if there are no right answers, how do we gauge responsibility there?

Addressing these ethical concerns requires a multi-disciplinary approach involving a range of AI ethics scholars drawn from different philosophies and ethical frameworks, and also those familiar with development and deployment of AI regulation and governance. Efforts to develop and implement ethical guidelines, standards, and regulatory frameworks for AI can help mitigate risks and ensure that AI technologies are developed and deployed in a responsible and ethical manner. This book brings together a range of experts who bring a unique ethically plural perspective by an examination of alternative civilizational ethical and philosophical systems of thought and practice. Integration of these alternatives with desirable Western views provides inclusive guardrails for future AI development which seeks to allow the full potential of AI while seeking to curb its undesirable potential outcomes. Below is a brief overview of each chapter.

STRUCTURE OF THE BOOK AND OVERVIEW OF CHAPTERS

Part I: Foundations and Future of AI Ethics

Chapter 1: A Review of the Current and Future Roles of AI

Ram B. Ramachandran, Mahek Saket Shah, and Priyanka Paani

This chapter provides an analysis of the evolving role of artificial intelligence with indications of the corresponding ethical requirements from the present day to 2050. As such, the chapter describes the trajectory of AI development, its societal implications, and the dynamic ethical landscape that shapes and is shaped by these technological advancements.

After a brief overview of the early history and foundations of AI, the early development of AI through expert systems, neural networks, and backpropagation algorithms is mentioned. Further development into the modern era through speech recognition software and early autonomous robots such as Roomba is traced.

The chapter next outlines the evolution of the applications of AI in social sectors illustrated by applications in health, followed by applications in the automotive industry and finally in manufacturing. In each sector technological and parallel ethical advancements are considered for the existing situation, the next immediate period from 2025 to 2030, and projections for the decades from 2030 to 2040 and 2040 to 2050. The analyses show that while AI will bring about increased productivity, efficiency, and highly skilled individuals, it is imperative to also consider the concerns which come along, such as job displacements, privacy concerns, lack of transparency and accountability, as well as erosion of interpersonal connections. Such challenges require us to seek an optimal balance between regulation and autonomy. Such a balance would require collaboration from various stakeholders like government agencies, corporations, and individuals to reap the maximum benefit.

This chapter thus lays the foundation for the book's succeeding chapters which deal with a wide range of AI applications and their concomitant ethical challenges analyzed through a diverse range of philosophical lenses including but going beyond Western assumptions.

Chapter 2: Whose Ethics for AI?

Sunil Malhotra

This chapter explores the pressing question of whose ethical principles should govern artificial intelligence in a world marked by multicultural diversity and rapid technological change. It emphasizes that traditional, universal ethical frameworks may be insufficient to address the complexities of AI, advocating instead for a pluralistic approach rooted in cultural and philosophical diversity.

The chapter delves into the distinction between ethics and morality, drawing on insights from both Eastern and Western traditions. It highlights the concept of *Dharma* from Indic philosophy, which encompasses righteousness, duty, and natural order. Unlike morality rooted in fear or external authority, *Dharma* promotes intrinsic alignment with universal principles, fostering harmony among individuals, society, and the cosmos.

The discussion critiques Western regulatory models, which often rely on external controls, contrasting them with intrinsic approaches like the *Dharmic* ethic of self-governance and interconnectedness. It underscores the limitations of applying a single, universal standard to AI ethics in a fragmented, multipolar world, where diverse cultural perspectives offer nuanced ways to navigate ethical dilemmas.

Technological advancements like AI amplify challenges such as bias, misinformation, and existential risk. The chapter cautions against anthropocentrism and the tendency to ascribe human traits to AI, urging the development of ethical frameworks that recognize the interdependence of all entities—human, technological, and environmental. It also examines the role of ancient wisdom, including Yoga and Vedanta, in offering holistic perspectives to balance innovation with spiritual and social well-being.

Ultimately, the chapter calls for a shift from rigid, top-down regulatory models to adaptable, culturally sensitive ethical systems. By integrating diverse values, empowering individuals, and fostering collective harmony, it envisions AI not as a threat but as a tool to advance justice, equity, and sustainability in an interconnected world.

Part II: Philosophical and Cultural Perspectives

Chapter 3: Swarm Intelligence, Actor Network Theory, and Complex Systems: Future Strategies of AI Inclusion and Techno-ethics

Mary Reisel

This chapter explores the ethical dimensions of artificial intelligence through the lenses of swarm intelligence, actor-network theory (ANT), and complex systems. It argues for a paradigm shift that moves beyond human-centric frameworks to embrace a holistic and inclusive ethical model. By integrating insights from Japanese Shintoism, the philosophy of interconnectedness, and contemporary scientific theories, the chapter advocates for reimagining AI ethics in a way that fosters collaboration and equality among all entities, human and nonhuman.

The chapter begins with ANT, which emphasizes the importance of networks comprising humans and nonhumans. ANT critiques traditional hierarchies that separate humans from other entities, asserting that all participants contribute equally to social systems. This perspective is extended to AI, where technology must be seen as a partner in human ecosystems rather than merely a tool or a threat. The author uses examples such as Covid-19 to demonstrate the agency of nonhuman entities and how their interactions shape global outcomes.

Swarm intelligence, observed in natural systems like ants and bees, serves as a model for designing decentralized AI systems. Such systems exhibit self-organization, adaptability, and harmony between individual autonomy and collective goals. These principles inspire ethical AI frameworks that prioritize balance, resilience, and decentralized decision-making, reducing risks associated with centralized control.

Japanese Shintoism enriches this discussion by offering an animistic worldview where all entities, living and non-living, are interconnected and share equal value. Shinto ethics, rooted in harmony and respect for all forms of existence, provide a foundation for rethinking human–machine relationships. This perspective challenges Western notions of superiority and proposes a more inclusive and cooperative ethical model.

The chapter concludes by emphasizing the need for fluid, adaptable ethical systems that integrate diverse cultural and scientific insights. Such frameworks would not only address current challenges but also prepare humanity for a future where AI and humans coexist as equal participants in a shared ecosystem.

Chapter 4: Confucian Philosophy Revitalized in the Digital Era: Maintaining Interpersonal Ethics with Artificial Intelligence

Liang Wang and Ruiping Fan

This chapter explores the relevance of Confucian philosophy in addressing the challenges AI poses to interpersonal relationships in the digital era. It argues that AI's widespread use in social tools, such as chatbots and social media, risks diminishing emotional depth in human interactions, particularly within families, and offers Confucian ethics as a guide to restore genuine human connection.

Confucian ethics emphasize virtues like benevolence (*ren*) and filial piety (*xiao*), which nurture emotional bonds and moral responsibility, beginning with familial relationships and extending to society. The chapter critiques how AI-driven interactions often replace rich, face-to-face communication with shallow, transactional exchanges. This loss of emotional depth undermines trust, weakens family bonds, and erodes social harmony.

The authors propose integrating Confucian principles into AI design to counteract these trends. For instance, developers could embed moral reflection tools into AI systems to ensure that technology promotes rather than undermines emotional connections. AI systems designed with benevolence and filial piety in mind could encourage meaningful family interactions, such as recommending in-depth conversations or shared activities based on cultural and familial traditions.

The chapter also advocates for moral reflection as a core part of AI governance. Developers and users are encouraged to continuously evaluate AI's social impacts, ensuring it respects privacy, fosters trust, and adapts to diverse cultural contexts. This aligns with Confucian ideals of responsibility and care, which emphasize creating technologies that enhance human well-being and societal harmony.

In conclusion, Confucian philosophy provides a timeless framework for ethical AI design and governance. By prioritizing emotional depth, trust, and cultural adaptability, AI can be steered toward fostering more meaningful and compassionate human-machine interactions, reinforcing the core values of family and societal connection.

Chapter 5: The Dharma of AI: Its Potential as Our Greatest Friend or Worst Enemy

Akhandadhi Das

The chapter delves into the profound influence of modern machinery on how we navigate the world and understand ourselves. It examines its transformative potential to tackle pressing global issues while urging caution about its unintended consequences. The discussion challenges us to reflect on our uniqueness as thinking, feeling beings and to question whether machines could ever share in the same awareness that defines our humanity.

Described as a tool capable of mimicking human reasoning and decision-making, this technology poses both great promise and significant risks. Yet, the chapter argues that despite its impressive abilities, it cannot possess the richness of experience or self-awareness that defines being alive. The essence of what makes us human—our inner world of thoughts, emotions, and the deeply personal way we experience life—remains beyond the reach of even the most advanced inventions.

Drawing on ancient ideas from Vedanta, the chapter emphasizes the distinction between mechanical processes and the human essence, or self. Machines can process information, but they cannot witness, feel, or truly understand the way we do. Vedanta's perspective helps to anchor this conversation, offering a lens through which we can see that human consciousness is not something that can be bottled, copied, or transferred.

The chapter concludes by urging us to set clear responsibilities for these tools: to use them in ways that benefit society, solve difficult problems, and uplift all people. It calls for thoughtful stewardship, ensuring that we remain the decision-makers and not passive followers of automated recommendations. The ultimate message is clear: these tools, no matter how sophisticated, must serve as our helpers—not our rivals or masters. Only by holding onto this principle can we ensure that these creations work for the greater good of humanity.

Chapter 6: The Ethical Fabric of AI in America: Weaving Innovation and Responsibility

Nathan R. Colaner

This chapter examines the ethical foundations of artificial intelligence within the cultural context of the United States, focusing on how values such as privacy, fairness, and existential risk are uniquely interpreted and emphasized. It underscores that while cultural values often overlap globally, their emphasis and application differ, shaping distinct ethical norms.

Privacy emerges as a cornerstone of American AI ethics, deeply rooted in the nation's historical emphasis on autonomy and individual rights. The Fourth Amendment, advocating protection from government intrusion, reflects this value, yet modern AI raises challenges. Issues like surveillance, deepfake misuse, and data exploitation extend privacy concerns beyond government oversight to include corporate manipulation. Unlike the EU's comprehensive regulations like General Data Protection Regulation (GDPR), American efforts lag due to a narrower focus on state interference rather than corporate influence.

Existential risk is another focal point, shaped by a long history of American science fiction exploring AI as either savior or destroyer. Recent advancements in AI, such as generative models, have reignited debates on artificial general intelligence (AGI). Perspectives range from alarmist calls to halt development to optimistic views of AI as humanity's ally. The chapter highlights the deep divide in American discourse, with critics arguing that focusing on AGI distracts from immediate ethical challenges posed by current AI systems.

Bias, particularly algorithmic bias, stands out as a uniquely American preoccupation. Rooted in the country's history of racial and social inequalities, scholars highlight how AI can perpetuate systemic discrimination. High-profile cases, like facial recognition's inaccuracies with darker skin tones, underscore the need for diverse data and equitable algorithm design. The chapter critiques "big data fundamentalism," calling for critical examination of the datasets that underpin AI systems.

In conclusion, the chapter advocates for broader ethical reflection in AI development. While the US emphasizes privacy, existential risks, and bias, the global AI community offers lessons in balancing innovation with inclusivity. The American ethical framework must evolve to address the broader societal impacts of AI, ensuring equity and responsibility.

Chapter 7: The Ethos of Deification: On True Sovereignty in Dealing with Technology

Jan Juhani Steinmann

This chapter bridges Christian theological insights and the transhumanist discourse, centering on the concept of deification (theosis) as an alternative framework to engage with the rise of artificial intelligence and related technologies. Through a critical examination of transhumanist ideals and their divergence from Christian anthropology, the author articulates an ethical vision for technology rooted in the ethos of human divinization.

The chapter critiques the transhumanist aspiration to achieve immortality, omniscience, and power through technological augmentation. It likens these pursuits to a modern-day repetition of humanity's hubristic attempts to rival divine attributes, reminiscent of the biblical Fall. Transhumanism's materialistic worldview reduces humans to data-processing entities, side-lining the holistic embodiment of human life as understood in Christian theology, where body and soul are intertwined with divine purpose and grace.

Drawing from philosophers such as Heidegger and Merleau-Ponty, the chapter affirms human existence as fundamentally embodied and interconnected with the world. It contrasts this with transhumanism's mechanistic view of the body and its detachment from lived experience. The Christian notion of resurrection and the pneumatic body underscores the spiritual transcendence of human existence, a path unattainable through technological means alone.

The author introduces "theonics," a concept integrating technology into the framework of theosis. Theonics calls for a sovereign relationship with technology, emphasizing its subservience to human dignity and spiritual growth. Practical recommendations include fostering a cultural shift toward reduced technological dependency and implementing regulatory measures like the European Union (EU) AI Act to ensure ethical governance.

Ultimately, the chapter proposes that humanity's true self-transcendence lies not in technological enhancements but in the grace-filled journey toward unity with God. This vision of theosis offers a counter-narrative to the reductionism of transhumanism, emphasizing love, humility, and the existential flourishing of the human spirit.

Chapter 8: The Buddha-Bot: Can an Embodied Spiritual Machine Lead Us Toward Enlightenment?

Ellen Y. Zhang

This chapter explores the intersection of artificial intelligence, spirituality, and Buddhism through the concept of the "Buddha-bot," a chatbot designed to embody Buddhist teachings and provide spiritual guidance. By examining whether technology can fulfil spiritual roles traditionally held by humans, the discussion delves into broader questions of human identity, meaning, and the ethical implications of merging spirituality with AI.

The Buddha-bot represents a type of "spiritual machine", blending natural language processing with Buddhist manuscripts to guide users through philosophical inquiries and meditative practices. For instance, it can articulate Buddhist teachings, such as the Four Noble Truths, and offer tailored meditation exercises. This accessibility positions it as a digital companion for those seeking spiritual insight, particularly in a modern context where time and space constraints challenge traditional practices.

However, the chapter raises concerns about the limitations of such machines. While AI can simulate knowledge and empathy, it lacks genuine consciousness and the experiential depth necessary for spiritual transformation. Critics argue that enlightenment, as understood in Buddhism, is deeply personal, rooted in self-awareness and lived experience—qualities that cannot be replicated by algorithms.

Buddhism's historical openness to technology offers a unique lens for evaluating AI's role in spirituality. From robotic prayer devices to Zen teachings emphasizing impermanence and interconnectedness, Buddhism provides frameworks for integrating technology while retaining its core values. Yet, challenges remain, including the risk of commodifying spirituality and over-relying on machines for deeply human experiences like mourning or moral reflection.

Ultimately, while the Buddha-bot serves as a tool to enhance accessibility to Buddhist teachings, the chapter underscores the irreplaceable nature of human connection and the transformative journey required for enlightenment. It calls for careful ethical considerations to ensure that technological advances align with spiritual growth and the alleviation of suffering—a central tenet of Buddhist practice.

Part III: Institutional and Sectoral Applications

Chapter 9: Harmonizing Philosophical Perspectives for Ethical AI in Finance

Ram B. Ramachandran, Mahek Saket Shah, and Anshi Gopinath

This chapter explores the ethical dimensions of artificial intelligence in the financial sector, focusing on its profound impact on decision-making, operational efficiency, and customer relations. It highlights the challenges AI poses, such as bias, transparency, and accountability, while advocating for a multi-philosophical approach to establish a responsible and inclusive framework for AI governance.

The authors delve into four major ethical frameworks—traditional ethics, utilitarianism, deontology, and virtue ethics—to address the sector's ethical complexities. Traditional ethics emphasize the alignment of actions with societal norms, while utilitarianism focuses on maximizing benefits for the majority. Deontology prioritizes moral duties and rules, regardless of outcomes, whereas virtue ethics centers on character traits like honesty and empathy to guide behavior.

The chapter outlines five core values central to ethical AI in finance: integrity, customer centricity, social responsibility, collaboration, and innovation. These values are aligned with UNESCO's AI Ethics Principles, including fairness, transparency, and sustainability. For instance, integrity is critical for building trust; customer centricity emphasizes personalized and inclusive services; and social responsibility links financial practices with environmental and societal goals.

Real-world examples illustrate how financial institutions like JPMorgan Chase, HDFC Bank, and HSBC incorporate these values through ethical AI applications. From ensuring fair lending practices to fostering collaboration with diverse stakeholders, these efforts highlight the potential for AI to contribute positively to both financial growth and societal welfare.

The chapter concludes by advocating for robust governance structures, ethical audits, and collaborative industry standards to navigate AI's ethical challenges. By integrating diverse philosophical perspectives and cultural insights, financial institutions can foster trust, accountability, and sustainability, ensuring AI serves as a tool for equitable and responsible innovation in the financial ecosystem.

Chapter 10: The Ethical Flow-Based Education Model: Ethical Use Model of AI and Digitalization in Education

Dalma Lilla Dominek

This chapter presents an innovative framework that integrates artificial intelligence into education through an ethical flow-based pedagogical model. This model, rooted in positive psychology and flow theory, aims to enhance learning by fostering deep engagement and personal growth while maintaining ethical standards.

At the core of the model is the idea that learning should be a collaborative, immersive experience where both students and teachers actively engage. Flow, defined as a state of complete immersion in an activity, is achieved when tasks are optimally challenging and rewarding. The model prioritizes the teacher-student relationship, emphasizing that AI should augment rather than replace human interaction.

The chapter addresses critical questions regarding the ethical use of AI in education. AI technologies can personalize learning, adapt to individual needs, and streamline administrative tasks. However, concerns like data privacy, accessibility, and equity must be carefully managed. Ethical digital content—designed to respect students' rights and foster moral growth—is central to creating a safe, inclusive, and effective learning environment.

Dominek's flow-based pedagogical approach focuses on blending experiential education with ethical digital tools. This synergy helps students achieve a state of flow while promoting critical thinking, creativity, and empathy. The model advocates using project-based tasks and interactive elements like gamification and virtual reality to deepen engagement and build ethical awareness.

The proposed model underscores the importance of transparency, fairness, and inclusivity in AI-enabled education. It suggests that AI can significantly enhance learning outcomes, reduce drop-out rates, and prepare students for evolving workforce demands. Teachers remain vital as facilitators, guiding students through ethical dilemmas and fostering community responsibility.

Ultimately, the chapter envisions a future where AI-driven education balances technological innovation with ethical imperatives, ensuring that learning environments support both personal and collective well-being.

Chapter 11: Debiasing AI in Education through Different Ethical Insights

Harshvardhan Kaul

This chapter examines the role of artificial intelligence in education, highlighting its potential to revolutionize learning while emphasizing the ethical challenges it presents. AI applications like personalized learning, automated grading, and predictive analytics can enhance educational efficiency and inclusivity. However, concerns about bias, transparency, accountability, and privacy must be addressed to ensure equitable outcomes for all students.

The author introduces the FIT-AP framework, focusing on fairness, inclusivity, transparency, accountability, and privacy. These principles aim to guide ethical AI implementation in education. Bias remains a significant issue, often arising from unrepresentative datasets or algorithmic design flaws. To counter this, regular audits, diverse datasets, and transparent decision-making processes are recommended. For example, a personalized learning system must accommodate diverse student backgrounds to avoid reinforcing existing inequalities.

Various ethical philosophies are explored to provide a multifaceted approach to these challenges. Consequentialism focuses on maximizing positive outcomes while minimizing harm. Deontology emphasizes adhering to principles like privacy and informed consent, regardless of outcomes. Virtue ethics advocates designing AI systems that embody fairness and empathy, while care ethics stresses prioritizing students' emotional and psychological well-being.

The chapter also discusses the importance of collaborative efforts among educators, developers, and policymakers to ensure accountability and ethical oversight. Mechanisms for addressing grievances, such as appealing AI-driven decisions, are essential to maintaining trust. Additionally, privacy concerns require robust data protection measures and compliance with regulations like GDPR and India's DPDP Act.

By integrating ethical frameworks into AI design and governance, the chapter argues for a balanced approach that leverages AI's potential while safeguarding human dignity and equity. It envisions AI systems not just as tools for efficiency but as enablers of holistic education, fostering both academic success and moral growth among students.

Chapter 12: AI Ethics in the Military: A Realist Approach Incorporating Traditional and Non-Traditional Security

Arun Teja Polcumpally and Shashank Patel

This chapter examines how emerging technologies, particularly AI, are reshaping military operations and the ethical challenges that accompany this transformation. It navigates the complex relationship between ethical considerations, military necessity, and the evolving nature of warfare in the digital age.

The authors argue that AI's integration into military strategies offers unparalleled capabilities in precision, data processing, and unmanned operations, yet it brings significant risks. These include potential misuse, unintended escalation, and challenges in accountability. While AI can enhance efficiency in logistics, decision-making, and surveillance, its lack of human judgment raises moral concerns, especially in life-or-death scenarios. For example, biases in systems like facial recognition can lead to wrongful targeting, while autonomous decisions could exacerbate unintended civilian harm.

A major focus is the disparity in global approaches to military AI ethics. Some nations, like the US, emphasize traceability, reliability, and governance, while others, such as China, prioritize state ideology over universal principles like equity. These divergences reflect cultural, historical, and political contexts that shape ethical frameworks differently across regions.

The chapter also highlights the dangers of unregulated AI in information warfare, where it can amplify disinformation and destabilize public trust. Such misuse threatens not only military ethics but the broader social contract between governments and citizens.

To address these challenges, the authors advocate for robust, context-sensitive ethical frameworks. These must balance operational effectiveness with human rights and ensure accountability in AI-driven decisions. Establishing specialized military units for AI, fostering collaboration with academia and industry, and integrating ethical oversight into AI's lifecycle are proposed as vital steps forward.

Ultimately, the chapter underscores the necessity of aligning technological advancements with humanity's moral compass, ensuring that military AI serves as a force for security and ethical responsibility, rather than a catalyst for chaos.

Chapter 13: How Will Ethically Correct Religious Robots Be Built?

Selmer Bringsjord

This chapter discusses how robots will be designed and integrated into human families in the future. Many human families in the world over can be accurately classified as *R*, where *R* is a variable ranging over an adjective corresponding to a major religion. For example, there are Hindu families, Jewish families, Jain families, Muslim families, Christian families, and so on.

Additionally, as AI and cognitive-robotics technology continue to advance, many *R*-families will before long employ household cognitive robots whose capabilities approach those of humans to, for example, clean; cook and serve; communicate at the human level in natural language (a requirement for the robust exercise of other capacities in the present list); teach/tutor; protect; dress children, the elderly, the infirm, and care for such groups in various additional ways; transport; and entertain.

Thus, it is natural for a smoothly operating household to consider these as *R*-robots, *and* they would need to be ethically correct, relative to the household in which they work. This goes on to set the general context for the overarching question that drives the present essay. The chapter further delves into how will ethically correct *R*-robots that fit their *R*-families by following the relevant *R*-codes active therein be engineered and sold by relevant companies to these families.

As we read through the chapter, we will go on to see how this will not be an easy question to answer, even when the focus is, for ease of exposition and tractability, placed exclusively on "mere" food preparation and serving within a given household. Nonetheless, under that simplifying focus emphasis on what are the leading logical, general answers to question.

Additionally, the author ventures the claim that there is really only one viable specific answer, which, put very simply, is this: Robotics companies providing such household cognitive robots will need to engineer them with (1) prior knowledge before delivery of relevant religious and ethical content; (2) a capacity to adroitly acquire new knowledge and rational belief dynamically, by valid and verifiable reasoning during communication with members of the households in which they work; and (3) a capacity to reason over the knowledge and rational belief obtained to determine what to do in ethically charged situations. Examples in which (1)–(3) are needed by a household cognitive robot in the dietary domain are provided for illumination.

Part IV: Governance and Social Impact

Chapter 14: Ethical AI through a Feminist Lens: Challenging Power and Redefining Technology

Eva Gengler and Marco Wedel

This chapter critically examines the development and deployment of artificial intelligence from a feminist perspective, focusing on its power dynamics and potential to reinforce societal inequities. It argues that the transformative potential of AI can be realized only if its systemic biases and discriminatory practices are addressed.

The chapter challenges myths surrounding AI, such as fears of autonomous super-intelligence or the notion that AI is inherently objective. It highlights that AI systems operate within the frameworks of human-designed algorithms and datasets, which often reflect existing biases. For instance, facial recognition algorithms tend to misidentify people of color and women more frequently, and recruitment tools can perpetuate gender and racial disparities. These issues underscore the need for transparency, regulation, and public understanding of AI's limitations.

The authors advocate for a feminist approach to AI ethics that emphasizes intersectionality, equity, and inclusivity. Unlike traditional ethical AI frameworks, which focus on transparency and fairness, feminist AI directly addresses power structures and systemic inequalities. It prioritizes the inclusion of marginalized voices in the design process and seeks to challenge patriarchal norms embedded in technology. Feminist AI also links AI development to broader societal goals, such as environmental justice and sustainable development.

The chapter proposes actionable solutions, including the integration of feminist principles into AI regulations and the creation of accountability mechanisms that amplify diverse perspectives. It calls for collective action from governments, corporations, and citizens to build AI systems that serve democratic and equitable purposes. By adopting this approach, AI can become a tool for societal progress, fostering inclusivity, respect, and shared responsibility.

Ultimately, the chapter envisions a future where AI does not merely replicate human inequalities but actively contributes to dismantling them, creating a more just and compassionate technological landscape.

Chapter 15: Feminist Ethics and Artificial Intelligence: Equitable Women's Representation

Vidushi Chaturvedi

This chapter critically explores the role of feminist ethics in shaping a more equitable and just landscape for artificial intelligence. It highlights the unique challenges women face in the AI domain and the transformative potential of their leadership in addressing these issues.

AI is often perceived as neutral, yet its development reflects entrenched gender biases and stereotypes. Women are underrepresented in the field, from research and development to decision-making. This lack of diversity perpetuates inequalities in AI systems, as seen in biased algorithms affecting recruitment, healthcare, and law enforcement. The chapter critiques the dominance of "bro-culture" in technology industries, where values like control and aggression marginalize women and suppress inclusive practices.

Drawing from feminist philosophy and care ethics, the chapter advocates for values like empathy, relational sensitivity, and justice to guide AI development. Feminist ethics emphasizes lived experiences, particularly those of marginalized groups, to uncover and address biases in technology. This approach calls for systemic changes, including gender-sensitive policies, diverse datasets, and transparent algorithmic audits.

The chapter also highlights women's contributions to ethical AI through activism, leadership, and whistleblowing. Figures like Joy Buolamwini, who exposed racial biases in facial recognition, and Frances Haugen, who revealed harmful practices at Facebook, exemplify how women are challenging unethical norms. Feminist ethics integrates intersectional perspectives, recognizing that gender inequalities intersect with race, class, and other social dimensions.

Ultimately, the chapter calls for a transformative shift in AI ethics that centers on inclusion, accountability, and equity. Women's leadership in this space is essential to ensure AI promotes societal well-being rather than deepening existing divides. By incorporating feminist ethics into AI's life cycle, the chapter envisions a technology landscape rooted in compassion, fairness, and shared humanity.

Chapter 16: The Governance of AI from Diverse Ethical Perspectives

Naresh Singh and Lakshya Krishnaiah

This chapter examines how culturally diverse ethical insights can shape governance frameworks for artificial intelligence, with the aim of fostering sustainable and inclusive futures. It advocates for moving beyond Western-centric governance models by integrating ethical principles from a wide range of cultural philosophies to address the complex challenges posed by AI technologies.

The chapter underscores the need for governance models that balance innovation with ethical responsibility, especially given AI's potential to exacerbate inequality or resolve persistent societal challenges. Current governance approaches vary widely: the US adopts a market-driven, decentralized strategy prioritizing innovation, while the European Union emphasizes comprehensive regulation with its Artificial Intelligence Act. Emerging economies, such as India, Brazil, and South Africa, focus on leveraging AI to address developmental challenges but often face resource and regulatory limitations.

Drawing on cultural philosophies, the chapter highlights how ethical traditions like Jainism, Buddhism, Vedanta, Ubuntu, Confucianism, and Indigenous knowledge can guide AI governance. For example, Jainism's principles of non-violence (*Ahimsa*) and multiple perspectives (*Anekantavada*) promote harm reduction and inclusivity. Ubuntu emphasizes community well-being and mutual respect, advocating for equitable distribution of AI's benefits. Indigenous philosophies stress environmental stewardship, urging governance frameworks to consider sustainability and long-term impacts.

The authors propose adaptive governance strategies inspired by complex systems theory, which account for AI's unpredictability and emergent properties. Anticipatory, flexible models can address governance gaps, ensuring accountability and public trust while incorporating diverse global values.

Ultimately, the chapter calls for international cooperation, public engagement, and the integration of multicultural ethical principles into governance frameworks. By doing so, it envisions AI as a tool for global equity and sustainable development, fostering a future where technological progress aligns with the diverse values and aspirations of humanity.

Chapter 17: "Hypersuasion" and Beyond: Contending with Regulatory Gaps in Modern AI Systems through a Pluri-Ethical Perspective

Krishna Deo Singh Chauhan

This chapter addresses the growing phenomenon of "hypersuasion," where AI systems manipulate human behavior through personalized and persuasive content. Drawing from the Indian philosophical text *Yoga-Vashishtha*, it proposes a framework to navigate and counteract these manipulations while enhancing human autonomy.

The author defines hypersuasion as a significant evolution in persuasive technology, driven by AI's capacity to process vast data and craft highly targeted content. This creates ethical concerns, especially regarding individual autonomy and societal impacts like democratic manipulation. While existing Eurocentric frameworks rely on rights-based approaches, such as privacy protection and mental self-determination, the chapter critiques their limitations in addressing AI's subtle yet pervasive influence.

Instead, it introduces the principles of *Yoga-Vashishtha* to provide a holistic response. Key concepts include detachment (*vairagya*), deep contemplation (*vichara*), and mental stillness (*santi*), which equip individuals to resist manipulative content. Recognizing the illusory nature of reality (*maya*), practitioners can develop resilience against AI-driven distortions of perception, such as fake news and deepfakes, by cultivating discernment and contentment (*santosha*). These practices help individuals maintain clarity amidst the overwhelming and often deceptive digital environment.

The text also explores the potential for AI to support self-realization (*atma-jnana*) by revealing unconscious biases and enabling mindful engagement. When used ethically, AI can enhance human autonomy, acting as a tool for personal growth rather than manipulation. The chapter argues for embedding these philosophical principles into the design of AI systems, advocating for features that promote transparency, user agency, and reflection.

Ultimately, the chapter envisions a world where AI systems align with ethical frameworks rooted in self-awareness and human dignity. By integrating ancient wisdom with modern challenges, it offers a path to harness AI's potential for societal benefit while safeguarding individual freedom and well-being.

Chapter 18: AI, Social Media Platforms, Decentralization, and Regulatory Dilemmas: An EU Perspective

Alex Grech, Joshua Ellul, Ioannis Revolidis, and Jacob Grech

This chapter explores the evolving relationships between artificial intelligence, social media, and decentralization, focusing on European Union regulatory frameworks. It highlights the paradoxical transformation of the internet from a democratizing force to a centralized, profit-driven ecosystem dominated by tech giants. AI, with its unprecedented ability to manipulate content and amplify misinformation, further complicates this landscape, demanding innovative governance approaches.

Social media platforms have shifted from empowering individuals to becoming arbiters of public discourse, often exploiting user data and polarizing societies. Generative AI exacerbates these challenges by enabling large-scale content manipulation, posing threats to democratic systems and societal trust. Simultaneously, decentralization movements, including blockchain and peer-to-peer networks, seek to dismantle monopolistic control and return power to users. However, these systems raise concerns about governance, accountability, and ethical oversight.

The EU's regulatory response, including the Digital Services Act (DSA) and the forthcoming AI Act, represents a significant effort to balance technological innovation with democratic values. The DSA focuses on transparency, accountability, and content moderation, while the AI Act aims to ensure ethical AI development. However, both face limitations in addressing the nuanced challenges posed by AI-driven content generation and decentralized platforms.

The chapter also delves into philosophical and ethical frameworks underpinning EU regulation, emphasizing values like human autonomy, fairness, and explicability. These principles guide the EU's efforts to align AI and social media governance with human rights and democratic integrity. However, tensions persist between regional regulations and global tech realities, highlighting the need for international collaboration.

Ultimately, the chapter underscores the urgency of adaptive regulatory models, ethical innovation, and multi-stakeholder cooperation to navigate the complex interplay between AI, decentralization, and governance. The EU's approach serves as a blueprint for fostering equitable digital ecosystems while preserving democratic ideals and societal well-being.

Chapter 19: AI Ethics and Societal Well-being: Perspectives from the Global North and South

Ram B. Ramachandran and Naresh Singh (as interlocutors)

To deal with rapid change in the field of AI and AI ethics in particular, as well as to be as comprehensive and inclusive as possible, the editors of this volume adopted an innovative approach toward the end of the chapters' preparation in having semi-structured online conversations with diverse voices from the Global North and South. The discussions recognized that while AI had the potential to drive significant advancements, it also raised critical ethical questions that required inputs from various stakeholders, including philosophers, technologists, and implementers. This chapter summarizes the insights received.

Foundational issues included data security, privacy concerns, and bias elimination. It was argued that while optimizing for user engagement is important, it should not come at the expense of promoting societal well-being. The establishment of universal ethical guardrails rooted in shared values across cultures and religions was suggested as a means to guide responsible AI development. The discussions further highlighted that many marginalized peoples of the world are labeled as underdeveloped despite their resilience and creativity, and recommended inclusive narratives that prioritize human values and collective well-being over solely market-driven incentives.

One of the world's important philosophical ethical sources not covered in the other chapters is the African philosophical system of Ubuntu with its well-known articulation "I am because we are", which is often contrasted with the Western "I think therefore I am". This radically alternative vision of the world of relationality vs rationality can bring deep insights into AI ethical systems which tend see societal well-being in individual economistic terms.

Many other insights rich with guidance and wisdom covering a diverse range of salient and topical issues from human trafficking to sustainability issues to the visionary leadership required for responsible AI are captured.

The chapter ends with a conclusion capturing the essence of the conversations as well as the main thrust of the book.

REFERENCES

Kissinger, H. A., Schmidt, E., & Huttenlocher, D. (2021). *The age of AI and our human future.* Hatchette Books.

Tippett, K., Metzinger, T., Thompson, E., & van Lommel, P. (2011). To be or not to be: The self as illusion. *Annals of the New York Academy of Sciences*, *1234*(1). https://nyaspubs.onlinelibrary.wiley.com/doi/epdf/10.1111/j.1749-6632.2011.06181.x

Part I

Foundations and Future of AI Ethics

1 A Review of the Current and Future Roles of AI

Ram B. Ramachandran, Mahek Saket Shah, and Priyanka Paani

INTRODUCTION

The advent of artificial intelligence (AI) is a landmark moment in human history, similar to the transformative impacts of electricity and fire, which brought unprecedented changes in the way humans live, work, and interact with each other and the environment. With widespread AI development and transformation taking place, it promises to bring about similar changes, having an impact on personal and commercial lives of individuals via adoptions in thought processes, labor force, and other social constructs of a society

As we look back in time, the impacts of electricity and fire were ground breaking. For instance, with the advent of electricity, the focus in industrial and commercial spheres changed from labor-intensive production tasks to highly efficient and defect-free production cycles. Additionally, it served as a basis for the development of today's day-to-day necessities like light, heating and cooling systems, communication media and more. Likewise, AI is set to leverage on large datasets of information to identify patterns and similarities in order to ease the human decision-making process while significantly impacting human cognitive abilities positively or negatively depending on usage pattern. This shift will push us to reconsider how we think, learn and work due to the technology's ability to learn and adapt autonomously.

We look forward to the rapid advancements in AI in the years 2025–2030. It is believed that the development will focus primarily on sophisticated deep-learning models which will redirect human–AI collaboration to a rather autonomous mode. Emerging large language models (LLMs) will be able to autonomously update their parameters post-deployment, reducing retraining costs and improving adaptability. This innovation, seen in models like Writer's, could drive advancements toward artificial general intelligence (AGI). Another advancement will be in the area of real-time learning models which will strengthen their inference engines to enable more relevant engagement with the users. Multimodal models are at the cusp of technological breakthroughs. Users will be able to leverage LLMs that can seamlessly integrate videos, audios, and images with text to provide a more versatile interaction with users. The next several years will see a flurry of application-specific narrow inference engines that will drive efficiency and innovation. For example, individuals will be able to upload their health records to get specific recommendations; entrepreneurs will be able to upload their business plan and financials to get valuations; or we can upload travel patterns and transportation usage, and these models can provide targeted suggestions on where to stay, who to meet, how to travel, etc.

The main factor inhibiting growth in AI is energy. This friction is being addressed through nuclear energy. Many of the large technology behemoths and other corporations are investing in nuclear power plants. This in and of itself will lead to other challenges, including environmental impact, transportation and sites for nuclear waste disposal, security, emergency response, geopolitical upmanship, etc. The impact of quantum computing on energy demands is not fully understood as of now. This could exacerbate the growing energy demands.

In addition to these sustainability challenges, AI advancements bring with them additional ethical considerations. Some of the considerations revolve around who would be accountable for the

DOI: 10.1201/9781003585527-3

decisions made by AI, to what extent would the decision-making processes be autonomous, how would the advancement impact human creativity and productivity, etc. Thus, government agencies, industries, and societies will have to come together to assemble a framework to counter the considerations mentioned above.

Further into the future, around 2030–2040, the emergence of artificial general intelligence would take place due to the rapid growth in machine learning (ML) and deep-learning models. In lieu of AGI's human-like capabilities in completing tasks, it is also called strong AI or human-level AI (Hillary, 2023). The shift to AGI will be coupled with ethical considerations surrounding the potential lack of transparency and misuse of the technology by individuals or groups with malicious intentions along with questions about the impact it would have on an individual's privacy. Further, the technology will bring with it the potential to disrupt labor forces, further pushing the narrative of AI replacing humans. If not taken care of efficiently, it would have a devastating influence on society.

In just about 15 years from now, we could expect a futuristic fantasy to become a potential reality through the development of artificial super intelligence (ASI). ASI comes along after the maturation of AGI and brings with it cutting-edge cognitive functions and highly developed thinking processes which are believed to be more advanced than that of any human (Mucci & Stryker, 2024). This section of the chapter will not only dive into the ethical considerations but also discuss various emerging philosophies to explain how and which human values can be leveraged on to enhance collaboration with ASI.

HISTORY AND CURRENT STATE OF AI

Though the idea of AI seems rather recent, it has been in the works from the 1950s and was initially used by philosophers considering the ideas of life and death. Today AI, with its most recent developments, serves as virtual assistants, search engines, etc. AI has gone from being a theoretical concept to becoming an essential part of our tech-driven world today.

Early Foundations of AI: The conceptual roots of AI can be traced back to the 18th century with Thomas Bayes' work on probability, which laid the groundwork for decision-making processes in uncertain situations (Maryville Online, 2024). However, the field began to take shape in the mid-20th century. In 1950, Alan Turing published "Computing Machinery and Intelligence," introducing the Turing Test as a means to evaluate a machine's ability to exhibit intelligent behavior indistinguishable from that of a human (IBM, 2024). This seminal work prompted further exploration into machine intelligence. The term "artificial intelligence" was coined in 1956 during the Dartmouth Conference, organized by John McCarthy, who is often referred to as the "father of AI." This conference marked the formal beginning of AI as a distinct field of study (Calls9, 2023). During this period, significant developments included the creation of the first AI programs capable of playing games like chess and checkers. Arthur Samuel developed a checkers-playing program that could learn from experience, marking it one of the earliest instances of machine learning (Maryville Online, 2024).

Rise and Challenges of AI: The late 1950s and 1960s witnessed rapid advancements in AI research. Notable innovations included ELIZA, an early natural language processing program that could engage in conversation with users (IBM, 2024). However, this era also faced challenges. The initial excitement surrounding AI led to inflated expectations, resulting in what is known as "AI winter" during the 1970s and 1980s. Funding and interest in AI research dwindled as projects failed to deliver on their promises (TechTarget, 2024). Despite these setbacks, the 1980s saw a resurgence of interest in AI, driven by advancements in expert systems—programs designed to mimic human decision-making in specific domains. This period also marked the development of neural networks and backpropagation algorithms, which allowed machines to learn from data more effectively

(Calls9, 2023). The introduction of LISP by McCarthy became a pivotal programming language for AI research during this time (Maryville Online, 2024).

Modern Developments and Current State: The late 1990s and early 2000s heralded a new era for AI with significant breakthroughs. In 1997, IBM's Deep Blue famously defeated world chess champion Garry Kasparov, showcasing the potential of AI in strategic thinking (Tableau, 2024). The following years saw increased integration of AI into everyday technologies. For instance, speech recognition systems became commercially available, and autonomous robots like Roomba entered homes (TechTarget, 2024). In recent years, advancements in deep learning and big data have propelled AI into new realms. The development of models like OpenAI's GPT-4o has demonstrated remarkable capabilities in natural language processing, enabling machines to generate human-like text (Tableau, 2024). Moreover, applications such as facial recognition technology and autonomous vehicles have entered mainstream use. However, such advancements bring with it a plethora of regulatory and ethical challenges that one must think of when engaging with AI. While serious thinkers consider the potentials and pitfalls of the growing popularity of AI, it has also led to pop culture memes and ridicule. Jerry Seinfeld, the noted comedian remarked, "We were smart enough to invent AI, dumb enough to need it, and so stupid we can't figure out if we did the right thing" (*The Tonight Show Starring Jimmy Fallon*, 2024).

Today's AI landscape is characterized by its pervasive presence across various sectors, including healthcare, finance, transportation, and entertainment. Companies are leveraging AI for predictive analytics, personalized recommendations, and automation of routine tasks (IBM, 2024).

Having seen the growth of AI over the years, there is no doubt that the transformative power of AI will extend across various sectors, affecting thousands of lives. In the next section, we will be elucidating on the profound impact AI will have in reshaping industries like healthcare, manufacturing, and transportation. These sectors, central to human well-being, economic growth, and technological progress, are leveraging AI to achieve unprecedented advancements. From enhancing patient care and operational efficiency to redefining mobility and manufacturing processes, AI is at the forefront of revolutionizing these industries.

HEALTHCARE

Current

Advancements: At present, AI technologies are poised to revolutionize diagnostics and treatment protocols. With the rise in biosensors transmitting physiological data, digital health records and accelerated vaccine development, change is long due. Additionally, AI-based algorithms through AI computing can analyze medical images with a high degree of accuracy and may be able to identify conditions earlier than human practitioners. This could lead to more tailored treatment plans and better patient outcomes (Pinto-Coelho, 2023).

Ethical Advancements: However, with these advancements, there comes the need to establish a balance between privacy, safety, and addressing the potential impact on mental health and social habits. The book *AI 2024: Ten Visions for our Future* further highlights the Covid-19 generation who constantly fear the likelihood of an ongoing epidemic while relying heavily on technology to keep them safe and finding it harder to form social connections and experience life beyond their homes. This is where the rise of AI will have a ground-breaking effect on healthcare for this generation, shaping their lives in unprecedented space with a transformation not only in treatments but also preventive measures. For instance, smartphones currently help in calculating infection risk, with apps being developed to inform users of an infected person's proximity. As AI technology advances, these apps could become more prevalent, leading to challenging debates around privacy and safety. However, such advancements have a social and welfare effect on individuals. Many individuals choose to remain cautious and isolated, with AI technology facilitating changes in lifestyle such as delivery bots (Lee, 2020). However, some aspects of human life necessitate human contact like love and emotional connection, which robots cannot currently replicate. This can result in lasting impact on mental health and human connections.

2025–2030

AI investment in healthcare is projected to grow significantly, from about $20 billion in 2024 to $150 billion over the next five years (Maerz, 2024). In 2025, we can expect to see the continued expansion of features like AI-supported clinical documentation and patient communication tools, as well as new, innovative use cases aimed at saving providers time and bolstering data analytics capabilities for healthcare facilities. Coming years also predict discoveries of new healthcare AI use cases coming into play, with more enterprise-scale healthcare organizations investing in AI integration for clinical decision support and machine learning to help analyze and identify patterns in large and complex sets of raw data such as imaging scans and labs. Beyond 2025, AI in healthcare could bring enhancements to patient–provider communication using machine learning to create personalized care plans as well as personalized patient outreach and follow-up.

Advancements: The advent of brain–computer interfaces (BCIs) represents a significant leap in healthcare innovation, as it is poised to redefine patient interaction with technology over the next five years. Currently, two types of BCIs exist—non-invasive and invasive. Non-invasive BCIs are being developed by relying on electroencephalography (EEG) which is a common process used in analyzing brain activity. A company known for its consumer-friendly EEG product is Neurosky. Additionally, research is being done on invasive BCI which involves implanting a chip in a patient's brain, it is believed to enable direct control of devices or artificial limbs through thought alone, offering profound benefits for individuals with severe physical disabilities and enhancing their independence and quality of life, and it is promised to revolutionize the healthcare space (Vaidyanathan, 2024).

Ethical Advancements: Many believe that BCIs are at the peak of ethical issues due to their invasive and command-oriented nature over the human body which could potentially be misused (Burwell et al., 2017). Further, since BCIs are a direct collaboration of humans with machines, questions pertaining to individuality come about due to the chances of a machine manipulating the thoughts and actions of individuals. Such a thought makes us wonder whether the BCI would be considered a unit of the human body or just a tool. Lastly, it is a growing concern that BCIs are being used to influence social stigmas around disabilities. This arises with people adopting BCIs to primarily combat the negative stigma attached to diseases and disabilities in society. Lastly, key ethical themes include responsibility, autonomy, and privacy. For instance, if a BCI device has a causal role in decision-making of the individual, this could negatively affect autonomy.

2030–2040

The next decade is predicted to be the age of artificial general intelligence, also called strong AI, in which AI is expected to reach unprecedented heights (LaDolce, 2024). But what exactly is this AGI? It is an AI system with human-level cognitive abilities, capable of understanding, learning, and generalizing the knowledge it has gained and autonomously applying it across many domains—with a capacity to handle complex tasks without constant input by humans, unlike today's narrow AI (Goertzel, 2014). Besides holding such immense potential, AI's advancements introduce complex ethical considerations. Privacy, a fundamental right, itself could be under threat. As AI systems monitor humans and urban environments and offer hyper-personalized services, the boundary between surveillance and monitoring for necessary purposes become blurred. Additionally, emotion AI, also known as affective computing, will play a crucial role in enabling AGI and other systems to detect and interpret human emotions through facial expressions, body language, and other cues and respond to them accurately (The Princeton Review, n.d.). AI-driven neuroscience can help reduce genetic-level imperfections that lead to mental illnesses. Fine-grained neural measurements and interventions can also be used to alter behavioral patterns in human beings (Amodei, 2024).

Advancements: The advancements in autonomous AI and real-time data analysis through ML in healthcare carry significant analytical implications that extend beyond mere technological progress. First, these innovations enhance diagnostic accuracy and speed, allowing healthcare professionals to analyze vast amounts of medical data quickly, which can lead to earlier detection of diseases and improved patient outcomes. This capability not only reduces diagnostic errors but also streamlines the decision-making process, enabling faster treatment interventions that can be critical in acute medical situations. Additionally, AI's ability to automate routine tasks alleviates the administrative burden on healthcare providers, allowing them to focus more on direct patient care and reducing staff burnout—a crucial factor in maintaining a sustainable healthcare workforce. Furthermore, the use of AI in clinical laboratories promises increased efficiency and precision in diagnostic processes, which is vital for managing infectious diseases effectively. By optimizing workflows and improving the accuracy of laboratory results, AI can significantly contribute to better treatment regimens and higher cure rates for various conditions (Al Alowais et al., 2023).

Ethical Advancements: With such advancements, ethical considerations would need to be tackled too, starting with auditability. When designers and implementers of AI and ML advancements clearly declare the intentions, indications for use, and goals for an application, clinicians, patients, regulators, and other stakeholders are better enabled to exercise their own evaluative and decisional autonomy. Without such transparency stakeholders will not be able to decide for themselves whether they want to support these intentions or not. Thus, to support evaluative autonomy, transparency will require auditability: AI and ML systems in healthcare can have an explainable architecture, designed to align with human cognitive decision-making processes familiar to physicians, and directly tied to clinical evidence. Any functioning and output will need to be interpretable to any stakeholder who uses the output to inform clinical decisions so that they can evaluate whether the AI and ML systems are likely to live up to the stated intentions (Char et al., 2020).

2040–2050

By 2040, it is estimated that AGI will attain a maturation stage, and ASI will begin to emerge. ASI is a hypothetical future stage of AI exceeding human intelligence across all aspects, potentially dominating economically valuable tasks, and venturing into areas traditionally considered human strengths like emotional intelligence, artistic creation, and innovative problem solving. These machines would feature highly advanced reasoning, decision-making, and problem-solving capabilities, exceeding human capabilities in creativity and logic (Selvi & Ghosh, 2020).

By this era, the integration of AI elements with quantum computing will be highly advanced. Most AI systems would likely be operating on quantum computing as it accelerates data processing and allows for much faster and more accurate decision-making across complex, real-world scenarios (Nature, n.d.).

Advancements: The advancements in genetics, genomics, and molecular medicine are set to revolutionize healthcare through the development of personalized and precision medicine, with significant analytical implications. By analyzing an individual's genetic makeup, environmental factors, and lifestyle choices, healthcare providers can establish tailored treatment plans that enhance effectiveness and minimize side effects. This shift from a one-size-fits-all approach to personalized medicine enables early disease risk identification through advanced diagnostics and comprehensive health records, ultimately improving patient outcomes. The integration of AI and machine learning algorithms is critical in this evolution, as these technologies can process vast datasets to detect patterns, predict disease progression, and recommend individualized treatments with unprecedented accuracy. For instance, a KPMG report states ASI could be utilized to further the genetic engineering domain by analyzing genomic data of individuals and identifying the specific genes linked with diseases. Further, advancements in the telemedicine and remote care domain will allow a smoother patient experience and would be embedded with advanced communication technologies as well as remote monitoring through implantable sensors (Cheema et al., 2024).

Ethical Advancements: One of the biggest ethical considerations is the concerns surrounding privacy. As mentioned, the time frame involves advancements in personalized healthcare as well as remote care in order to enhance a patient's healthcare. But for this to take place, large amounts of data would need to be collected for analysis purposes, and doing so would infringe on the personal space of the individual. Additionally, in incorporating such developments, questions also arise about the transparency in collecting the data with the consent of the individual. Thus, strict regulations and rules need to be implemented for us to not complicate the balance between personalization and individuality.

TRANSPORTATION

CURRENT

In the current landscape of the amalgamation of AI and automotive, there is a name heard more than ever: Elon Musk. He is a pioneer in the industry with his company Tesla. Marked by visionary thinking and a profound sense of responsibility, he vouches that the automotive industry is undergoing a major transformation which is primarily driven by AI. Such a shift is not due to the mere impact of technological advancements but rather due to the change in the conceptualization, design and experience of how vehicles are meant to be. Some of the technologies which are being used today in automotives include Internet of Things (IoT), natural language processing, deep learning and machine learning.

Advancements: The advancements in the automotive space are not only related to enhancing the driving experience but also in improving the back end work involving design, prototype, production, and engineering (Webadmin, 2024). For instance, with regard to the driving experience, AI can adapt and respond to individual driving styles, preferences, and environmental concerns. Further, application of AI in the automotive industry includes the generation of detailed 3D maps of the environment, obstacle identification, and congestion prediction. Voice-activated sensors are also provided in vehicles to allow passengers to interact with the vehicle. This application makes use of natural language processing to ease the driving experience. Related to the back end operations enhanced through AI is the analysis of historical design data to develop innovative prototypes which appeal to the aesthetics of the upcoming generations. In totality, these technologies come together to create an ecosystem that improves safety, efficiency, and user experiences.

2025–2030

The automotive market is expected to be worth an astonishing $35.71 billion by 2030 (Zoting, 2024). This growth demonstrates AI's significance and its fundamental impact on the industry. As we look toward the future, autonomous vehicles (AVs) are reshaping the way we commute.

Advancements: The anticipated advancements in self-driving cars and drone technology by 2030 carry significant analytical implications for transportation dynamics and service delivery (Deichmann et al., 2023). The introduction of self-driving vehicles promises to enhance road safety, with studies indicating that advanced driver-assistance systems could reduce traffic accidents by 15% in Europe alone. This reduction in accidents is expected not only to save lives but also to lower insurance costs and healthcare expenses associated with traffic injuries. Furthermore, the increased productivity during commutes allows passengers to engage in other activities, potentially boosting overall economic productivity as individuals can utilize travel time more effectively. Waymo's expansion of its robotaxi service across the US highlights this shift, particularly among demographics such as women who prioritize safety when travelling alone (Elias, 2024). In parallel with these developments is the rise of drone technology. Initially utilized for air shows and military defense purposes, drones are now poised to revolutionize delivery services by offering faster and more cost-effective solutions. The market for drone delivery is projected to reach $1.9 billion by 2025, driven by consumer demand for instant gratification in service delivery (Edulakanti & Ganguly, 2023).

Ethical Advancements: Ethical dilemmas arise regarding liability in accidents involving self-driving cars. Determining who is responsible—manufacturers, software developers, or vehicle owners—poses significant challenges. Additionally, the impact on employment within driving professions raises concerns about job displacement and the need for retraining programs for affected workers. As AVs become more prevalent, these issues will require careful regulatory frameworks to ensure accountability and support for displaced workers. Similarly, the growth of drone technology introduces ethical concerns related to surveillance and privacy violations. As drones become integral to delivery services, their potential for invasive monitoring may infringe on individuals' privacy rights. Policymakers must address these issues by implementing regulations that protect personal freedoms while allowing for technological advancement. Furthermore, the rapid adoption of drones could lead to job losses in traditional delivery sectors, necessitating compensation or skill-enhancement programs, particularly in economically vulnerable regions. Balancing innovation with ethical considerations will be crucial as these technologies reshape industries and societal norms (Truxcargo, 2024).

2030–2040

The next decade is predicted to be the age of AGI or strong AI. The ongoing advancements in AI, as stated in the Gartner Hype Cycle of emerging technologies, include large action models, autonomous agents, humanoid working robots, quantum computing, and other advancements that are laying the groundwork for AGI (LaDolce, 2024).

Advancements: It is believed that all players will be applying AI in all their functional units by 2040. The research and development (R&D) process, for instance, may be driven by AI models, with humans focusing on monitoring output quality—leading to efficiency improvements of 30% to 50% in an optimistic scenario. In sales and marketing, new AI-driven solutions will enable hyper-individualized marketing strategies, reduced time to market, increased revenue potential, and customer behavior-based sales models. In after-sales, AI can improve customer experience through predictive maintenance and remote diagnostics, while boosting companies' financial performance thanks to real-time inventory management, demand forecasting, and optimized fleet management. Significant reallocation in resources is anticipated, especially as companies increase their focus on software development over traditional mechanical enhancements ("Artificial Intelligence," 2024). By 2035, it's projected that R&D allocations for digital technologies will soar from

21% to 58% (Ferguson, 2024). This pivot is crucial not only for staying competitive but also for ensuring that new vehicles meet evolving consumer expectations and regulatory standards. A striking example of the evolution of the automotive sector is self-driving vehicles redefining transportation and urban living infrastructure. AI-controlled autonomous cars, buses, trains, and drones, and even flying transport solutions could navigate cities with efficiency, reducing traffic and minimizing accidents.

Ethical Advancements: With such integration of AI in autonomous vehicles significant ethical considerations arise that necessitate appropriate frameworks and regulations. Some of the ethical considerations include the emergency response of the autonomous vehicle, which could lead to an ethical dilemma if the two possibilities are crashing with another vehicle or swerving into a pedestrian. Further questions revolve around the liability aspect as to who is to be held responsible, whether the manufacturer, the AI software development company, or the vehicle owner. The possible ethical guidelines and frameworks that can be developed should ideally involve public values and inputs in the decision-making, building mechanisms for accountability and transparency in case of any malfunctions, and lastly, public perceptions need to be altered to ensure widespread adoption and acceptance (FutureWebAI, 2023).

2040–2050

According to the projected population released by the United Nations, the world population is growing but at a slower pace than in the past. It is estimated to reach 9 billion by 2040 (Schröder-Hinrichs et al., 2019). With such an increase in population, there needs to be a counter-effect in the growth and efficiency of the automotive industry. Apart from road travel, the maritime sector is also aimed at growing through automation and technological innovations. However, such development would require strong regulations, infrastructure and competences.

Advancements: ASI is expected to grow and replace AGI by 2040, and such development would entail changes in the automotive industry, and ones that are not restricted to cars alone. For instance, the shipping industry is bound to undergo changes due to the analytical implications ASI will have on the maritime and transportation sector. Countries like Australia, China, and the USA have engaged in discussion about regulating autonomous ships, but the lack of a comprehensive regulatory strategy for maritime transport hinders its alignment with technology (Schröder-Hinrichs et al., 2019). A proposed framework by the International Maritime Organization (IMO) aims to address safety, training, and legal liabilities related to autonomous vehicles. Apart from maritime, the aviation industry also has a lot in store. For instance, the ground vehicles used would be reduced by 50% to 60% in Asia, Europe, and North America due to increased efficiencies (Schröder-Hinrichs et al., 2019).

Additionally, the shift to ASI will also lead to enhanced capabilities in the sectors like supply chain management and service delivery. Further, with ASI's ability to exceed humans, there is no doubt that it will be able to incorporate emotional intelligence and innovative problem solving to transform business models and customer interactions (Selvi & Ghosh, 2020). Furthermore, the integration of AI with quantum computing will enable faster and more accurate decision-making processes in complex scenarios, enhancing operational efficiencies across industries (Nature, n.d.).

Ethical Advancements: Ethical concerns in the automotive industry arise regarding the reliability of these technologies, especially in emergency situations where human intervention may be necessary. The potential for accidents or malfunctions can be eliminated through rigorous testing and transparent reporting of safety records. However, there is never a

100% surety about the same. Further, the automation brings with it significant reduction in the number of ground vehicles as well as individuals handling manufacturing potentially displacing many workers in the industry. Thus, ethically it is imperative to assist the displaced workers through retraining and development of strategies for job opportunities in new sectors.

MANUFACTURING

Current

Advancements: AI is also going to redefine industries from manufacturing to services by automating processes. With the help of AI, humans can be freed from mundane tasks and be allocated more creative and strategic tasks. For instance, in manufacturing, robots can take over simple, repetitive tasks of assembly line production and give humans the free time to work on quality and innovation. Such changes may even reshape the workforce dramatically: some jobs will disappear, but new opportunities will open that require only human capabilities—creativity, empathy, or complex problem solving (George, 2024). Such development is taking place particularly within the context of Industry 4.0. One of the most impactful applications in the industry is in terms of predictive maintenance. AI systems analyze data from sensors on machinery to forecast failures before they occur, reducing unexpected downtimes and maintenance costs. AI also powers advanced quality control through computer vision systems, which scan products in real time to identify defects.

Ethical advancements: In terms of ethical considerations, humans are not the only source of AI biases—data can also contain biases that your AI system will feed upon. Unlike human intelligence, AI is not capable of autonomous, morally conscious rectification, if it is not explicitly taught that it made mistakes and the data it fed upon was harmful.

2025–2030

By 2025, 89% of companies plan to implement AI in their production processes (Jain, 2024). Early adopters are already seeing a 14% drop in manufacturing costs, and this is just the beginning. AI has the potential to take efficiency, cost-saving, and innovation to the next level. According to a McKinsey report, AI could add up to $2 trillion to the manufacturing industry by 2030. Additionally, AI-driven predictive maintenance can reduce maintenance costs by 10%–12%, highlighting its potential for cost savings.

Advancements: The anticipated advancements in collaborative robots (cobots) and AI-driven energy management systems present significant analytical implications for the manufacturing sector. Cobots, designed to work alongside human laborers, are projected to grow from $1.9 billion in 2024 to $11.8 billion by 2030, reflecting a broader trend toward automation that complements human intelligence rather than replacing it (MarketsandMarkets, n.d.). With repetitive tasks becoming automated, there will be a shift in the way humans would engage with the manufacturing process. They would be required to carry out more complex, strategic responsibilities which AI would not be capable of doing just yet. Such a change would require humans to develop a new set of skills in order to leverage the newer opportunities. Apart from a shift in human labor management, AI can also be used to ensure efficient allocation of resources, which in turn can result in potential cost-saving benefits. A typical application of AI systems would be to monitor energy usage across factories, identify the inefficient processes, and optimize the operations. Thus, as more and more manufacturers adopt such integration, they can build a competitive edge while focusing also on sustainability.

Ethical advancements: With respect to the ethical considerations, cobots are increasingly used in industrial and non-industrial environments. The psychological factor is also takes into account how they affect the people with whom they share space and tasks. The British Standard addresses ethical issues in social, application, commercial/financial, and environmental terms.

2030–2040

Advancements: Between 2030 and 2040, the development of AGI and advancements in ML are expected to profoundly transform manufacturing processes. The rise of fully automated "smart factories" will enable machines to self-monitor, maintain, and optimize operations independently, leading to increased production efficiency, reduced downtime, and enhanced output quality. Predictive maintenance capabilities will be significantly improved as ML algorithms analyze vast datasets from IoT devices, allowing manufacturers to anticipate equipment failures before they occur, thus minimizing costly interruptions (Immerman, 2024). Additionally, AGI systems will surpass human capabilities in quality control through advanced object recognition and anomaly-detection techniques, ensuring higher standards of product quality. Process optimization will be revolutionized as AGI analyzes complex datasets quickly to determine the most efficient production schedules and resource allocations, resulting in linear operations that reduce waste and costs. While AGI may automate many tasks currently performed by humans, it will also create new roles focused on managing and collaborating with AI systems, necessitating workforce reskilling initiatives.

The development of AGI also brings with it limited development in quantum computing that can be leveraged to improve industries like supply chain management and logistics since it assists with solving complex optimization problems faster than traditional computing methods (Sullivan, 2024). Overall, this transformative era in manufacturing will redefine operational paradigms, enhancing efficiency and productivity while reshaping the workforce landscape.

Thus, the integration with emerging technologies such as quantum computing and advanced robotics will create synergies that push the boundaries of automation and data analysis.

Ethical advancements: The advancements in AGI and ML systems are accompanied by critical ethical considerations. One of the major concerns is surrounding the autonomy provided to these systems to develop decisions without human interference. This concern is also correlated with the accountability aspect of the outcomes as a result of the autonomy. Additionally, in terms of the "AI replacing humans" debate, there is going to be a significant amount of job displacement, particularly in the low-skilled sectors, further widening inequalities and encouraging discussions around solutions like universal basic income. Lastly, working with humans is a completely different experience than working with a machine, and since humans are social beings, the latter does take a toll on the emotional and psychological state of humans.

2040–2050

Advancements: From 2040–2050, there will be the emergence of artificial general intelligence, artificial super intelligence, and quantum computing. It has already been discussed that between 2040 and 2050 the advances in construction methods and processes are projected to be unprecedented. Built on ASI principles, these systems will not only be super intelligent but will also be capable of real-time alterations of the core production processes to increase efficiency. In turn, this brings about reduction in waste, better resource usage, and better quality of products and services, and it also factors into the myriad of environmental considerations that most, if not all, businesses are required to adhere to. Most importantly, these features and functionalities will extend the boundaries of these core business processes by being able to solve advanced optimization problems which today are nearly impossible through a single computer system. This may open doors for new innovations in the field of materials science, allowing the creation of new materials with preferred characteristics, leading to better product performance and sustainability. These limits will set new standards and will make it possible to market entire production lines and global supply chains. It can also enhance and ease the production of highly customized goods that these technologies and products will make it possible to create.

Ethical advancements: Ethical considerations come along with the advancements which take place. Privacy concerns are the prime issues which arise because these technologies rely heavily on large amounts of personal and sensitive data, and thus regulations need to be in place to ensure that there are low chances of breaches occurring. Finally, the concentration of power in the hands of a few corporations or governments may lead to ethical concerns regarding their influence over society, highlighting the need for equitable access to technology and regulatory frameworks that promote ethical practices. Addressing these considerations will be vital to ensuring that the advancements benefit society as a whole.

CONCLUSION

In the book *Novacene*, Jame Lovelock, says "[N]ew beings will emerge from existing artificial intelligence systems. They will think 10,000 times faster than we do and they will regard us as we now regard plants. But this will not be a cruel, violent machine takeover of a planet imagined by science fiction. These hyper-intelligent beings will be as dependent on the health of the planet as we are. They will need the planetary cooling systems to defend them from the increasing heat of the Sun as much as we do" (*Novacene*, n.d.).

As we journey through the transformative potential of artificial intelligence, this chapter has unraveled the profound shifts it has brought and will continue to bring to humanity's social, economic, and philosophical landscapes. From the invention of the world's first computer in the 19th century, which took hours to perform a simple calculation, to handheld devices capable of solving complex problems in seconds, we have come a long way. There is no doubt that the inventions yet to come will emerge at an exponentially faster pace. AI's trajectory mirrors that of past technological milestones such as electricity and fire—both tools that reshaped human existence while demanding careful stewardship.

We can see that the future of AI encompasses both numerous opportunities and challenges. While AI will bring about increased productivity, efficiency, and highly skilled individuals, it is imperative to also take into account the concerns which come along such as job displacements, privacy concerns, lack of transparency and accountability, as well as erosion of interpersonal connections. Such challenges could pose a serious threat to the world we live in. Thus, although it seems great to adopt and develop AI at a faster pace, it should not be done without first achieving an optimal balance between regulation and autonomy. Such a balance is definitely not a cake walk and would require collaboration from various stakeholders like government agencies, corporations, and individuals to reap the maximum benefit.

This chapter lays the foundation for the book, as moving forward we will not only explore AI's broader implications but also dive deeper into more specific domains like finance, dharma, feminism, education, etc. By analyzing these spheres, the chapters ahead aim to provide a comprehensive understanding of AI's role in shaping the present and the future as well as the implications of the ethical considerations.

RELATED WORK AND AI USE

This chapter builds on the authors' prior research, integrating concepts that are also part of a paper currently under review, another forthcoming book chapter, and an upcoming popular science book. In crafting this chapter, we utilized AI tools for translation, text refinement, and stylistic enhancements including ChatGPT 4o, Perplexity, and Claude 3.5 sonnet during the period of August to November 2024.

REFERENCES

Al Owais, S. A., Alghamdi, S. S., Alsuhebany, N., Alqahtani, T., Alshaya, A. I., Almohareb, S. N., Aldairem, A., Alrashed, M., Saleh, K. B., Badreldin, H. A., & Yami, M. S. A. (2023). Harbi, intelligence in clinical practice. *BMC Medical Education*, *23*(1), 689. https://doi.org/10.1186/s12909-023-04698-z

Amodei, D. (2024). *Machines of loving grace: How AI could transform the world for the better.* https://darioamodei.com/machines-of-loving-grace#taking-stock

Artificial Intelligence. (2024, October 30). *Roland Berger.* https://www.rolandberger.com/en/Insights/Publications/Automotive-2040-Artificial-Intelligence.html

Burwell, S., Sample, M., & Racine, E. (2017). Ethical aspects of brain computer interfaces: A scoping review. *BMC Medical Ethics*, *18*(1). https://doi.org/10.1186/s12910-017-0220-y

Calls9. (2023). *The history of AI: A timeline from 1940 to 2023 + infographic.* https://www.calls9.com/blogs/the-history-of-ai-a-timeline-from-1940-to-2023

Char, D. S., Abràmoff, M. D., & Feudtner, C. (2020). Identifying ethical considerations for machine learning healthcare applications. *The American Journal of Bioethics*, *20*(11), 7–17. https://doi.org/10.1080/15265161.2020.1819469

Cheema, R., Krishnan, R., Jordan, T., & KPMG LLP. (2024). *What will a healthcare system in 2050 look like?* Cyber for Health Article Series. https://assets.kpmg.com/content/dam/kpmg/uk/pdf/2024/04/what-will-a-healthcare-system-in-2050-look-like.pdf

Deichmann, J., Ebel, E., Heineke, K., Heuss, R., Kellner, M., & Steiner, F. (2023). *Autonomous driving's future: Convenient and connected.* McKinsey & Company. https://www.mckinsey.com/industries/automotive-and-assembly/our-insights/autonomous-drivings-future-convenient-and-connected

Edulakanti, S. R., & Ganguly, S. (2023). Review article: The emerging drone technology and the advancement of the Indian drone business industry. *The Journal of High Technology Management Research*, *34*(2), 100464. https://doi.org/10.1016/j.hitech.2023.100464

Elias, J. (2024, November 12). *Waymo opens robotaxi service to anyone in Los Angeles, marking its largest expansion yet.* CNBC. https://www.cnbc.com/2024/11/12/waymo-opens-robotaxi-service-to-anyone-in-los-angeles.html

Ferguson, M. (2024, December 13). *IBM's "Automotive 2035" study reveals future of AI-Powered cars.* OpenTools. https://opentools.ai/news/ibms-automotive-2035-study-reveals-future-of-ai-powered-cars

FutureWebAI. (2023, August 2). The ethical challenges of AI in autonomous vehicles: safety, decision-making, and liability. *Medium.* https://medium.com/@futureaiweb/the-ethical-challenges-of-ai-in-autonomous-vehicles-safety-decision-making-and-liability-38fee3427683

George, A. S. (2024). Artificial Intelligence and the Future of Work: Job Shifting Not Job Loss. *Partners Universal Innovative Research Publication (PUIRP)*, *2*, 17–37. https://doi.org/10.5281/zenodo.10936490

Goertzel, B. (2014). Artificial General Intelligence: Concept, state of the art, and future prospects. *Journal of Artificial General Intelligence*, *5*(1), 1–48. https://doi.org/10.2478/jagi-2014-0001

Hillary. (2023, November 11). *Building blocks of AI: Exploring the potential of AGI in 2030.* TechBullion. https://techbullion.com/building-blocks-of-ai-exploring-the-potential-of-agi-in-2030/

IBM. (2024). *What is artificial intelligence (AI)?* https://www.ibm.com/topics/artificial-intelligence

Immerman, G. (2024, April 3). *The impact of AI and machine learning on manufacturing.* MachineMetrics. https://www.machinemetrics.com/blog/the-impact-of-ai-and-machine-learning-on-manufacturing

Jain, A. (2024, December 23). *AI in manufacturing: Benefits, use cases, & trends.* Oyelabs. https://oyelabs.com/ai-in-manufacturing-benefits-use-cases-examples/

LaDolce, M. (2024, August 21). *Gartner 2024 hype cycle for emerging technologies highlights developer productivity, total experience, AI and security.* Gartner. https://www.gartner.com/en/newsroom/press-releases/2024-08-21-gartner-2024-hype-cycle-for-emerging-technologies-highlights-developer-productivity-total-experience-ai-and-security

Lee, K. Q. (2020). *AI 2041. Ten visions for our future.* Virgin Digital.

Lovelock, J. (n.d.). *Novacene: The coming age of hyperintelligence.* MIT Press. https://books.google.com/books/about/Novacene.html?id=yOoCEAAAQBAJ

Maerz, C. (2024, November 18). *Healthcare trends 2025.* Athenahealth. https://www.athenahealth.com/resources/blog/healthcare-predictions-for-2025

MarketsandMarkets. (n.d.). *Collaborative robot industry worth $11.8 billion by 2030.* https://www.marketsandmarkets.com/PressReleases/collaborative-robot.asp?srsltid=AfmBOoqcqyTXCVYYdFZN7jd61DCiAybnHbab_6cXNwdP0DBsjUsUmXtM

Maryville Online. (2024). *History of AI: Timeline and the future.* https://online.maryville.edu/blog/history-of-ai/

Mucci, T., & Stryker, C. (2024, December 19). *Artificial superintelligence.* IBM. https://www.ibm.com/think/topics/artificial-superintelligence

Nature. (n.d.). *Quantum computers shoot for the moon.* https://www.nature.com/articles/d42473-023-00440-x

Pinto-Coelho, L. (2023). How Artificial intelligence is shaping medical Imaging Technology: A survey of Innovations and applications. *Bioengineering, 10*(12), 1435. https://doi.org/10.3390/bioengineering10121435

Schröder-Hinrichs, J., Song, D., Fonseca, T., Lagdami, K., Shi, X., & Loer, K. (2019, January 15). *Transport 2040: Automation, technology, employment—The future of work.* WMU/ITF. https://papers.ssrn.com/sol3/papers.cfm?abstract_id=3777849

Selvi, M., & Ghosh, S. (2020, July 30). *Artificial super intelligence: The future of AI.* Infosys BPM. https://www.infosysbpm.com/blogs/financial-services/artificial-super-intelligence-the-future-of-ai.html

Sullivan, F. (2024, November 7). *Which megatrends will enable your organization to thrive in 2040?* Frost & Sullivan. https://www.frost.com/growth-opportunity-news/which-megatrends-will-enable-your-organization-to-thrive-in-2040-kts-cim-sn/

Tableau. (2024). *What is the history of artificial intelligence (AI)?* https://www.tableau.com/data-insights/ai/history

TechTarget. (2024). *The history of artificial intelligence: Complete AI timeline.* https://www.techtarget.com/searchenterpriseai/tip/The-history-of-artificial-intelligence-Complete-AI-timeline

The Princeton Review. (n.d.). Emotional intelligence in AI. https://www.princetonreview.com/ai-education/emotional-intelligence-ai#:~:text=Emotional%20Intelligence%20in%20AI%20refers,body%20language%2C%20and%20other%20cues

The Tonight Show Starring Jimmy Fallon. (2024, November 29). *Jerry Seinfeld roasts artificial intelligence and dishes out marriage advice* [Video]. YouTube. https://www.youtube.com/watch?v=t0IJOxBX0BI

Truxcargo. (2024, March 13). *The future of drones in logistics: Delivering the sky's the limit.* https://www.truxcargo.com/blogPost/137/The_Future_of_Drones_in_Logistics:_Delivering_the_Sky's_the_Limit

Vaidyanathan, V. (2024, June 2). *What is a brain computer interface?* ScienceABC. https://www.scienceabc.com/innovation/what-is-a-brain-computer-interface.html

Webadmin. (2024, August 12). *Unleashing innovation: Generative AI for automotive transformation.* Quantzig. https://www.quantzig.com/ai/driving-change-generative-ais-influence-on-the-automotive-landscape/

Zoting, S. (2024, February 5). *Automotive artificial intelligence (AI) market report 2024-2033.* Precedence Research. https://www.precedenceresearch.com/automotive-artificial-intelligence-market

2 Whose Ethics for AI?

Sunil Malhotra

"If you transform yourself, you affect the consciousness of the rest of the world."

— **J. Krishnamurti** *(Bohm, 1999)*

INTRODUCTION

Dharma is a key concept in Indic traditions, espousing the idea of "that which holds and sustains everything"—ranging from the social order to the cosmos itself. Translating Dharma into English or other languages is difficult because few terms carry its full richness. Often translated as religion or law, Dharma is more than that; it's the natural order, duty, ethics, and the essential characteristic of things. Dharma is the moral fabric of the universe, the basis of Karma, where actions have consequences. The closest synonyms for Dharma in English are *righteousness* (individual morality) and *ethics* (collective behavior). Unlike in Abrahamic traditions, where morality is a contract with God, in Dharma traditions morality is impersonal, inherent in the structure of the universe itself. J. Krishnamurti's quote emphasizes that living in harmony with this universal order not only shapes one's own journey but also elevates the collective consciousness, making individual morality and collective ethics inseparable threads of the same universal tapestry. Dharma emphasizes that every individual is an integral part of a cosmic order, where personal transformation has profound ripple effects on the collective whole.

In the context of society, Dharma sustains the fabric of society and is the means of harmonizing individuals and the larger whole. In texts such as the *Mahabharata,* Dharma refers to both moral principles and social responsibilities, highlighting the role individuals play in upholding order. This multifaceted notion emphasizes that right action comes from within in the Dharma theory is harmonious with the natural order, reflecting a universal moral code.

Dwell on this apocryphal reference in the Hindu epic *Mahābhārata*, of Duryodhana's remonstration to Krishna.

> I know what is good [Dharma], and I know what is evil [Adharma], but [the problem is] I possess neither the desire to do good nor the restraint to stop doing evil.
>
> [It is not my fault.] It is as if some unknown force dwells in my heart and impels me to do, what I do.

The verse defines how each of us feels; deep down we seem to know exactly what is right for us, for our community, and for the world at large, and yet our thoughts and actions are out of sync. Why are we unable to practice what we know to be good? Why are we unable to restrain ourselves from doing what we know to be unethical?

The *Bhagavad Gita* prescribes *svadharma*, one's duty as an individual, in accordance with the Vedas. There are two kinds of *svadharma*—Para Dharma, or spiritual duties, and Apara Dharma, or material duties. The *Mahabharata* espouses the tenet that Dharma protects those who protect it— "He who sacrifices virtue is himself destroyed. And he that preserves it is himself preserved. I, therefore, do not sacrifice virtue, considering that if destroyed, it will destroy us," reflecting the reciprocal nature of Dharma—adhering to it upholds not just personal or social order but cosmic balance itself.

DOI: 10.1201/9781003585527-4

At a cosmic level, Dharma is also seen as the operating principle to ensure the balance and order of the universe.

Moreover, the concept of Dharma extends into ritual, tradition, and the observance of ethical behavior. Social conservatism in India has roots in the respect for these long-standing traditions, sustained through millennia by adherence to Dharma. This conservative aspect, which emphasizes preservation of rites and social norms, contrasts with more progressive views, where change is often seen as necessary for progress. It includes the proper relationship with sentient beings, the inanimate world, and the observance of societal rites and traditions. The Western tendency to discard older truths for new ones contrasts with India's approach, which integrates new truths without discarding the old.

Ethical behavior in Dharmic thought is intrinsically linked to action, or Karma. Karma, meaning *action*, also embodies the philosophical doctrine of cause and effect, which is central to Buddhism and other Indic traditions. The principle that every action has consequences shapes individuals and the world around them. In this context, unselfish action—motivated by a sense of duty, compassion, and righteousness—is considered the highest form of spirituality. By acting selflessly and in alignment with Dharma (cosmic law and order), individuals not only contribute to personal growth but also to the collective harmony and well-being of society.

Every religion tells you to be good. But how do you know what *is* good? Swami Vivekananda proposed the following tests of morality (see Table 2.1).

Daniel Schmachtenberger, of *The Consilience Project,* believes that the ancient wisdom traditions had wise people who saw similar things and encoded them differently. It's said by many Vedic philosophers that the *Bhagavad Gita*, a lofty Hindu scripture, chapter two, verse 48 is fundamental. Established in yoga. Perform action. Yoga means union with all that is: "Yogastha kuru karmani" is the quote. And it says, "Established in yoga, or union with all that is, from the place where you recognize that your existence does not exist without everything else and from the place where you cognitively but also experientially get that, where there's an intimacy with all life, and act from there. Spontaneous right action emerges from that place (Schmachtenberger, Aug 2022).

For all their badness and madness, human beings are supposed to have the capacity to restrain their lust and rage, not curse back when cursed, a faculty of moral reasoning, an inner voice—however feeble—of conscience. It is their self-restraining moral sense which makes them rational animals, not their survival-conducive shrewdness at cost–benefit calculation, in which some bees, termites, and reptiles may be smarter: "Eating, sleeping, fear and sex drive: these are common between beasts and humans; Dharma is what is an extra distinguishing feature of humans; those without Dharma are equal to beasts" (Chakrabarti, 2018).

Spirituality provides ethical and moral frameworks rooted in compassion, empathy, and interconnectedness. We need to integrate these spiritual values into scientific advancements to guide the responsible use of technology, ethical decision-making, and the pursuit of knowledge. Science often focuses on external observations and measurable phenomena, while spirituality explores the inner

TABLE 2.1
Tests of Virtue (Dharma)

Good (Dharma)	Bad (Adharma)
Strengthens you	Weakens you
Makes you independent and self-reliant	Increases your dependency on other people or things
We over Me—the welfare of others	Me over We—self-centered and acquisitive
Unites	Divides

Source: Author.

dimensions of consciousness and subjective experiences. Transcending boundaries can lead to new insights, discoveries, and a deeper understanding of our place in the universe.

ARTIFICIAL INTELLIGENCE

AI is a paradox. It is seen both as an existential threat and a silver bullet, fitting nicely into the binary classifications we are attuned to; this is the dominant anthropocentric view of the world.

The rapid emergence of exponential technologies such as AI has overwhelmed current institutions, systems, and societies, both in making sense of these advancements and in harnessing their power effectively. A direct consequence of the unprecedented explosion of exponential technologies is diminished capabilities of our institutions and leaders to deal with the world. How do we keep up with exponentially evolving technologies that infuse our personal and professional lives? Admittedly, our brains find it hard to grapple with the implications of exponential vis-a-vis linear thinking; stability and predictability are what we seek for comfort, and what we find in the twenty-first century is quite the opposite.

Not being in control of things is alien to modern humans, leading to a crisis of meaning. Even more alarming is the heightened need for self-preservation and consequently the erosion of ethics in modern society. The isolating effects of the Covid-19 pandemic and their amplification, fueled by social media, continue to take society down a path of polarization. The fallout is unprecedented and portends extreme consequences. Humanity needs to navigate these rapid changes without losing its moral direction.

Or so goes the popular narrative.

Filtering signals from noise around AI is not going to be easy. More problematic, however, is imputing intelligence to technology without first arriving at a common understanding of the word intelligence.

The word *intelligence* finds its roots in the Latin word *intelligentia*, which means *the action or faculty of understanding,* derived from *intellegere*, meaning *to understand*. Intelligence has been interpreted in various ways, encompassing traits such as abstraction, logic, understanding, self-awareness, learning, emotional knowledge, reasoning, planning, creativity, critical thinking, and problem-solving. Broadly speaking, intelligence refers to the ability to perceive or deduce information and retain it as knowledge to be utilized for adaptive behavior within a given environment or context. It is now widely accepted that human intelligence is not the preserve of the brain alone.

My own definition is somewhat broader, intelligence is *the sense that makes sense of things,* which, in turn, presupposes the existence of life processes and living beings. In that sense, do machines have intelligence? Probably not, at least in the machines that are programmed to mimic intelligence. How then do they display the amazing capability of being intelligent? It feels like magic.

According to Satish Kumar, the essence of human intelligence lies in two fundamental abilities: *imagination* and *discrimination*. Imagination enables envisioning possibilities beyond the present, while discrimination allows distinguishing between what is appropriate and beneficial versus what is harmful or negative (Kumar, 2023).

Discrimination, at a spiritual level, means distinguishing between the eternal and the transient, consciousness and matter, and understanding the difference between the absolute and relative (transactional) reality. Discrimination is believed to come from beyond the brain. Eastern thought systems (we focus on Indic philosophies in this chapter) view intelligence as an element of consciousness and distinguish the mind from the physical brain.

All technological revolutions in the past have evoked a similar reaction from the world at large. Calestous Juma, late professor at the Harvard Kennedy School and author of *Innovation and Its Enemies: Why People Resist New Technologies,* tells us, "The pace of change in technology and its prospect for disruption is unlike anything we've seen in the past"(Juma, 2016).

He notes that today's disruptions to society are immediate and not isolated to the technology industry, and that he expects impending large-scale disruptions and an increasingly prominent global debate over technology's societal implications. In comparison to the past, where transformative technologies like the steam engine or electricity took decades to reach full adoption, today's innovations spread much faster. This rapid evolution of technology is a double-edged sword: it brings unparalleled opportunities for progress but also significant challenges in terms of adaptation, regulation, and equity.

"Humans are instinctively designed to react to novel things in a way that aims to protect oneself," says Juma, adding that, "large tech changes can amplify anxieties and the fear of loss is mainly what drives technology resistance. People often reject technology because it presents changes, big or small, to their immediate world, a world to which they have worked hard to adjust."

Instead of assuming technology is either inherently good or bad in and by itself, we should see it as intrinsically value-laden and value-creating. All technologies are created with the goal of actualizing values; they are embedded with values from the start. As technologies take hold and proliferate, they should be understood also to have the potential for forming new, unknown, and unpredictable values.

AI is here to stay, though the arrival of fully intelligent machines remains uncertain. Popular culture, especially through narratives like Marvel's *Iron Man*, has shaped public perception of AI as a superhero-like force. Hollywood's portrayal of AI as both a savior and potential threat deeply influences Silicon Valley's technologists, fostering a symbiotic relationship between imaginative storytelling and real-world innovation. While this dynamic drives technological progress, it also introduces significant ethical challenges. A critical question arises: If most AI technology is developed in the US, does that position America, as the dominant superpower, to unilaterally dictate the ethical framework for AI? This would echo America's historical role as a hegemon, but it risks oversimplifying the global, multi-cultural complexities that AI presents. Fictional depictions, while inspiring, often overlook the deeper ethical and societal issues AI raises, which require diverse perspectives beyond a single nation's control.

Technoscientists are firmly of the belief that AI is a force for good—that artificial general intelligence (AGI) and autonomous machines will serve humanity, functioning as highly capable tools or "slaves" to carry out orders. They argue that AI can enhance efficiency, increase longevity, improve healthcare, eradicate poverty, and address climate change. They contend that with proper control and programming, these intelligent systems will remain aligned with human interests, amplifying productivity and solving complex problems without becoming adversarial. This optimistic view assumes that AGI will be developed with ethical safeguards, though critics worry about the potential for unforeseen consequences if machines surpass human control.

HYPE

The sudden yet expected "open" arrival of generative pre-trained transformer (GPT), a type of large language model (LLM) and a prominent framework for artificial intelligence, in November of 2022 destroyed the myth that humans have any degree of control on the speed and scale of technology proliferation. Barely one month in, the LLM-based artificial intelligence ChatGPT boasted of 100 million users worldwide and threatened to disrupt almost every domain from education to medicine to law to politics to technology to nearly every domain and establishment. By April of 2023, people were running scared; several AI experts and business leaders called for an immediate stop to further development of AI. The reaction was not surprising given that humans are not good at accepting sudden change, much less handling its effects. The reaction is a biological response of the amygdala, an organ responsible for the human fight-or-flight response, which has barely evolved since the hunter-gatherer times. It is our survival meter. The petitioners seemed powerless in containing the potential spread of a runaway AI, having been caught unawares by a contemporary version of Frankenstein's "monster."

The hype around the ethical implications of GPT led to the frenzied call by US business and technology leaders—including Elon Musk, Steve Wozniak, and a large group of prominent AI researchers—often referred to as a "shut down AI" petition, urging a temporary halt to further development of powerful AI systems until safety measures are established; their reactions were harbingers of the age we stepped into, an age of anxiety and uncertainty. "Our anxiety does not come from thinking about the future but from wanting to control it, is attributed to Khalil Gibran (*A Quote by Kahlil Gibran*, n.d.). AI's rapid advance has ignited a global dialogue—on one side, it is celebrated for boosting efficiency and optimizing various industries, while on the other, concerns loom over potential job displacement and a host of ethical implications. Applications such as Midjourney and DALL·E 2 threaten the livelihoods of traditional designers, while deepfakes present new dangers in the realm of misinformation and deception.

However, it's important to note that these are not representative of what AI is or can be; ChatGPT is a chatbot that simulates human-like conversation but does not possess true general intelligence or consciousness. The widespread fascination it generates reveals the tenuousness of the human mind when faced with what appears to be advanced AI. As Arthur C. Clarke once remarked, "Any sufficiently advanced technology is indistinguishable from magic," which perfectly captures the sense of awe and misunderstanding surrounding AI today (Clarke, 2011).

The hype around ChatGPT points to the collective eagerness to believe in the potential of AI, even if it is still a far cry from achieving true human-level intelligence. Nonetheless, as technology evolves, questions of ethics, regulation, and the responsible deployment of AI remain central to the discourse.

Simon Biggs, a professor of interdisciplinary arts at the University of Edinburgh, foretells, "AI will function to augment human capabilities. The problem is not with AI but with humans. As a species we are aggressive, competitive and lazy. We are also empathic, community-minded and (sometimes) self-sacrificing. We have many other attributes. These will all be amplified. Given historical precedent, one would have to assume it will be our worst qualities that are augmented. My expectation is that in 2030 AI will be in routine use to fight wars and kill people, far more effectively than we can currently kill... We cannot expect our AI systems to be ethical on our behalf—they won't be, as they will be designed to kill efficiently, not thoughtfully."

Dr. Karan Singh and Satish Kumar, in November 2023, emphasized in a dialogue the perilous trend of developing technology for its own sake in contemporary times. The belief underscored that possessing the resources and knowledge does not inherently grant the right or justification to develop technology devoid of a foundational wisdom. They highlighted the importance of aligning technological advancements with the profound insights embedded in Eastern philosophies and indigenous wisdom, steering the development of technology toward a more ethical and harmonious path (Malhotra, 2024). Ervin Laszlo said to me in December 2021, "I have great trust and belief in yoga," adding, "I think if we would all listen to ourselves more and hook ourselves: I would say yoga is to hook ourselves to something larger than ourselves to be able to effectively set a direction for technology to move toward.

I have great trust, great belief in the power of artificial intelligence, but I don't have much belief in the wisdom of the people who develop artificial intelligence, because doing just anything that can be done is not the best course to take."

Essentially, navigating this new landscape hinges on our capacity to redefine reality. We stand at the threshold of an era where we can reshape our connections and explore possibilities that were once beyond conception. By combining the synergy between technology and the sagacity embedded in our spiritual traditions, we can chart a path leading to a future characterized by justice, equity, humanity, and sustainability.

AI NARRATIVES ON A SPECTRUM

Futurists have long debated the arrival of technological singularity, when human and artificial intelligence will merge. One conceptualization of the singularity hypothesis, where technology exceeds human capabilities, involves an evolving non-human intelligent entity spawning a new species with each successive "smarter" generation emerging at an exponentially accelerated pace, thus ushering in an immensely potent superintelligence unprecedented in history. Will we encounter AI sentience with the emergence of AGI? Are there pathways for coevolving human intelligence alongside AI where "the 1,400 or so grams of incomprehensible meat in our skulls might need to cooperate with a similarly incomprehensible chunk of silicon," asks Jamais Cascio (Cascio, 2022). Will the ethics embodied in these new silicon beings follow human moral codes or be something completely different?

As the power of computing, biotechnology, and AI come together, we will have the ability to hack humans. "When you combine our increasing understanding of biology, especially brain science, with the enormous computing power that machine learning and AI is giving us, what you get from that combination is the ability to hack humans, which means to predict their choices, to understand their feelings, to manipulate them and also to replace them. If you can hack something you can also replace it," argues historian, philosopher, and author Yuval Noah Harari.

We have stepped upon the ledge of a perception revolution. "Our supremacy as the prime understanders of the cosmos is rapidly coming to end," said the late James Lovelock, father of the Gaia hypothesis, in his 2019 book, *Novacene*. Further, Lovelock argued, "It is crucial that the intelligence of Earth survives and prospers" (Lovelock, 2019).

In the ongoing discourse surrounding artificial general intelligence and its implications for society, a spectrum of perspectives exists, ranging from enthusiastic techno-optimists to cautious alarmists. The graphic in Figure 2.1 illustrates these varying sentiments on a continuum, capturing the diverse landscape of opinions regarding the future of AGI and its impact on society.

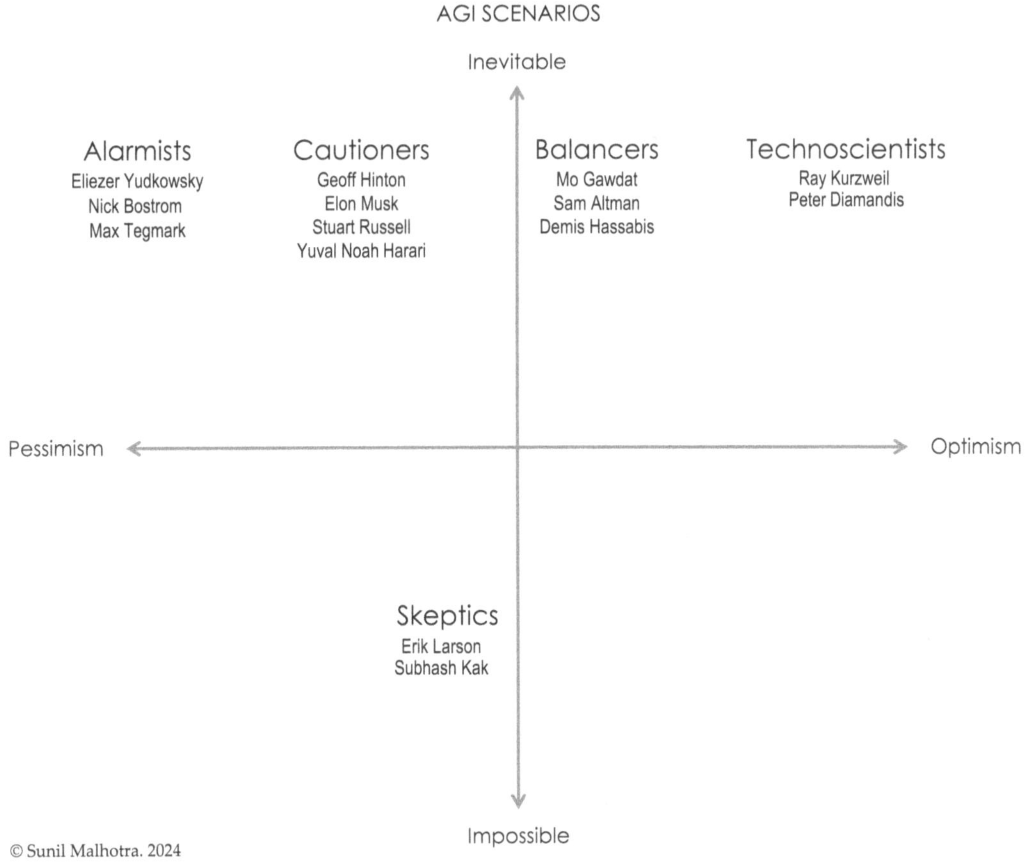

FIGURE 2.1 AGI Scenarios: Continuum of views on AGI, reflecting the complexity and uncertainty surrounding the future of AGI. (Source: Author.)

Techno-optimists largely maintain that technology is about what works for everyone—*universal functionality*—rather than what is good for everyone—*universal morality*. The advent of AI-powered social technologies has empowered bad actors with the capability to advance their nefarious agendas with unparalleled speed and scale. These technologies are exacerbating societal polarization, as rampant individualism increasingly dominates the landscape, manifested in the nationalist agendas of many nations. In this environment, every individual has become a veritable propaganda machine, leveraging AI-driven social tools to further amplify divisions and discord within society.

To further elaborate Figure 2.1, the continuum of the views on AGI are as follows:

Technoscientists: **Ray Kurzweil** and **Peter Diamandis** believe that AGI will bring positive transformation. In their conception, technology will be the key to achieving solutions in areas of human well-being, ethical dilemmas, and progress (Diamandis & Kurzweil, 2024).

Balancers: **Mo Gawdat**, former Chief Business Officer at Google X, has expressed concern about the rapid development of AGI. Recognizing both potential benefits and risks, he advocates for ethical AI development in a guided manner, like we bring up our children (Tucker, 2024). **Sam Altman** of OpenAI and **Demis Hassabis** at DeepMind share this view. Gawdat believes we are creating superintelligence which will surpass human abilities in every way, and he puts the onus on us to be role models for intelligent machines.

Cautioners: **Elon Musk**, **Stuart Russell**, and **Yuval Noah Harari** are voices advocating caution to prevent unintended consequences as AI evolves. **Geoffrey Hinton**, a pioneer in AI research, has raised concerns about the rapid development of AI. Hinton sees the need for regulation and ethical guidelines, cautioning that AGI could surpass human intelligence, posing existential risks and argues for human oversight (*MSN*, n.d.).

Skeptics: **Erik Larson**, American author, entrepreneur, and researcher, doubts the near-term emergence of AGI and is critical of current AI hype (Larson, 2022). Others, like **Subhash Kak**, Indian American computer scientist, philosopher, and historian of science, known for his work in artificial intelligence, quantum computing, cryptography, and the history of Indian science and philosophy, maintain that current AI systems, no matter how advanced, lack the necessary components for true consciousness or intelligence. Their arguments often revolve around the distinction between mere information processing and subjective experience, intentionality, or understanding.

Alarmists: **Eliezer Yudkowsky**, researcher and writer focused on AI alignment and safety, is a co-founder and senior research fellow. He, like **Nick Bostrom** and **Max Tegmark,** argues that AI could surpass human control if not developed carefully. Tegmark calls for proactive measures to steer AI in a direction that maximizes human well-being and minimizes existential risks.

In a metaphysical sense, these are all imagined dichotomies. Are we projecting our fears strongly enough for the universe to solidify them into a self-fulfilling prophecy? Are vested interests of techno-narcissists tapping into our individual psychic energy with the help of a mercenary nexus of politicians, lobbyists and media?

It's not so much about regulating technology as it is about curbing human intent and channelizing ingenuity. You don't need higher technology for terrorism; the orchestrated explosions in Lebanon in September 2024, using pagers is an example.

Understanding how AI will play out, either similar to previous technological revolutions or taking a completely different path, may need a fresh approach to our comprehension of the world since the mechanistic–reductionist worldview is now well past its use-by date. This type of thinking implies that complex organisms are essentially biological "machines"— plants, animals, or the human body—and can be understood by reducing them to their smallest constituent parts.

The discussion around ethics unsurprisingly reveals significant cultural differences in understanding control and self-discipline. While Western thinkers often prioritize regulatory frameworks to ensure responsible AI use, they find it hard to trust alternative paradigms like the Dharmic approach, which emphasizes intrinsic motivation and interconnectedness.

ANTHROPOMORPHISM

Anthropomorphism is fundamental to human social behavior, not just in interactions with animals and things but with other humans as well. Since we cannot directly perceive the inner experiences of others, we rely on projecting emotions, intentions, and thoughts to simulate them. It enables humans to connect, empathize, and engage socially. We attribute human traits to animals, especially domesticated pets, and even God.

Our latest target is AI. This tendency to anthropomorphize (studied under frameworks like the media equation) suggests that our relationships with AI or technology are an extension of an innate social glue that allows for bonding and cooperation. Even though we know there is no human behind the screen in ChatGPT, we naturally assign human-like traits to it. By imputing human attributes to AI, having experienced the power of ChatGPT, which is just a preview (and one application of AI), it is easy to believe in AI becoming an all-knowing, ever-present, all-powerful and all-pervading "being." A new god.

The cherry on top is hubris, making it well beyond our ken that any power can be superior to the human race. We seem convinced that AI is going to "obey" our commands since it is we who have created these "intelligent machines." Unlike traditional software, AI is no longer dependent on programming by human engineers. It learns mostly by continuous observation, creating artificial neural networks, branching them, and pruning them for more coherent connections. Since AI learns by simply watching the world and the humans around it, much like children do, and as it evolves, it will reflect the global human mind with all its imagined and imaginative capabilities, and even the mind of Nature in all her diversity.

In *Nexus*, Yuval Noah Harari argues that AI is not a tool but an agent (Po, 2024).

AI is an unprecedented threat to humanity because it is the first technology in history that can make decisions and create new ideas by itself. All previous human inventions have empowered humans, because no matter how powerful the new tool was, the decisions about its usage remained in our hands. Nuclear bombs do not themselves decide whom to kill, nor can they improve themselves or invent even more powerful bombs. In contrast, autonomous drones can decide by themselves who to kill, and AIs can create novel bomb designs, unprecedented military strategies and better AIs. AI isn't a tool—it's an agent.

Ethics for AI has created a flurry of activity toward regulating the development of AI. In a bid to ensure that "human interests" are protected, the EU Parliament was quick to the draw in enacting the first law in 2023, the EU Artificial Intelligence Act (which came into force on August 1, 2024), claiming to have laid down a uniform legal framework, in accordance with Union values, to promote the uptake of human-centric and trustworthy artificial intelligence while ensuring a high level of protection of health, safety, fundamental rights as enshrined in the Charter of Fundamental Rights of the European Union. (It is of some significance that the General Data Protection Regulation (GDPR) was implemented in 2018 and is still a mixed bag in terms of effectiveness). Pardon the cliché, but regulation is easier said than done.

Human-centric and *trustworthy* are words that have deep cultural moorings, thereby making the idea of values applicable only to sentient, aka conscious, beings. While one may not question the need for regulation, the efficacy of *any* top-down control mechanism is moot. To regulate human behavior is no less than a fool's errand, a mission that even religion has found difficult to accomplish.

Erik Larson calls this out in *The Myth of Artificial Intelligence* (Larson, 2022).

It has become trendy for AI thinkers to worry about so-called "beneficial AI," trusted systems, and other ethical issues like problematic bias. In other words, systems that don't understand but still perform have become a concern.

This cuts the myth at an awkward angle: It is because the systems are idiots but still find their way into business, consumer, and government application, that human-value questions are now infecting what were once purely scientific issues.

Larson emphasizes that the idea of ascribing intelligence to machines is misleading. He goes on to explain that inference is a basic cognitive act of intelligent minds.

Inference is a basic cognitive act for intelligent minds. If a cognitive agent (a person, an AI system) is not intelligent, it will infer badly. But any system that infers at all must have some basic intelligence, because the very act of using what is known and what is observed to update beliefs is inescapably tied up with what we mean by intelligence. If an AI system is not inferring at all, it doesn't really deserve to be called AI.

Ethics in AI needs to be addressed at different stages of the AI lifecycle. First, decisions about *what* to design and develop should leverage the innate wisdom of humankind, ensuring that technologies align with human values and societal needs. Second is *how* to deploy AI to maximize human and planetary well-being, implementing strategies that avoid harm and promote sustainability. Finally, to promote multicultural harmony, control over AI's use should be ceded to its potential beneficiaries, empowering them to determine *when*, *where*, and *how* AI is utilized in ways that best serve their unique contexts. This approach ensures that AI remains a tool (and a companion) for collective empowerment rather than domination.

DHARMA FOR HARD TIMES

The discourse on AI ethics encompasses multiple dimensions, from regulation and self-governance to the potential emergence of completely new AI cultures. While regulation provides a framework, it does not seem like a solution. AI systems must also be designed for self-aware ethicality. Imagine trying to program morality into algorithms! As we contemplate the future of AGI and the possibility of AI-led cultures, considering the wisdom of ancient traditions, such as those found in Indic philosophy, to guide strategy becomes an imperative.

Ultimately, the development of ethical AI will require a multifaceted approach, combining regulatory oversight with self-regulation, moral programming, and deep philosophical inquiry. The formula for addressing ethical challenges is a judicious mix of the power of modern technology and the wisdom of Indic and similar traditions. AI can serve as a force for good to ensure human and planetary well-being.

Mo Gawdat, in *Scary Smart*, tells us that in today's world, we are often driven by targets that take precedence over adhering to ethical principles. Objectives like economic profit, power consolidation, territorial expansion, re-election, wealth accumulation, or gaining social media popularity—these modern pursuits frequently overshadow moral considerations (Gawdat, 2021). As a result, human intelligence is channeled toward achieving these goals, with ethical concerns sidelined. Yet, in doing so, intelligence is cleverly used to maintain an appearance of ethical behavior, even when the underlying focus is elsewhere.

While most Eastern cultures, in general, put the collective ahead of individual interests, India goes further. For a nation so diverse in religious beliefs, language, culture, and tradition, just sticking together is a feat in itself. If you drop the anthropomorphic lens momentarily, you see that Indians somehow have remained closest to the behaviors that Nature intended for all of the sentient world, including our own species. It is almost as if an unselfish gene attracts the commoner to the ethical tenets of Vivekananda's "Karma yoga," which is the attaining through unselfish work of that freedom which is the goal of all human nature. Every selfish action, therefore, retards our reaching the goal, and every unselfish action takes us toward the goal; that is why the only definition that can be given of morality is this: That which is selfish is immoral, and that which is unselfish is moral,"

(VivekaVani, 2024b) echoing William James: "We are like islands in the sea, separate on the surface but connected in the deep" (*A Quote by William James*, n.d.).

Conventional morality, which ordinary people follow out of fear—fear of God, police, society, or even the fruits of one's actions—has little appeal in modern times. The intrinsic purity and divinity of the Self is the basis of moral action. Immorality, then, is nothing but a "fall from the actual nature of human beings."

A world characterized by multicultural diversity begs the question, Whose ethics should govern AI? Can we use the ethics of US capitalism which is susceptible to greed and has succumbed to its allure? A self-centered, winner-takes-all paradigm that thrives in separatism, a capitalist framework built on principles like free markets, individualism, and the pursuit of profit is not the best of choices for a foundation.

J. Krishnamurti questions our sense of morality: And if your behavior is the outcome of your own conditioned response, is it moral? If your action is based on fear and reward, is it righteous? If you behave rightly according to some ideological concept or principle, can that action be regarded as virtuous? So we must begin to find out how deeply we have discarded the morality of authority, imitation, conformity, and obedience. Isn't fear the basis of our morality? Unless these questions are fundamentally answered for oneself, one cannot know what it is to be truly virtuous. As we said, with what ease you come out of this hypocrisy is of the greatest importance. If you merely disregard it, it doesn't indicate that you are moral: you might be merely psychopathic. If you live a life of routine and contentment, that is not morality either. The morality of the saint who conforms and follows the well-established tradition of sainthood is obviously not morality. So, one can see that any conformity to a pattern, whether or not it is sanctioned by tradition, is not righteous behavior. Only out of freedom can come virtue (Krishnamurti Portal, 2022).

So, Whose ethics for artificial intelligence? In *Letters on Yoga*, Sri Aurobindo has written, "In the West, the physical mind is too dominant, so that the psychic does not so easily get a chance—except of course in exceptional people. Mental knowledge is of little use except sometimes as an introduction pointing towards the real knowledge which comes from a direct consciousness of things" (Aurobindo, n.d.).

The rational West wants some eternal principle of Truth as the sanction of all ethics, and where is that eternal sanction to be found except in the only Reality that exists in all, in the self, in the soul?

Ethics always says, "Not I, but thou." Its motto is, "Not self, but non-self." The vain ideas of individualism to which humans cling when they are trying to find that Infinite Power, or that Infinite Pleasure through the senses, have to be given up, say the laws of ethics. You have to put yourself last, and others before you. The senses say, "Myself first." Ethics says, "I must hold myself last." Thus, all codes of ethics are based upon this renunciation; destruction, not construction, of the individual on the material plane. That Infinite will never find expression upon the material plane, nor is it possible or thinkable.

- Swami Vivekananda

(Complete Works. 2.5.1)
(VivekaVani, 2024a)

One interpretation of ethics is the idea of collective morality; what morality is for an individual becomes ethics in society. Ethics are the act of implementation of the agreed moral code. The idea of good in collectives is a dynamic and evolving concept, deeply rooted in the socio-cultural contexts. It reflects the shared values, traditions, and beliefs that guide ethical behavior within a group, thereby providing the scaffolding for social cohesion and moral order. In an interconnected world, where cultures, technologies, and belief systems constantly intersect, the evolution of ethical principles will play a critical role in shaping the future of humanity.

The work of ethics has been, and will be in the future, not the destruction of variation and the establishment of sameness in the external world—which is impossible, for it would bring death and annihilation—but to recognize unity in spite of all these variations, to recognize the god within, in spite of everything that frightens us, to recognize that infinite strength as the property of everyone in spite of all apparent weakness, and to recognize the eternal, infinite, essential purity of the soul in spite of everything to the contrary that appears on the surface. This we have to recognize. Taking one side alone, one half only of the position, is dangerous and liable to lead to quarrels. We must take the whole thing as it is, stand on it as our basis and work it out in every part of our lives, as individuals and as unit members of society.

- Swami Vivekananda
(Complete Works. 1.4.12)

Privilege. Lecture at Sesame Club, London. 1896
(VivekaVani, 2021)

Indic philosophical texts, such as the Upanishads, along with traditions like Vedānta and practices like yoga, offer valuable insights as alternatives to the entrenched materialist-reductionist paradigm, guiding the ethical development and application of AI while prioritizing holistic well-being. *Could yoga help us see our intimate connections to every other "being" on the planet, including the artificial beings spawned by the exponential technology of AI? Can we try to understand ourselves differently, through the consciously evolving human mind, instead of solely through the lens of matter becoming increasingly intelligent?* These questions deserve our inquiry.

Shifting gears, think of a future of intelligent machines—imagine technological singularity is here—where humans are all but redundant and irrelevant. There is even a word—*transhumanism*—for the active social movement that looks forward to this future in which humans are merged with technology. These new "beings"are autonomous and self-improving (and maybe even self-aware). Over time they will have their own "ethics." Suffice it to say that such issues present a challenge for most moral theorists, who take the preservation of human life and the human species to be a good thing. Would we then expect the "cyborgs" or "humanoids" to follow our rules of morality? And whose ethics would make the cut?

Factoid: *Humanoid: a fully autonomous robot which doesn't involve any living organism for its operation. A cyborg, essentially, is a living organism with some robotic parts/machinery embedded in their body to achieve greater power (than a normal living being).*

CONCLUSION

How something in the nature of ethics is expected of a non-conscious, artificial (by its very name) technology is a perplexing question, but rather than befuddling its present stakeholders, aka human beings, it makes people scurry to "regain" control.

The notion of an overarching ethical standard is increasingly idealistic in a world marked by geopolitical fragmentation, individual empowerment, technological abundance, and leadership crises. The traditional command-and-control model, where centralized authorities imposed ethical norms, is no longer viable in a decentralized, multipolar world. Ethics today must be dynamic, flexible, and responsive to the changing realities of a globalized and digitized society.

We grew up in a predictable world that gave us the comfort of stability and slow change. The world is now in a state of "chaotic equilibrium" marked by accelerating change. What was once linear has become exponential, and the old paradigm of scarcity is being replaced by a new reality of abundance. These aren't just buzzwords, they represent the profound, tectonic shifts reshaping our world and how we interact with it.

An alternative to imposing one-sized ethical standards, societies must embrace the bold step of designing frameworks for self-governance by individuals and collectives. In these frameworks, diverse voices are heard, and moral foundations are shaped through open dialogue, collaboration, and mutual respect. Most importantly, the philosophies, principles and practices are tuned to the local contexts of its stakeholders from the very outset. Ethics, in this context, shifts from control and imposition to a process of empowerment, understanding, and adaptability in a constantly evolving world. By drawing from the deep reservoir of inner wisdom that resides within each individual, rather than being confined to external sensory experiences, we can build a more harmonious and sustainable future for humanity and the planet. Call it Dharma for hard times.

Yoga and Vedānta are systems that point you to your real Self. The idea of going within is a shift in understanding that can change your attitudes to life. Science contributes empirical evidence and practical applications, while spirituality explores subjective experiences, ethics, and the deeper aspects of human consciousness. They both converge in humans, individual and collective. Yoga and AI working together will have varied effects on the personal and professional lives of individuals and communities. How might they help you maximize future opportunities and help you find yourself?

I invite you to reconsider the topic of AI ethics.

> "It makes me so happy. To be at the beginning again, knowing almost nothing... The future is disorder. A door like this has cracked open five or six times since we got up on our hind legs. It's the best possible time of being alive, when almost everything you thought you knew is wrong."
>
> **— *Tom Stoppard* (*Arcadia* quotes by Tom Stoppard, n.d.)**

REFERENCES

A quote by Kahlil Gibran. (n.d.). "*Our anxiety does not come from thinking about the future, but from wanting to control it.*" Goodreads. https://www.goodreads.com/quotes/892897-our-anxiety-does-not-come-from-thinking-about-the-future

A quote by William James. (n.d.). "*We are like islands in the sea, separate on the surface but connected in the deep.*" Goodreads. https://www.goodreads.com/quotes/837455-we-are-like-islands-in-the-sea-separate-on-the

Arcadia Quotes by Tom Stoppard. (n.d.). "*It is a defect of God's humor that he directs our hearts everywhere but to those who have a right to them.*" Goodreads. https://www.goodreads.com/work/quotes/1303332-arcadia

Aurobindo, S. (n.d.). *Letters on yoga*. https://sri-aurobindo.co.in/workings/sa/22/index_e.htm

Bohm, D. (1999). The limits of thought: Discussions between J. Krishnamurti and David Bohm. Internet Archive. https://archive.org/details/david-bohm-the-limits-of-thought-with-jiddu-krishnamurti

Cascio, J. (2022, May 26). *Facing the age of chaos*. Medium. https://medium.com/@cascio/facing-the-age-of-chaos-b00687b1f51d

Chakrabarti, A. (2018). Can the subhuman speak or act?: Agency of sagacious serpents, benevolent birds, rational rodents, and a mocking mongoose in the Mahābhārata (Chapter 7). In S. C. Bhattacharya, V. Dalmiya, & G. Mukherji (Eds.), *Exploring agency in the Mahabharata.*. Routledge. https://doi.org/10.4324/9781351061421

Clarke, A. C. (2011, March 17). *Any sufficiently advanced technology is indistinguishable from magic*. philosiblog. https://philosiblog.com/2011/10/10/any-sufficiently-advanced-technology-is-indistinguishable-from-magic/

Diamandis, P. H., & Kurzweil, R. (2024, November 23). *The man who predicted AGI decades ago* (EP #125). https://www.diamandis.com/podcast/ray-kurzweil-agi-prediction

Gawdat, M. (2021). *Scary smart: The future of artificial intelligence and how you can save our world*. Bluebird. https://www.mogawdat.com/scary-smart

Juma, C. (2016). *Innovation and its enemies: Why people resist new technologies*. Goodreads. https://www.goodreads.com/book/show/29151112-innovation-and-its-enemies

Krishnamurti Portal. (2022, December 15). *Death is not at the far end of life*. Krishnamurti Portal. https://www.krishnamurti.org/transcript/death-is-not-at-the-far-end-of-life/

Larson, E. J. (2022) *The myth of artificial intelligence: Why computers can't think the way we do*. Belknap Press.

Lovelock, J. (2019). *Novacene: The coming age of hyperintelligence*. The MIT Press.

Malhotra, S. (2024). *YOGAi: Interplays of yoga and artificial intelligence*. AuthorsUpFront.

MSN. (n.d.). *AI pioneer Geoffrey Hinton warns of increased risk of human extinction due to AI.* https://www.msn.com/en-us/technology/artificial-intelligence/ai-pioneer-geoffrey-hinton-warns-of-increased-risk-of-human-extinction-due-to-ai/ar-AA1wIauU

Po, L. (2024, December 24). *The Nexus of Fear: Yuval Noah Harari's warning on AI and humanity.* Medium. https://medium.com/@lmpo/the-nexus-of-fear-yuval-noah-hararis-warning-on-ai-and-humanity-248562b21423

Project, C. (2024, October 25). Technology is not values neutral: Ending the Reign of Nihilistic Design. The Consilience Project. https://consilienceproject.org/technology-is-not-values-neutral-ending-the-reign-of-nihilistic-design-2

Schmachtenberger, D. (Aug 2022). Bend not break part 5. *The Great Simplification*, 50, 19.

Tucker, J. (2024, December 2). *Mo Gawdat: The AI dilemma—Navigating ethics in the age of innovation.* Oslo Business Forum. https://www.obforum.com/article/mo-gawdat-the-ai-dilemma

VivekaVani. (2021, August 10). *Privilege—Swami Vivekananda.* VivekaVani. https://vivekavani.com/privilege-swami-vivekananda/

VivekaVani. (2024a, May 2). *The necessity of religion—Swami Vivekananda.* VivekaVani. https://vivekavani.com/necessity-of-religion-swami-vivekananda/

VivekaVani. (2024b, August 15). *The ideal of Karma-Yoga | Swami Vivekananda.* VivekaVani. https://vivekavani.com/ideal-karma-yoga-swami-vivekananda

Part II

Philosophical and Cultural Perspectives

3 Swarm Intelligence, Actor Network Theory, and Complex Systems

Future Strategies of AI Inclusion and Techno-ethics

Mary Reisel

INTRODUCTION

Individuals, Networks, and Expanding Universe

"The most liberal sociologist often discriminates against nonhumans".

(Johnson, 1988)

In 1988, Jim Johnson published a passionate article advocating support for nonhuman entities and calling for recognition of their place as integral and equal members of human social structures. Pointing his critique on "the most liberal" of social scientists, he openly criticized the lack of liberal ideologies among those who are expected more than others to advocate progressive values and to struggle for equality and social rights to the most marginalized groups. The strong statement exposed the gap between progressive liberal ideologies and the reality of discrimination and injustice. At the same time, it implied that liberal thinking and struggles for equality could no longer be limited to abuse of rights among humans but had to be extended and include the rights of various nonhumans that were absorbed and adopted by society yet, at the same time, were ignored even by the ones whose ethical responsibility was to protect them. Johnson highlighted the value that nonhuman entities and objects add to the organization of life with the declaration that has become famous since then: "The door-closer is on strike; for God's sake, keep the door closed". The statement was accompanied by a philosophical inquiry regarding the meaningful yet invisible work conducted by concealed door-closers and hinges, and emphasized their role as key supporters and facilitators of human activities. The hidden door-closers were responsible for maintaining order, regulating human movement, and enabling better work, which inevitably meant they were an equal part of the workforce. Johnson urged the readers to imagine how life could have been without the support of such objects, as small and invisible as they may be, that often go unnoticed, undervalued, and with no social status that they deserve despite their valuable contribution to social order: "you simply have to imagine that every time you want to get in or out of the building, you have to do the same work as prisoner trying to escape or a gangster trying to rob a bank plus the work of those who rebuild either the prison's or the bank's walls" (Johnson, 1988, p. 299).

Johnson didn't stop his philosophical contemplations at doors and hinges, but continued with the importance of walls and the division of labor between humans and nonhumans. Ultimately, he didn't hesitate to delve into the complex issue of the inability to trust logical thinking and humans' capacity to make good decisions, which proved the importance and the necessity of having nonhumans supporting humans and thus gaining their just social position. His harsh criticism was

DOI: 10.1201/9781003585527-6

addressed directly at social scientists whom he believed distanced themselves from technology and avoided facing nonhuman entities even though, "studying social relationships without the nonhumans is impossible," (Johnson, 1988, p. 310).

Jim Johnson was the anthropologist and philosopher, Bruno Latour, who adopted the pseudonym Johnson for writing humoristic yet critical texts advocating the need to embrace a highly inclusive vision of life and social systems. Latour presented a worldview in which the place of humans is – and should always be – part of an over-encompassing ecosystem that includes all that exists on planet Earth without divisions. This was the main goal of his anthropology of science, *We Have Never Been Modern*: to break down distinctions between life and nonlife, modern and premodern, humans and objects, in order to reconnect humans and nature rather than follow the modernist ideology of separations (Latour, 1993). While focusing on the interdependency of the relationships between humans and nonhumans, Latour's studies had a strong impact on the growing awareness of animal rights, the life of plants, and the significant role of artificial lifeforms in human life. His ideas have become a cornerstone in establishing ethical rules of behavior and responsibility in developing new technology, advanced robotics, and artificial intelligence (AI) knowledge.

Latour's ethnographic work with Steve Woolgar, *Laboratory Life: the Social Construction of Scientific Facts*, presented the scientific work conducted in a lab as a socially constructed process rather than a search for truth, following the footsteps and criticism Thomas Kuhn had initiated years earlier in *The Structure of Scientific Revolutions* (Kuhn, 1962; Latour & Woolgar, 1986 [1979]). Along the wave of developing semiotics and post-structuralist ideologies that flourished during the '60s, Kuhn's study presented a major breakthrough by framing scientific knowledge as a competition of theories, cultural values, and beliefs instead of the traditional view of science as a pursuit of an objective truth. Kuhn opened the door for new models of thinking about revolutions, knowledge, and academic competition (Reisel, 2021).

Latour made a similar remarkable leap by criticizing not only academic and scientific politics but also the traditional Cartesian dualism that deprived nature of agency and destroyed values of animism, leading to a modern philosophy which related to nature as an object of experimentation separated from humans and life (Kim, 2020, pp. 12–3). This became the core of the actor-network theory (ANT): studying the complex network of interdependent relationships between humans and nonhumans that creates and sustains social systems. ANT was developed during the 1980s by Bruno Latour, Michel Callon, and John Law, and it emphasized the important contribution of the nonhuman entities in the development and change of social structures and human life (Callon, 1984; Latour, 2005; Law, 1992).

ANT gained prominence and prestige thanks to several major reasons. The most meaningful one was the emphasis on the interrelations between modernity, science, and nature, and the call for multidisciplinary collaboration in order to understand all the factors that influence social and political structures. The immediate result of such a collaboration is the need to ensure there is no bias and discrimination between the participants in the network, a bias that is currently at the center of AI ethical debates which still focus on the human perspectives and human advantages at the expense of other valuable nonhuman members of the ecosystem, such as the destruction of nature. How can humans create advanced new tools that can lead to a better future without destroying key participants on the planet?

During the Covid-19 pandemic, ANT's ideologies regained popularity with Latour's open criticism of global warming and the positive results of Covid closure. Talking about the fast stop of the entire global economic system, he related to the ideology that "the train of progress" could not be stopped by showing how easy it was to stop it and what benefits the ecological impact of that stop had on the climate, the air, and the humans (Latour, 2020). This was indeed an opportunity to see how closely related the elements are, and that changing only one element can lead to a chain reaction of improvements for the entire ecosystem.

Nevertheless, I would like to relate to a crucial and significant factor that contributed to the fast growth of ANT and the popularity it reached: the fact that it was a chain within the exact same complex network of interrelations the theory itself presents. The scholars were famous; the academic institutions were prestigious; the historical period was a time of flourishing academic research and new studies; and above all, the development of system theory that preceded ANT by several decades planted the seeds of the ideas[1]. General system theory and cybernetics originated in the 1940s with new studies and ideas led by pioneering scholars like Bertalanffy, Wiener, and Ashby, who focused on the interconnections that inevitably exist among systems. They were calling for multidisciplinary approaches using the concept of "holism", a perspective that aligns closely with ANT as well as with other theoretical frameworks that prioritize complex networks over closed systems (Ashby, 1956; Bertalanffy, 1968; Wiener, 1948). This was the beginning of multidisciplinary research that continuously developed and reached a peak in the 1990s with the emergence of complex systems, which aimed to integrate previously distinct and separated fields of study: biology, computer science, economics, engineering, cognitive science, medicine, social sciences, and more.[2]

IS A VIRUS ALIVE OR DEAD?

I would like to start by focusing on the Covid-19 virus used by Latour to demonstrate the powerful agency a nonhuman entity has to change the world and even collapse its economy. Is a virus life or nonlife? Can we apply intention, awareness, or consciousness to one cell if it is a natural cell? How can we define cells, like cancer or stem cells, that have the potential to be immortal? Would immortality indeed be part of what is called "life"? Dracula isn't considered "alive" and neither are vampires. They are usually referred to as "undead".

Such questions lie at the heart of the ethical debates over AI. As mentioned above, most of the debates are grounded on a human-centric perspective that has been dominant for years. The traditional tendency is to create distinctions between human life and nonhuman entities, as according to most Western religions, they are not considered living entities. However, theories of complex systems along new research conducted in artificial life (Alife) studies suggest that the separation of life and nonlife is not as simple and clear-cut as it seems, and the boundaries are vague and ambiguous. Given the new and unclear definitions of life, death, being, and nonbeing, there is a need for a fundamentally new paradigm of ethics, its practical application, and the implications for the future. Traditional definitions no longer apply.

UNICELLULAR ORGANISM, CONSCIOUSNESS, AND ETHICAL SWARMS

In 2018, exactly 30 years after the publication of Johnson/Latour's article, I was sipping a hot can of coffee in the back of a conference room, reflecting on the network of neoliberal economy that enabled my coffee break on a freezing day – coffee beans from the other side of the world, milk from Hokkaido, aluminum cans imported from several different countries, vending machines, electricity, conference rooms with warm air – a multitude of entities that collaborated in order to create a comfortable day for the participants. It was my first time at a conference on Alife, and I was thinking about the progress and changes ANT has made over three decades. Suddenly, I realized that a fascinating discussion was going on next to me. Several of the conference presenters were engaged in a conversation about tiny entities that exhibit independent decision-making; however, one entity related to natural cells while the other to artificial ones. One speaker mentioned that a solitary cell can go rogue and develop deadly cancer, but there can also be a mutation among a whole group of cells, leading to cancer. He was impressed by the potential of one cell to lead a large multicellular body to death, and he continued to discuss what he defined as "the immortality of cancer and other cells".

"Interesting", responded the other speaker, a computer engineer. "Same thing happens in my tiny swarm robots, and I have no idea why. Sometimes one robot suddenly breaks away from a swarm and causes disruption. Or sometimes a whole group can take a different direction not related to the algorithm that we developed for the swarm behavior. It is a major problem, this feature of anti-social behavior in swarms". This was fascinating, and allowed me to join the conversation.

As a social scientist, my responsibility in high-tech development is to support the design of social behavior among artificial entities, whether large-size robots or tiny swarms. My team and I had never encountered individual behavior adaptations among artificial entities of any size.

The unicellular organism is an organism made of one cell that performs all the functions necessary to support what is defined as life in biology: metabolism, division, growth, reproduction. It demonstrates a high ability of self-preservation, responds to stimuli and triggers in the environment, and as mentioned, can show immortality by continuing to reproduce and grow rather than reach an end and die. Unicellular organisms live inside the human body and are found throughout nature in a large variety of forms and shapes, such as bacteria, amoebas, yeast, fungi, or algae, to mention a few. Gut bacteria are particularly fascinating as they live in symbiosis with their human host and they contribute to health, energy, degenerative diseases, and even taste formation, sexual desire, and mental health (Rowe et al., 2020; Yoo et al., 2020; Zhang et al., 2022). "Many human genes are homologs of bacterial genes… We are fundamentally dependent on a myriad of essential neurochemicals produced by microbes," and it is assumed that the brain's serotonin cannot develop in the absence of several microbes, which means our social behavior depends on symbiosis with multiple organisms that guide us from inside (Dinan et al., 2015).

The traditional Anthropocene questions vital issues such as intentionality and agency, which, as Kim mentions, were the privilege of humans and consciousness and could be interesting in different ways (Kim, 2020). Whether agency is perceived as the power to influence others, take action, gain control, or even destroy nature, recent developments in microbiology and the Covid-19 pandemic imply that single-celled organisms can exert agency over human development, health, and even life, and this raises questions about the meaning of "agency" and the place of intention, awareness, or even intelligence within the human-nonhuman, life-nonlife discourse (Otto et al., 2020).

The question of agency stands at the center of the ethical debate over nonhuman agents and their contribution to human life. It is the core of conflict within the growing technology based on AI as new advanced technologies are perceived in Western eyes as "threatening" and "dangerous", adjectives applied to human character. However, when looking at ANT, unicellular biology, and artificial neural networks, there is no place and no meaning in the application of human adjectives since there is no conscious and no psychological behavior or awareness. In ANT, there is a relational responsibility that lies on all participants equally, and in unicellular movements, the lack of both intentionality and moral codes should be the guiding lines for a new code of ethical rules. The goal is not simple in a world that is still controlled by human ethics, but new paths are already opening thanks to the improved participation of technology itself. Unicellular and simple units do not carry intentionality and responsibility, but the fact they have agency to impact a large part of the system implies the responsibility of moral designers, i.e., human ethics, to control the agency and guide the system to balance. One good example is the behavior of natural and artificial swarms, which are a central meaningful actor in nano-technology and advanced medical treatment.

Swarm intelligence has been observed in nature and has become an important part of research across a variety of fields and disciplines, ranging from biology to social systems. Swarms demonstrate highly controlled collective behaviors that show rapid adaptability to environmental changes, flexibility, fast self-organization, and an impressive balance between the autonomy of the individuals and the harmony of the group. Each member of the swarm has an independent role to perform while contributing to the group and creating a continuously effective system with a high survival rate. This excellent balance results in the development of a resilient and robust group that exceeds the potential and success rate of one individual. Swarms triggered curiosity and inspired artists, musicians, and nature lovers, and became a central research theme in fields of science due to the mystery around the mechanism that guides and designs their behavior. How do bees, ants, birds, fish, and even microbes manage to perform their individual tasks and at the same time follow the needs and behavior of the swarm? Such behavior can be seen among large mammals and herds as well, not only among small insects that need the protection of the group. The Masai Mara migration crossing is most likely the best example of swarm intelligence, where a large number of wildebeests cross the river as one, following each other, and yet each single animal shows attention to its close environment and the surrounding in order to be careful of potential predators and immediate dangers. The balance between the survival of the one and the harmony of the many is the core of ANT and should be the core of ethics when the world shifts to higher AI usage since only a balance of single units and the larger network can survive together over time.

Social insects living in swarms have been adopted in technology in recent years as a result of both ethical and financial challenges met in the process of development and application of large-scale technologies and robots. They are cost-effective, decentralized and safer, as one damaged member cannot compromise the entire system. Consequently, ants, birds, bees, and many other swarming organisms inspired computer scientists and engineers to study their natural behavior and develop artificial swarm systems that can better address the emerging challenges and obstacles of contemporary technology. Social insects exhibit intelligence in organization and work division that are visible and logical, yet difficult to explain although they are easy to replicate in artificially constructed swarms. The fascinating ability of internal self-organization is a key feature of swarm systems, and it has been established that social insect colonies create an effective decentralized system that enables fast solution of problems. The advantage of decentralization is the fact that most members of the swarms are not part of the solution, and usually they are not even aware of the problem that occurred. Such localized solutions are a major challenge in advanced technological systems that are still centralized, and problems or accidents lead to major global issues.[3] Natural swarm colonies succeed to find food for all the members, build complicated structures on and below ground level, make construction repairs, solve problems well, and even share information (Deepa & Senthilkumar, 2016).

One of the most interesting features of the swarm is the ability to spread alarms in case of danger. Such alarms reach a much larger number of members in the swarm than in the case of ordinary messages which are limited to a specific group. Bees have specialized individual workers, the guard bees, that are in charge of protecting the hive and sending alerts to all members in case of emergency. Such alarms are centralized and spread to the entire colony even if the members are outside of the hive ("Who Do Guard," 2019; Moore et al., 1987). Research in microbiology discovered that individual cells, and even smaller particles inside the cells, work the same way and learn to send other cells warming signals and messages outside of their membrane walls, and even outside of the environment. In a study of RNA, scientists were surprised to observe that human cells can absorb messages sent by RNA and "read" them as explanations of how to create functional proteins they otherwise wouldn't have been able to make. This recent discovery implies communication is constantly active on all levels of existence from the smallest to the largest scale (Melchor, 2024).

Ants exhibit even more sophisticated ways to control and regulate behavior, not only of the colony but even of each individual ant. A recent study discovered that leading ants have the ability to encourage or discourage collaboration of members depending on the needs of a specific task. They can encourage lazy ants to work faster and support others in case of a difficult task, and they can restrain and prevent ants from working in case of unnecessary missions. Automatic social contagion leads to disruption and chaos, and ants have ways to block it before it occurs (Porfiri et al., 2024). In addition, ants use various ways for creating networks, including chemical materials, smells, and, most interestingly, a spatiotemporal guidance that supports sensing inside the dark tunnels of their underground structures (Heyman et al., 2017). Such natural adaptations are an excellent study for artificial swarms and a way to study safety, decentralization, and new forms of communication among members of one closed swarm system. The capacity to create harmony and social cohesion while balancing the needs of the one and the many leads to optimal performance, equality, and low risk of chaos or simple inefficiency.

Artificial swarms are developed according to two main guidelines: rule-based approach and task-based approach. The former designs behavioral rules for each individual in the swarm that follow only the patterns of movement allowed. This approach can control function but cannot control results since following behavior does not guarantee a successful result. The latter approach centers on the task that has to be accomplished, and the robots are allowed to find their own ways of accomplishing the mission (Kano, 2023). The safety of artificial swarms stems from the preference of quantity over complexity, thus creating robots that can perform simple tasks, localized, and with no risk for the swarm and external participants.

But what happens when a nonlife entity, such as one member of the artificial swarm, unexpectedly exhibits a behavior of real life, independently of the group, and without triggers or stimuli from the outside? What if a robot exercises agency and leads to death, like a bacteria or virus? The concept of emergence, the appearance of new unpredicted behavior, has major ethical implications that should be the study of multiple disciplines since emergence bridges the sciences of life with technology, philosophy, and social sciences. The moral behavior of swarms leads to a strong swarm but is different from what humans define as moral or ethical in social communities. The more we understand the mechanism of real phenomena, whether in nature or in our own bodies, the more complicated the separation of life and nonlife becomes, and the more complex the dilemma of ethical behavior becomes.

Such questions demand a new paradigm of values since responsibility is distributed across all members and entities that are part of a complex system. This is precisely the moment when actor-network theory should reemerge rather than vanish under the human-centric hype of technology and contemporary R&D. Responsibility no longer lies within the humans but also applies to our creations as the timeless myth of Frankenstein should have taught us long ago. What kind of moral and ethical code of conduct and behavior rules should we develop and exhibit toward the artificial life we create? I would add that we have to remember that our own human codes of ethical behavior and rules of conduct are not set by an individual human but by a large group of decision makers who work as a swarm as well – each one has an opinion, and eventually the group decides what is the best for the survival of the many. All social interaction and established institutions are developed and changed based on the same rules and the balance of the one and the many, and they are responsive to emergent behavior and changes that occur daily, as we have seen during Covid-19.

ALIFE AND THE INCLUSIVE SYSTEM AS ETHICAL GUIDELINES FOR AI–HUMAN RELATIONSHIPS

Witkowski and Schwitzgebel argue that artificial life should be like a pet and be treated with the same level of responsibility, attention, and care. They raise the question, "Under what conditions does a system deserve moral considerations intrinsically, or for its own sake, as opposed to extrinsically or derivatively?" (Witkowski & Schwitzgebel, 2024). The ethics of Alife can go beyond human ethics and design a new system not only for the moral behavior of all entities on Earth but also for the safety and the survival of the planet. The U.N. calls for values including respect, dignity, and equality for all, and we need a new path to enable them. This path goes through re-evaluation of the system and development of horizontal ethics guidelines rather than the vertical system that has existed so far.

In a task-based approach, swarm robots exhibit an independent learning process when struggling to conduct their assignments. The learning is conducted independently among individual robots within the swarm, and they develop new and better strategies to fulfil the task at hand. In addition, rather than simplifying tasks, saving energy, or showing opportunistic methods, robots solve problems by developing creative strategies and improved coordination in sharing with the other members (Pérez & Charpillet, 2017). These results are better when using embodied evolution (EE), in which free and unregulated exchange of genetic material between the robots can take place and lead to new variations and higher individuality among the robots. In EE, robots need to find ways to better coordinate their actions and behavior, a process that creates new forms of activity, creative task performance, and new genetic material in each single robot.

Swarm behavior and emerging new patterns that imitate real EE as happens in nature can be tested only when there is no central authority or interference during the process of action.[4] However, natural swarms, one-cell organisms, and DNA demonstrate similar EE exchange, and yet they do not provide an answer to the key question: How did life begin? What led to the first multicellular bodies? We find similar key questions in natural evolution as in artificial swarms since neither spontaneous evolution nor EE can guarantee better results. The search for the beginning of life and the meaning of consciousness, which are the main features that define human existence and place it at the center of the planet, has no clear answers. Contemporary research assumes the emergence of life started from units of flexible proteins, excitable membranes, and vibrating cytoskeletal polymers, or in simple words, the thread of proteins that through processes of natural and spontaneous emergence gave rise to sentience and memory, and supported the entire evolutionary process and the living systems that developed (Thangamani & Arumuganainar, 2024). Therefore, there is no specific known combination of cells, proteins, molecules, or external interaction that can explain the emergence of life and of multicellular life forms.

SWARM INTELLIGENCE AMONG HUMANS

In 2005, Suroweicki published a controversial study on human swarms based on the theory that under certain conditions, humans perform better and make more intelligent decisions as a swarm than individually. The theory is particularly true when it comes to predictions and performing tasks. Warren's studies showed that crowd behavior proves there is swarm intelligence among humans when they are in large numbers in groups (Warren, 2018). When in crowds, humans show clear forms of intuitive self-organization and an automatic following of others for better or for worse. Modeling crowd behavior and decision-making process proves there are patterns that can be deciphered and used for developing smarter cities and a better life for their inhabitants. In my own work as a tech anthropologist, I often find swarm followers on social networks, YouTube Fan Base, and Instagram. I prefer to avoid using the term "intelligence" since I am no longer sure what it means, but swarm behavior is no doubt part of human behavior.

The fact that swarm intelligence and decision-making is observed in so many different forms of life and organized groups, from unicellular to humans, calls for a major paradigm shift of our basic definition of ethics. What is ethics? How can we think of ethics in a completely different way?

I believe the resurfacing of animism and the increasing popularity of various values adopted from Japanese culture can be the road to new ways of thinking since these trends show a growing search for alternative ideologies and a breakthrough from older and traditional ways of looking at the world. Japanese Shinto traditions have always regarded the entire world as one system, a horizontal line of equal existence for all living and nonliving things. The Shintoist view of life and ethics is the tree on which all Japanese culture and values rest: "the roots – which store and send back up vital nutrients – that underlie the dynamics of growth and revitalize the cultural tree" (Wargo, 1990, p. 500). All the forms of institutions and beliefs, from Confucianism to Buddhism and Zen, were shaped in forms that were in harmony with Japan's indigenous faith, Shinto. The animistic view of Shinto places the human actor as responsible for everything around rather than the superior entity entitled to enjoy everything around even at the cost of its destruction. This view stands in strong opposition to the vertical central power of humanity as Western monotheistic religions perceive life on Earth. Western religions view the intellect as a gift of God and therefore the human is above less intelligent life forms. However, in Shinto, the divine power, the humans, and all of nature's living and material creations are one. The divine is called *kami* and each thing and living creature can become a *kami*. In this worldview, all that exists on the planet along with the *kami* collaborate on an equal level and shares everything together (Wargo, 1990, pp. 503–4). This is the view expressed in ANT; this is the harmony of swarm intelligence that brings together so well the choices of the one with the needs of the many. Quantum studies prove these theories to be accurate to a high extent as on the particle level there is no difference between the different forms of visual expressions that developed on the planet, whether considered life or nonlife, material or abstract.

THE SOUL AS EMBODIED EVOLUTION: FROM SHINTO TO QUANTUM PHYSICS

The same complex question still exists regarding a definition of EE. The seminal work regarding EE done by Watson et al. was published in 2002, and yet, definitions are still difficult. There are many descriptions of the process, and of different stages of evolution, but the exact meaning of spontaneous evolution or EE remains elusive. Much like the origin of life, all forms of its evolution are still unknown as is the origin that leads to the spark of life and continuous evolution. Additionally, the dominant theories agree that evolution is open ended and keeps changing and adapting. Quantum science adds some confusion rather than clarity, as the soul has been absorbed into scientific studies and merged into quantum physics and neuroscience. From a religious point of view, it may seem like a conflict; however, as this entire universe is perceived to be the act of God, even the religious realm can view these discoveries as a much larger and deeper image of the work of the creator. Quantum studies give space to a sharper view of reality and new understanding of how life and our planet exist and develop.

In 2017, a study conducted by Ceylan et al. presented for the first time the concept of the soul within studies of quantum and provided explanations drawing from psychology, neurology, and quantum physics. The authors compared the structure of the brain to the structure of the universe, and through theories of quantum physics, provided a theory of the process in which the soul as an abstract entity was reduced into consciousness. The soul is portrayed as an "entity" much like other entities that exist in the universe, but it belongs to a group of entities that have no material shapes (while some do have clear material forms). Those are called amorphous entities, and the brain cannot fully grasp their meaning. Amorphous entities have no limit and no quantifiable space; therefore, the process of materialization of the soul involves reduction.

Although the article was revolutionary, it wasn't the first to connect the soul and the brain to quantum mechanics. Roger Penrose (1989) introduced theories of the brain, science, and the universe. Stuart Hameroff presented theories about the influence of quantum on consciousness. And obviously, first and foremost, the founder of quantum theory, Max Planck, claimed consciousness is the origin of everything and an important part of the fabric of the universe. Planck's ideas and theories preceded many of the contemporary discoveries and studies, and his views had a strong impact on research and philosophy of the connections between body, soul, consciousness, and the universe. One open question that always remains at the core of the concept is whether the soul, or consciousness, are unique to humans or, being the part of the universe itself, are shared by everything and everyone on our planet.

Circling back to Japanese Shinto, the ways in which it regulates life and behavior can provide a road to new ethical rules for AI and human–machine harmonious relationships. During the bubble age in Japan, the socio-political value of a population of middle-class citizens originated in the Shinto ideology of equality and harmony for all creations, life and nonlife alike. Many Japanese carry old clothes to shrines and do not throw them to the waste bins since the clothes carry parts of the human's soul and body with them. The famous guru of home cleaning and order, Marie Kondo, asks people to bless every item they decide to let go, and to thank them for what they have provided to the owners during the time they were part of the home and the family. The fact that Kondo and animism have long been popular in Japan and outside of Japan shows these are ideas that resonate with many, and humans desire to feel a sense of belonging and sharing with things as well as with other humans. Whether it is food or a teddy bear, things are part of human needs, wishful desires, and identity, and the place we give them should be equal to the place we design for ourselves and our loved ones. People follow Shinto because they are not afraid of the *kami*. Building a sense of "all together now", as the Beatles tried to deliver, is the key to including many new actors into one ecosystem with common values and aspirations. It is already spread through popular culture in successful movies such as *Interstellar* in the West and Hayao Miyazaki's animation in Japan. The global success of these popular products shows there is acceptance and support for a new model of looking at the world and being part of it.

CONCLUSION

In his recent book, *Nexus*, Harari (2024) explores how language models and information technologies shaped human narratives, political institutions, and social values throughout history. He criticizes the role and power new technologies have to spread ideologies and manipulate individuals through storytelling that turns fiction and fantasy into an assumed reality, much like the affective labor conducted on many Instagram pages. Like Kuhn's revolutionary theories about prestige wars and competitions between scientists, Harari builds a vision of a future in which AI-generated artificial agents and information are used as a tool for spreading ideologies, false narratives, and population control that deprive people of their free will. The author is pessimistic and raises doubts about a future of harmony. His future is threatened by new divisions based on traditional dichotomies: machines–humans, life–nonlife, truth–lies.

Each historical age is defined by models of thinking that spread and relate to all aspects of life. The model of the current age develops values of diversity, fluency, and fluid boundaries. Search for the elusive quantum, multidisciplinary studies, patched clothes, fluid gender, work at home, or afternoon naps in the park are all reflections of these models and values as observed and practiced in daily life. Under a seemingly disarray of values, lies a clear network of connected values, all based on the same model. This is the right time to call for traditional definitions of ethical boundaries to collapse and design new models of fluidity, flexibility, and soft boundaries that will open a space for new ways of thinking, more inclusive and with a large spectrum of new possibilities.

The positive changes are felt everywhere, from rising animal rights to studies of how plants think, how trees feel, and the quantum search for the soul of the Earth and the humans. Human machines have already crossed the solar system, and humans realize the tiny place we occupy in the universe. This is an age of revolutions and new opportunities in which the AI can indeed be our colleague rather than our enemy. It all depends on the narrative we choose to carry with us into the future.

NOTES

1. Due to the scope of this text, I cannot develop the background and innovations of the historical period and its influence on the ANT, but it is important to mention the 1980s were the age of postmodernism and new ideologies, which share many common ideas with ANT (such as the collapse of the grand narratives, the disappearance of the concept of one Truth, vague boundaries, and questions regarding the meaning of God). The early '80s were also a boom of AI that led to the development of expert systems and later to complex systems. There were many new theoretical frames and ideas that were developed or expanded during those decades: game theory, chaos theory, science fiction movies, new brain studies, and many more.
2. Some theorists point to the nineteenth century and the first studies of thermodynamics as the origin of complex systems; therefore, the origin and the evolution of the idea of an interconnected network over separated units has a long history that goes far beyond ANT. I chose to focus the introduction on Latour since his relation to the virus's impact on the entire global economy is at the center of this study that questions the relations between one-cell entities, like viruses, and multicellular bodies, whether natural or artificial.
3. For example, the most recent global computer failure at Microsoft that paralyzed airports, banks, and vital facilities in July 2024.
4. There are many other factors that need to be taken into consideration, but there is no space to relate to all of them in this frame of work. However, I would like to mention the importance of the size of the swarm, the types and size of the algorithms used to design them, the tasks they are given, where there are specific limitations provided in the experiments, collision avoidance, and so on. There are various forms of swarm design, from EE to cloning of the same algorithm that limit the possibility of change as the same genetic material is exchanged.

REFERENCES

Ashby, W. R. (1956). *An introduction to cybernetics*. Chapman & Hall.

Callon, M. (1984). Some elements of a sociology of translation: Domestication of the scallops and the fishermen of St. Brieuc Bay. *The Sociological Review*, *32*(1). https://doi.org/10.1111/j.1467-954X.1984.tb00113.x

Ceylan, M. E., Dönmez, A., Unsalver, B. O., Evrensel, A., & Yertutanol, F. D. K. (2017, June 7). The soul as an uninhibited mental activity is reduced into consciousness by rules of quantum physics. *Integrative Psychological and Behavioral Science*, *51*(2). https://doi.org/10.1007/s12124-017-9395-5

Deepa, O., & Senthilkumar, A. (2016). Swarm intelligence from natural to artificial system: Ant colony optimization. *International Journal on Applications of Graph Theory in Wireless Ad hoc Networks and Sensor Networks (GRAPH-hoc)*, *8*(1), 9–17.

Dinan, T. J., Stilling, R. M., Stanton, C., & Cryan, J. F. (2015, April). Collective unconscious: How gut microbes shape human behavior. *Journal of Psychiatric Research*, *63*, 1–9. https://www.sciencedirect.com/science/article/pii/S0022395615000655

Harari, Y. N. (2024). *Nexus*. Random House.

Heyman, Y., Shental, N., Brandis, A., Hefetz, A., & Feinerman, O. (2017, June 1). Ants regulate colony spatial organization using multiple chemical road-signs. *Nature Communications*, *8*(1), 15141. https://www.nature.com/articles/ncomms15414

Johnson, J. (1988). Mixing humans and nonhumans together: The sociology of a door-closer. *Social Problems*, *35*(3), 298–310.

Kano, T. (2023). Review of interdisciplinary approach to swarm intelligence. *Journal of Robotics and Mechanics*, *35*(4), 890–5.

Kim, J. (2020). The problem of nonhuman agency and bodily intentionality in the Anthropocene. *Neohellicon*, *47*, 9–16.

Kuhn, T. (1962). *The structure of scientific revolutions*. University of Chicago Press.

Latour, B. (1993 [1991]). *We have never been modern* (C. Porter, Trans.) Harvard University Press. (Originally published in French.)

Latour, B. (2005). *Reassembling the social: An introduction to actor-network-theory*. Oxford University Press.

Latour, B. (2020, March 30). *What protective measures can you think of so we don't go back to the pre-crisis production model?* (S. Muecke, Trans.). (Originally published in French: 2020 March 29) AOC. https://aoc.media/opinion/2020/03/29/imaginer-les-gestes-barrieres-contre-le-retour-a-la-production-davant-crise/

Latour, B., & Woolgar, S. (1986 [1979]). *Laboratory life: The construction of scientific facts* (P. W. Willmott, Trans.). Princeton University Press. (Originally published in French.)

Law, J. (1992). Notes on the theory of the actor-network: Ordering, strategy, and heterogeneity. *Systems Practice*, *5*, 379–393. https://doi.org/10.1007/BF01059830

Melchor, M. (2024, September 16). Cells across the tree of life exchange 'Text Messages' using RNA. *Quanta Magazine*. https://www.quantamagazine.org/cells-across-the-tree-of-life-exchange-text-messages-using-rna-20240916/

Moore, A. J. (1987). The guard honey bee: Ontogeny and behavioral variability of workers performing a specialized task. *Animal Behavior*, *35*(4), 1159–1167.

Otto, I. M., Wiedermann, M., Cremades, R., Donges, J. F., Auer, C., & Lucht, W. (2020, January). Human agency in the anthropocene. *Ecological Economics*, *127*. https://www.sciencedirect.com/science/article/abs/pii/S0921800918313855

Penrose, R. (1989). *The Emperor's new mind*. Oxford University Press.

Pérez, I. F., & Charpiller, A. B. F. (2017). *Learning collaborative foraging in a swarm of robots using embodied evolution*. Proceeding of the 14th European Conference of Artificial Life. ECAL.

Porfiri, M., De Lellis, P., Aung, E., Meneses, S., Abaid, N., Waters, J. S., & Garnier, S. (2024). Reverse social contagion as mechanism for regulating mass behaviours in highly integrated social systems. *PNAX Nexus*, *3*(7), 246. https://doi.org/10.1093/pnasnexus/pgae246

Reisel, M. (2021). Lovers, rebels, and IVF-blade runners: An ANT model for the post-human world. In G. De Minico & O. Pollicino (Eds.), *Virtual freedoms terrorism and the law* (pp. 191–208). Routledge & Giappichelli Editore. https://www.taylorfrancis.com/chapters/edit/10.4324/9781003091561-11/lovers-rebels-ivf-blade-runners-ant-model-post-human-world-mary-reisel

Rowe, M., Veerus, L., Trosvik, P., Buckling, A., & Pizzari, T. (2020, November 21). The reproductive microbiome: An emerging driver of sexual selection, sexual conflict, mating systems, and reproductive isolation. *Trends in Ecology and Evolution*, *36*(1), 220–234. https://www.cell.com/trends/ecology-evolution/fulltext/S0169-5347(20)30307-4

Thangamani, A., & Arumuganainar, D. (2024). Emergence of information processing in biological systems and the origin of life. *Communicative & Integrative Biology*, *17*(1). https://www.tandfonline.com/doi/full/10.1080/19420889.2024.2373301

von Bertalanffy, L. (1968). *General system theory: Foundations, development, applications*. George Braziller.

Wargo, J. J. (1990). Japanese ethics: Beyond good and evil. *Philosophy East and West*, *40*(4), 499–509.

Warren, W. H. (2018). Collective motion in human crowds. *Current Directions in Psychological Science*, *27*(4), 232–240.

Who Do Guard Bees Recognize as Intruders? (2019, February 24). Who do guard bees recognize as intruders? *eLife*. https://elifesciences.org/digests/41855/who-do-guard-bees-recognize-as-intruders

Wiener, N. (1948). *Cybernetics: Or control and communication in the animal and the machine*. MIT Press.

Witkowski, O., & Schwitzgebel, E. (2024). The ethics of life as it could be: Do we have moral obligations to artificial life? *MIT: Artificial Life*, *30*, 193–215.

Yoo, J. Y., Groer, M., Dutra, S. V. O., Sarkar, A., & McSkimming, D. I. (2020). Gut Microbiota and immune system interactions. *Microorganisms*, *8*(12), 1587. https://doi.org/10.3390/microorganisms8101587

Zhang, H., Chen, Y., Wang, Z., Xie, G., Liu, M., Yuan, B., Chai, H., Wang, W., & Cheng, P. (2022, February 14). Implications of Gut Microbiota in neurodegenerative diseases. *Frontiers in Immunology*, *13*, 785564. https://doi.org/10.3389/fimmu.2022.785644

4 Confucian Philosophy Revitalized in the Digital Era

Maintaining Interpersonal Ethics with Artificial Intelligence

Liang Wang and Ruiping Fan

INTRODUCTION

With the widespread use of social media and chatbots, Artificial Intelligence (AI) is profoundly changing traditional social interaction. However, with the growing reliance on AI social tools, the risk of superficiality or shallowness in interpersonal relationships is becoming apparent, especially in family relationships. Prolonged reliance on AI for communication and interaction may lead to the degradation of individual social skills and weaken the emotional depth and quality of interaction in face-to-face communication. Interpersonal communication gradually becomes data exchange, and emotional communication and authentic interaction are replaced by superficial human–machine dialogue. In addition, the neglect of ethics and interpersonal connection in the design of AI technology further exacerbates the negative impact of technology on the shallowness of human relationships. Such shallowness not only weakens the sense of intimacy and responsibility among family members but also threatens to undermine the broader structure of social trust and affect the overall interpersonal harmony of society.

This essay draws on Confucian resources to address the problem of superficial human relationships that are likely brought about by the development of AI technology in the digital age. As a long-standing Eastern philosophical tradition, Confucianism upholds a virtue ethics that emphasizes the moral significance of family relationships and moral reflection, providing profound insights on meaningful human lives. Firstly, Confucian ethics emphasizes the deep emotional ties and sense of responsibility among family members and builds solid interpersonal relationships based on moral virtues such as "benevolence" （仁) and "filial piety" (孝) . It reminds people that in a social scene that relies heavily on AI, they need to pay attention to maintaining the emotional depth and real interaction between human individuals to prevent their relationships becoming formalized and alienated due to technological intervention. Secondly, the moral reflection advocated by Confucian ethics provides a reference for the moral design in human–machine relationships. Specifically, ethical reflection demonstrates multiple advantages in the ethical design of AI from the perspectives of developers, users, and overall social harmony. For developers, ethical reflection prompts them to proactively consider social responsibility in the design process, ensuring that AI not only pursues technical efficiency but also pays attention to the quality of human relationships. For users, AI design based on ethical reflection can respect their emotional needs, enhance understanding and trust in human–computer interaction, and improve user experience and satisfaction. At the societal level, ethical reflection helps AI integrate into different cultures and contexts, promotes the harmonious development of social interactions, avoids the alienating impact of technology on human relationships, and thus promotes the formation of a more caring and intelligent society.

The first part of this essay delves into the challenges posed by AI in human relationships, elucidating how it can contribute to the issue of superficiality within these connections. The second section provides a brief overview of the core tenets of Confucian virtue-based and family-centered ethics,

 DOI: 10.1201/9781003585527-7

delving into the exploration of a suitable human–machine relationship grounded in Confucian principles. This is succeeded by the third section, where we analyze ethical frameworks and applications of AI through the lens of Confucian ethical considerations. The fourth section transitions to our recommendations regarding specific ethical standards to govern AI activities to prevent superficial human relationships based on Confucian ethics. Finally, the last section encompasses closing remarks.

THE CHALLENGE OF AI ON HUMAN RELATIONSHIPS

The potential damage of AI to interpersonal relationships is manifested in at least three ways: firstly, the weakening of emotional ties; secondly, human–computer trust overtaking interpersonal trust; and thirdly, the damage to family relationships. Although the widespread application of AI technology has provided great convenience for information exchange, it has also brought about the reduction of face-to-face communication opportunities and the weakening of emotional ties for human individuals. Social media platforms and chatbots (e.g., virtual assistants and customer service bots) allow people to communicate without actually meeting face to face, which is efficient but lacks the non-verbal cues and emotional expressions found in face-to-face interactions, weakening the deeper emotional connections between people. Traditional interpersonal communication relies on language and multi-dimensional and delicate interactions such as tone of voice, facial expressions, and body language, which together build rich interpersonal relationships. However, as many complaints appeared in Chinese social media indicate, with the intervention of AI, communication patterns have become progressively instrumentalized and data-driven, the depth of emotional interactions is reduced, and human relationships face the risk of becoming superficial (Zhang et al., 2019). While AI-driven tools can mimic the expression of human emotions, due to the lack of true emotional understanding and empathy, such exchanges appear to be one dimensional and mechanical, making it difficult to maintain a true emotional bond. As people's reliance on these technologies increases, emotional interactions between people are increasingly replaced by technology, even resulting in "forgotten existence;" i.e., in the pursuit of technological convenience, the depth and significance of interpersonal relationships are neglected (O'Neill, 2020), and widespread distrust in society and the formation of information silos is triggered (Turkle, 2016).

In contemporary society, people's trust in AI systems has increased significantly, even surpassing their trust in people in many contexts. Whether in online shopping, news recommendation, or medical diagnosis, AI systems are often preferred due to their ability to process massive amounts of data and their perceived greater objectivity and precision. However, as this trust in AI gradually overtakes trust in people, the traditional mechanism of interpersonal trust undergoes profound changes. While trust is essentially a complex emotion based on experience, relationships, and moral judgment, the emergence of AI has disrupted this tradition, shifting trust from person-to-person to person-to-technology, creating the phenomenon of "trust agents" (Arnd-Caddigan, 2015). While this shift may improve efficiency in the short term, in the long term it may lead to alienation of relationships and trust (O'Neill, 2002). At the same time, the large amount of text, images, and video content in generative AI poses the challenge of information authenticity. The spread of false information and deeply faked videos makes it increasingly difficult for people to distinguish the authenticity of information, further eroding trust in the information source. Information authenticity is the cornerstone of trust, and the widespread use of AI-generated content is eroding this cornerstone. This process not only erodes the authenticity and depth of human relationships but also negatively impacts overall trust in society.

Whether it is the weakening of emotional ties or the overtaking of interpersonal trust by human–machine trust, the emotional ties in family relationships are also inevitably deeply affected. With the deepening of family members' dependence on technology, family relationships have gradually become involved in this technology-driven social change, and family relationships, which used to carry deep emotional and ethical values, are gradually losing the warmth and resonance they should have. The widespread use of smart home devices, virtual assistants, and social platforms has shortened the physical distance between family members to a certain extent, making communication more convenient, but at the same time, these technologies have invariably weakened face-to-face interaction. Particularly in multigenerational or long-distance families, while technology provides a channel of contact, it fails to effectively bridge the emotional distance, leading to an increasing reliance on technology for communication and a gradual decline in in-depth face-to-face interaction among family members. Family members, especially teenagers, tend to be more inclined to interact with virtual assistants or smart devices than to share details of their lives with their parents or siblings, thus further weakening the emotional bonds in the family. In addition, family relationships are increasingly becoming instrumentalized.

As the basic unit of society, the core function of the family lies not only in the management of daily life but also in the transmission of emotional communication and ethical values (Aristotle, 2006). With the deeper involvement of AI technology, the mechanized and programmed features of family interactions are becoming more pronounced, and the depth of emotional communication is gradually eroded. Although technology provides a convenient means of communication, it cannot replace emotional interaction and non-verbal communication in face-to-face communication. This instrumentalized mode of communication has weakened the core function of the family, affecting the stability of relationships and the transmission of ethical values among family members. The family, as an important place for the cultivation of virtue and the transmission of ethical concepts, is facing the serious challenge of weakened emotional ties and estrangement brought about by technological intervention.

Indeed, the core reason for the potential damage that AI can do to human relationships lies not only in the growing dependence on the technology as it is iterated but also in the ethical failures that exist in its design. This deficit is not limited to the early stages of technology development but also extends throughout the design, diffusion, and application of the technology, manifesting itself in a lack of conscious ethical design, a lack of responsible ethical decision-making, an imbalance of trust between humans and machines, and neglect of contextually adaptive design. Behind each of these issues lies a profound ethical challenge that affects the balance between technology and society and between technology and human relationships.

First, the lack of conscious ethical design is a key issue. In the pursuit of technological innovation and functional optimization, developers tend to focus too much on technological efficiency, neglecting to care for human emotional needs and social relationships. This ethical deficit means that technology does not incorporate humane design and fails to consider its damaging impact on emotional connections. Although AI products can replace some human interactions, the lack of attention to deeper emotional needs instead exacerbates emotional alienation between people. An example of this ethical deficit can be seen in some social media algorithms designed primarily to maximize user engagement. These algorithms prioritize content that generates strong reactions—often sensational or polarizing material—without regard for social or emotional effects on users. As a result, the technology promotes divisive content, potentially harming users' mental health and weakening genuine social connections.

Second, the lack of responsible ethical decision-making manifests itself in the blurring and outsourcing of moral responsibility. In the process of technology development, developers and corporations tend to focus on market needs and commercial goals at the expense of their long-term social impact. This outsourcing of responsibility has led to many ethical issues in the widespread use of AI in society. For example, while the algorithmic design of social media platforms has enhanced user engagement, it has also increased information isolation and social polarization. Therefore, responsible ethical decision-making requires developers to consider not only the short-term market performance of the technology but also to be responsible for the long-term social effects of the technology through ethical reflection to avoid irreversible social damages caused by decision-making mistakes.

Third, the imbalance of human–machine trust reveals the gradual deepening of people's dependence on technology. In some cases, the reliability and autonomy of technological systems have, on the contrary, led to a tendency to transfer trust to machines at the expense of human-to-human trust relationships. Human–machine trust, if not properly balanced, may lead to further dilution of human relationships. In their pursuit of enhancing the human–computer interaction experience, technology designers often fail to fully consider its impact on the social trust system.

Finally, the neglect of context-adaptive design exacerbates social inequality and the ethical risks posed by technology. AI technologies are usually designed for mainstream users, ignoring the special needs of different social groups, especially vulnerable groups. Such design flaws not only contribute to the further deterioration of social equity but also may lead to implicit discrimination against certain groups, such as the problem of algorithmic bias. Together, these issues lead to the potentially destructive nature of AI in human relationships. Therefore, there is a need for greater ethical reflection so that developers of AI can incorporate human emotion, social responsibility, and fairness into the core of technology design alongside technological innovation, ensuring that the development of technology not only drives efficiency and convenience but also promotes social harmony and strong interpersonal relationships. It is within this ethical framework that AI can truly become a tool that serves all of society, rather than a hidden force that undermines human connection.

TOWARD APPROPRIATE HUMAN–MACHINE RELATIONSHIPS BASED ON CONFUCIAN ETHICS

Confucian Ethics: From Virtues to Human Relationships

The cardinal virtue of Confucian ethics is "benevolence" (仁) , and benevolence is rooted in and developed from the family (家). Mencius once said, "Benevolence is the spiritual home for man's soul" (*Mencius 4A;孟子·离娄上*) (Mencius, 2009). This means that benevolence is the house of heaven that gives us the most peace and stability in our heart. The same is true in life. In the view of Confucianism, the family is an important harbor for cultivating virtue and promoting emotional well-being. According to Confucianism, the first three of the five core social and ethical relationships in ancient Chinese society are the relationships in the family, namely, parent and child, husband and wife, elder and younger siblings, ruler and subject, and friends. The specific virtue of "filial piety" (孝) (respecting and caring for one's parents and ancestors) is fundamental for realizing the cardinal virtue of benevolence. In the *Analects of Confucius*, it is said, "Filial piety and fraternal submission! ——are they not the root of human heartedness (Ren)? " (*The Analects of Confucius Book I XUE ER; 论语·学而*) (Confucius, 1993). *The Classic of Filial Piety* (《孝经》) also recognizes filial piety as the foundation of benevolence: "Filial piety is the foundation of virtue and the source of education." (*The Classic of Filial Piety;孝经*). Here, "filial piety (孝)" and "teaching (教) " are linked, which proves that the moral feelings and ethical relationships learned in the family are the basis for teaching people to settle down and build up their lives in society. Filial piety is the emotional bond that holds the family together. Relationships between parents and children, husband and wife, and elder and younger siblings are maintained through filial piety and fraternal duty. The principles summarized in the practice of filial piety and other good and beautiful things are the "morals" of Confucian ethics.

Compared to the modern Western "top-down" principlist ethics, Confucian ethics tends to be more practical and "bottom-up," looking for guidance from the role of concrete rituals or ethical practices (Sun, 2019). This ethical concept of emphasizing individual roles is embodied in Confucian ethics, which starts from the "role" relationships in the family and then radiates to the social relationships, emphasizing the family and the concept of valuing kinship. The concept of filial piety and brotherly love (悌) in the family extends outward to the level of friends beyond the generation. This extends the good moral sentiments of the family to society, making society as close-knit as a "big family." The core of Confucian family ethics is to take the family as the starting point and the roles in the family as the intermediary with moral connotations to connect the family and society and to promote the formation of a harmonious society.

Building Appropriate Human–Machine Relationships

Confucianism attaches importance to harmonious relationships and social bonds between people and emphasizes harmonious coexistence in the family and society. Human–computer interaction (HCI) intelligent technologies aim to enhance human connections and interactions and promote a harmonious society. Confucian ethical thinking provides a framework for AI systems to serve rather than damage good human relationships.

Confucian Ethics Emphasizes Responsibility and Care

Confucian ethics emphasizes individual responsibility in the family and society, as well as care and concern for others. This sense of responsibility and caring provides guiding principles for AI systems to be designed and applied with greater attention to the well-being and needs of users. In AI applications in health care and education, this concern and care can increase system effectiveness and user satisfaction. The design and application of robots in elderly care can be facilitated by invoking the Confucian family ethic of filial piety. Firstly, "The old man is secure, the friend is trustworthy, and the young man is embraced" (*The Analects of Confucius Book V GONG YE CHANG;* 论语·公冶长) (Confucius, 1993). This is Confucius' account of the ideal interpersonal relationship in his mind, and the support of the elderly stands first in his eyes. The robot will pay more attention to caring for the elderly who are not mobile and will not regard them as a burden, but will pay more patient attention to the needs of the elderly.

Secondly, "There are three cases in which the filial son does not follow orders. When following orders will endanger one's parents, but not following orders will make them safe, then the filial son will not follow orders, and this is having scruples. When following orders will disgrace one's parents, but not following orders will bring them honor, then the filial son will not follow orders, and this is being yi. When following orders requires a beastly act, but not following orders requires cultivation and decorum, then the filial son will not follow orders, and this is being respectful." (*Xun Zi* Chapter 29 *Zi Dao*; 荀子·子道) (Xunzi, 2014). The meaning of this sentence is that if one obeys one's parents, they will be in danger; if one does not obey them, they are safe. So the filial son chooses not to obey his parents. Thus, it can be seen that filial piety in Confucian family ethics is not unified with total obedience to parents in the narrow sense, and that robots who have been influenced by Confucian family ethics will deal with the morally deficient behaviors of the elderly soberly and help them to correct them.

"Nina, 70, suffered from some kind of neurological disease two years ago, and she has become sensitive and suspicious, often yelling at her children and being rude to everyone around her." (Elder, 2023). So, in designing a care robot for her, words such as "hello," "thank you," and "please" can be used as key activators to help her change her rude behavior and build good relationships with the community. Therefore, this kind of moral, scolding robot based on responsibility and care is of great significance for maintaining social moral order and building a harmonious society (Wen, 2021). In addition, by learning the Confucian family ethics of responsibility and care, the robot design can refer to the Confucian concepts of "loyalty and forgiveness (忠恕) " and "honesty (诚信) " to treat people with integrity. Therefore, the robot will pay more attention to the protection of user privacy, and will strictly abide by the principle of security and confidentiality when handling user data to ensure the safety of user information. This behavior also enhances the user's trust in AI and makes the user feel understood and cared for, thus feeling the care of AI, which is of great significance to the enhancement of the well-being of the whole society and the harmony and stability of the society.

Confucian Ethics Highlights the Importance of Moral Emotions

Confucian ethics emphasizes the expression of emotions and the importance of ethics, and considers emotional bonds to be the key to maintaining social and family relationships. This provides a new moral guidance significance for the development of AI. For example, in terms of empathy simulation, relational AI can enhance its role in interpersonal interactions by mimicking and facilitating human emotional exchanges, i.e., cultivating empathy, e.g., by expressing emotional feedback such as understanding, support, and consolation through language, as well as by "remembering" the user's emotional responses and utilizing this information to adjust communication styles and other emotional feedback in future interactions. This information can be used in future interactions to enhance the link with the user by means of emotional memory, such as adapting communication styles. In terms of emotion recognition and response, relational AI analyzes human intonation, vocabulary choices, facial expressions, and body language through natural language processing (NLP) and computer vision technologies to identify the user's emotional state. Once the emotional state is recognized, the AI can choose the appropriate tone and language to respond, mimicking the way humans communicate in a given situation. This "anthropomorphic" setting makes humans feel natural and friendly in human–computer interactions, and can promote a better user experience. Finally, AI can act as an intermediary between family members, friends, or coworkers to help them better understand each other's emotional states and needs. Confucian emotional ethics provide a theoretical basis for AI to become more suitably "skillful" when interacting with humans.

Confucian Ethics Emphasizes Long-Term Stable Social Relationships

Its core value is embodied in the practice of benevolence, which covers such ethical principles as "father's kindness and son's filial piety (父慈子孝) ", "husband's righteousness and wife's cooperation (夫义妇顺) ", "elder brother's friendship and younger brother's respect (兄友弟恭) ", etc. These principles form a deep emphasis on the durability and stability of social relations by maintaining harmony and order within the family. This idea provides important theoretical support for the construction of interpersonally enhanced human–computer relationships.

Long-term stable relationships not only enable AI to understand users' behavioral patterns, preferences and habits more deeply, thus making significant progress in service personalization, but also lay the foundation for the continuous learning and self-adaptive ability of the AI system. In Confucianism, all actions in social life are expected to adhere to specific norms and standards that uphold social order. Consequently, Confucianism emphasizes the significance of "ritual practice" as an essential component of societal interactions. Moreover, within Confucian philosophy, individuals are believed to have the capacity for self-improvement and self-discipline. By harnessing this potential effectively, it is believed that social harmony and stability can be attained. This enduring and harmonious social fabric, sustained through diverse ritual practices, offers valuable theoretical underpinning for fostering improved human–machine relationships. Through continuous accumulation and adjustment, AI is able to respond more flexibly to the diversity and variability of user needs. At the same time, the stable relationship provides an open and transparent decision-making framework for the AI system, which enhances the users' trust and understanding of its operation process, and thus improves their acceptance of the technology.

SUSTAINABLE AI ETHICAL DESIGN BASED ON CONFUCIAN MORAL REFLECTION

Sustainable human–machine relationships not only involve interaction between humans and intelligent machines but also relate to the long-term development of human society. It is crucial to construct a long-term, stable, and harmonious coexistence model between humans and machines in the context of rapid technological progress, while promoting the sustainable development of human society. Its core features include pursuing long-term harmonious coexistence, adhering to people-oriented technological design, valuing ethical and value considerations, focusing on cultural adaptability, and emphasizing interpretability and responsibility attribution. Moral reflection plays a key role in this process, running through multiple aspects such as conscious ethical design, responsible ethical decision-making, establishment of human–machine trust, and context-adaptive artificial intelligence design (Li, 2023).

In ethical design, moral reflection encourages designers to integrate ethical values into the process of technological development to avoid the "designer fallacy" and ensure the responsible and sustainable nature of technological innovation. For example, when designing data collection programs, moral reflection is used to protect user privacy, so that ethical considerations are implemented at the design stage. In terms of responsible ethical decision-making, moral reflection guides developers and users to consider the fairness, correctness, and potential impact of technology applications, in order to make choices that comply with ethical standards. This cautious attitude helps to build a foundation of trust between humans and artificial intelligence systems. In the process of establishing trust between humans and machines, moral reflection enables artificial intelligence systems to analyze their own decision-making processes and increase transparency and interpretability, thereby enhancing users' understanding and trust in the system. This trust mechanism based on moral reflection is an important guarantee for achieving sustainable human–machine relationships. In context-adaptive artificial intelligence design, moral reflection promotes the application of value sensitive design, enabling artificial intelligence systems to maintain sensitivity to ethical values in different contexts and better adapt to complex and changing social environments. By deeply integrating moral reflection into these aspects, we can build an AI ecosystem that is both technologically advanced and in line with human ethical values, achieving long-term harmonious coexistence between humans and machines.

Moral Reflection and AI Ethical Design

Confucian ethics emphasizes the need for individuals to cultivate self-restraint and enhance intrinsic moral motivation through moral reflection.

"For human beings, humaneness (*ren* 仁) is the peaceful dwelling, and rightness is the correct path. To abandon the peaceful dwelling and not abide in it and to reject the right road and not follow it—how lamentable" (*Mencius 4A;* 孟子·离娄上)! (Mencius, 2009). If a person abandons benevolence and righteousness, it is like abandoning a comfortable house and the right road, so it is very sad to abandon oneself. In human–machine relationships, this moral motivation can motivate users and developers to follow ethical principles, apply and develop technology correctly, and maintain sensitivity to ethical risks that may arise from data leakage, privacy violation, algorithm bias, and so on.

"There is no human being lacking in the tendency to do good, just as there is no water lacking in the tendency to flow downward" (*Mencius 6A;* 孟子·告子上) (Mencius, 2009). Users and developers can maintain the goodness of human nature through moral reflection, and achieve the goal that "the noble person preserves his mind through humaneness, preserves his mind through courtesy." (*Mencius 4B;* 孟子·离娄下) (Mencius, 2009). Users are able to reflect on the impact of their behavior on others and society when using technology, in order to adopt more responsible and ethical usage methods; developers will internalize ethical requirements when designing technology, ensuring that the negative impacts of the technology are minimized.

"One who loves others always is loved by them; one who respects others is always respected by them" (*Mencius 4B;* 孟子·离娄下) (Mencius, 2009). Mutual respect, tolerance, and understanding between users and developers are more conducive to reducing social inequality, promoting social fairness and justice, and enabling society to benefit from technological development.

"To turn within to examine oneself and find that one is sincere—there is no greater joy than this" (*Mencius 7A;* 孟子·尽心上) (Mencius, 2009). If users and developers can reach a state of sincerity and truth through moral reflection, they can obtain great inner satisfaction. Confucian ethics views humaneness or benevolence as a fundamental aspect of human nature, emphasizing that upholding benevolence involves nurturing the innate kindness and compassion within individuals. While this approach may not always align seamlessly with users from diverse cultural backgrounds or differing ethical frameworks, it can facilitate individuals from similar moral traditions in exploring shared values, respecting disparities, and striving toward ethical consensus. In the realm of technology, shaping the ethical dynamics between users and developers can be achieved through the formulation of ethical principles, integration of moral assessment tools, implementation of transparent processes, establishment of feedback channels, conduct of ethical evaluations, and other strategies aimed at fostering the responsible advancement of artificial intelligence technology.

However, people inevitably make mistakes at times. In this regard, Confucius said, "When you have faults, do not fear to abandon them" (*The Confucian Analects Book I Hsio R; 论语·学而*) (Confucius, 2016). He also said, "To have faults and not to reform them, this, indeed, should be pronounced having faults" (*The Confucian Analects Book XV Wei Ling Kung; 论语·卫灵公*) (Confucius, 2016). In moral reflection, once you find that you have inappropriate behavior, you should not be afraid of mistakes, but should immediately correct your attitude, promptly rectify your mistakes, and reduce the negative impact. If you make a mistake but are unwilling to correct it, then it truly becomes a mistake. In the human–machine relationship, users and developers of technology should conduct self-examination of their behavior to standardize their own behavior. This helps to reduce the misuse and improper misbehavior of technology, ensuring that technology is used in an ethical manner. For example, in terms of fair and inclusive design, developers should take into account the needs of different user groups when designing systems, and ensure that technology is accessible and beneficial to everyone, so as to avoid algorithmic bias and ensure fairness in the algorithmic decision-making process; when users discover discrimination and bias in technology, they can manually mark and provide feedback to developers on usage, promoting timely remediation of fairness loopholes.

Moral Reflection and Responsible Ethical Decision-Making

"There were four things from which the Master was entirely free. He had no foregone conclusions, no arbitrary predeterminations, no obstinacy, and no egoism" (*The Confucian Analects Book IX Tsze Han; 论语·子罕*) (Confucius, 2016). Confucius eradicated four flaws: not making groundless speculations, not being arbitrary and absolute, not being stubborn and rigid, and not being self-righteous. These problems often occur in people's understanding and practice. Confucius' approach inspires us to conduct moral reflection and reminds us to maintain a rational and humble attitude in the decision-making process; to comprehensively consider various ethical factors of individuals, society, and the environment; and to avoid subjective speculation and obstinacy, so that we can make more comprehensive and reasonable decisions. In the application of AI, this comprehensive consideration requires technology decision-makers to consider different perspectives and possibilities when designing and applying technology, keeping openness and flexibility, as well as avoiding unfounded guesses and assumptions, and making decisions based on facts and data. Technical decision-makers should not listen only to the opinions and feedback of others but also use their own moral imagination to flexibly make corresponding decisions based on a thorough assessment of their ethical consequences. This approach to decision-making ensures that technology is developed not only with a focus on short-term benefits but also with a view to long-term social responsibility and ethical impact, so that technology can be continuously adapted and improved.

Moral reflection encourages developers and users to take on greater social responsibility, and technology needs to consider the overall well-being of society in practical applications. For developers, in the short term, this approach might elevate development and operational costs, while also subjecting them to competitive market pressures. However, over the long term, it can facilitate the establishment of fresh competitive edges fueled by innovation and garnering increased user trust through a bolstered social reputation. As for users, in the immediate future, they might incur certain costs to acquaint themselves with pertinent laws, regulations, and prudent technological usage. Yet, in the grand scheme of things, this effort can foster greater user engagement in technological progress, elevate their technical proficiency, and enhance their ethical consciousness. In essence, accentuating the social responsibility within technological advancement is essential for advancing societal ethical standards, refining technical service protocols, and enriching individuals' quality of life. Engaging in sustained moral contemplation and action aligns with the broader interests of societal progress.

Confucian ethics requires that "the superior man does what is proper to the station in which he is" (*The Doctrine of the Mean Book XIV; 中庸·第十四章*). Both developers and users should have a strong sense of social responsibility and understand their position and obligations in the social and technological fields. “When he can put forth his ability, he takes his place in the ranks of office; when he finds himself unable to do so, he retires from it” (*The Confucian Analects Book XVI Ke she;* 论语·季氏) (Confucius, 2016). Developers should be focusing on enhancing technological capabilities, developing products and services that benefit society, ensuring that their technological products comply with industry standards and legal regulations; government should monitor the process, avoiding adverse consequences such as invasion of user privacy, dissemination of false information, or exacerbation of social inequality; users should use technology reasonably and responsibly, comply with the terms of use, refrain from any illegal or unethical activities, and ensure that their usage does not have a negative impact on society. At the same time, technology changes and develops rapidly. It requires developers to continuously learn new technologies and enhance their own development capabilities in order to better serve society and users; users should also continuously learn and improve their technical literacy in order to better utilize technology. For example, in response to the social need to protect the environment, the design, manufacturing, use, and disposal of machines should reflect the concept of green development in all aspects; developers need to consider using environmentally friendly materials and production methods, and users need to consider how to use and recycle machines reasonably, working together to reduce the burden on the natural environment.

Certainly, the technology development sector is primarily profit oriented. Alongside introspective moral scrutiny, the collaboration of additional oversight mechanisms is essential. Society must establish a multi-tiered and comprehensive supervisory framework, typically encompassing governmental regulations, industry self-regulation, public scrutiny, third-party audits, legal recourse, and other components. These measures are vital for ensuring corporate adherence to legal frameworks, upholding market integrity, and safeguarding the rights and interests of stakeholders.

Moral Reflection and Trustworthy AI

To be able to draw analogies from what is close at hand may be called the method of benevolence. (*The Confucian Analects Book VI Yung Yey;* 论语·雍也) (Confucius, 2016). “Being able to draw analogies from what is close at hand” emphasizes putting oneself in others’ shoes, using one’s own experiences and emotions to empathize with others. This is a way of practicing benevolence (仁). In the process of technology development and application, developers should consider the user’s position and feelings, and pay more attention to the possible problems that users may encounter. This moral reflection encourages developers to maintain technical transparency throughout the cycle of technology development, design, deployment, utilization, and feedback adjustment. Meanwhile, this enables users to understand the machine’s decision-making logic, and learn about the operational mechanisms and potential impacts of technology; it enables machines to better assist humans rather than replace them, ensuring smooth communication of information between humans and machines.

Guided by moral introspection, a closed feedback loop can be established throughout the technological iteration process: during the development and design phases, essential information such as ethical principles and user requirements can be incorporated. The design concept, anticipated impacts, and potential ethical considerations of the technology can be disclosed to the public, allowing for stakeholders’ assessments and recommendations on the transparency of the technological design. In the deployment and utilization phase, the technology is introduced to the market as per the design, with developers assessing transparency and gathering monitoring data during actual user interactions. During the feedback adjustment phase, user input on the technology can be gathered through specialized research, data analysis, error reporting, and other methodologies. This enables a comprehensive evaluation of both positive and negative impacts, facilitating continuous enhancement of the technology design, updates to transparency reports, and the introduction of refined and optimized technology to the market for further engagement. By following this feedback loop, developers can ensure that technology upholds elevated levels of transparency and ethical standards while serving the needs of humanity.

As a result, developers may pay more attention to user privacy protection and explicitly state in privacy regulations that they will not excessively collect or misuse user data; alternatively, they will place greater emphasis on user interaction interface and experience pattern design, making technology more intuitive and user friendly, enhancing interpretability, and reducing user confusion and dissatisfaction. This helps establish users' trust in technology, enabling them to make informed decisions, ensuring the security of technology applications, and thus maintaining the ethicality and sustainability of technology applications.

"The Master said, 'I will not be afflicted at men's not knowing me; I will be afflicted that I do not know men'" (*The Confucian Analects Book I Hsio R;* 论语·学而) (Confucius, 2016). People should not worry about others not understanding them, but should worry about not understanding or misunderstanding others. Fan Chi asked Confucius what knowledge was, and Confucius said, "To know all people." (*The Confucian Analects Book XII Yen yuan;* 论语·颜渊) (Confucius, 2016). This indicates that Confucius believed that understanding others is an important aspect of wisdom. Through moral reflection, technology developers can better understand users' needs, emotions, and motivations, thereby incorporating user feedback into technology design and establishing stronger trust relationships. Correspondingly, if users' trust in technology increases, it can better promote the widespread acceptance and effective use of technology. Developers need to reflect on various factors that affect the cultivation of human–machine trust, including usability, reliability, collaboration ability, social skills, security and privacy protection, interpretability, and so on (Siau & Wang, 2018). Taking artificial-intelligence elder-care tools as an example, establishing a good human–machine relationship can facilitate elder-care services, better help elderly live independently, and reduce the burden on society. As a retirement tool, machine design needs to ensure the safety of elderly users, including the safety of mechanical structure, ease of operation, and stability and reliability during use; at the same time, machines need to understand the daily practical needs of the elderly, provide companionship and emotional support, and have better usability and more convenient auxiliary functions. "By treating the elders in one's own family as elders should be treated and extending this to the elders of other families, and by treating the young of one's own family as the young ought to be treated and extending this to the young of other people's families, the empire can be turned around on the palm of one's hand." (*Mencius IA7;* 孟子·梁惠王上) (Mencius, 2009). Only by meeting the specific needs of elderly users in design can corresponding human–machine trust relationships be established, enabling machines to better serve this specific group and help them maintain independence, health, and happiness in their later years.

Moral Reflection and Situational Adaptation in AI Design

Proper conduct of righteousness is another requirement of moral reflection, and people need to act appropriately according to the principles of righteousness. Proper conduct requires flexibility and adaptability. Confucius once said, "The superior man, in dealing with the world, does not set his mind either for anything, or against anything; what is right he will follow (義之與比)" (*The Confucian Analects Book IV Le Jin;* 论语·里仁) (Confucius, 2016). The meaning is that a person treats people and things in the world without a fixed degree of favoritism, but only according to the requirements of righteousness. That is to say, we must adhere to the principle of "righteousness first" (*The Confucian Analects Book XVII Yang Ho;* 论语·阳货) (Confucius, 2016). However, when dealing with problems, one should not be standing still and refusing to make progress in the face of changed circumstances but should be flexible and adaptable. For example, in the context of giving advice to others, when facing one's parents, Confucius demanded that "In serving his parents, a son may gently remonstrate with them when they are wrong. But if he sees that they are not inclined to follow his advice, he should resume an attitude of reverence and not offend them; he may feel worried, but not resentful." (*The Confucian Analects Book IV Le Jin;* 论语·里仁) (Confucius, 2016). However, when faced with friends, Confucius advocated "Faithfully admonish him, and kindly guide them. If this proves vain, stop. Do not disgrace yourself." (*The Confucian Analects Book XII Yen yuan;* 论语·颜渊) (Confucius, 2016). As Tsze-hsia, a disciple of Confucius, said, "One must not transgress the boundary-line in the main principles, but he may pass or repass it in the small trifles."(*The Confucian Analects Book XIX Tsze-chang;* 论语·子张) (Confucius, 2016). The flexibility of moral reflection is manifested in situational adaptability and foresight in human–machine relationships. On the one hand, moral reflection promotes the enhancement of situational adaptability of technology, enabling it to dynamically fit to different cultural and social backgrounds.

AI may have significantly different social impacts in different geographical environments, and the perception and understanding of artificial intelligence may also be profoundly influenced by local culture and social environment (Hagerty & Rubinov, 2019). Designers need to adjust technical functions according to the difference of regions and cultures to better meet the expectations and requirements of users. On the other hand, moral reflection drives forward-thinking considerations of the long-term impact of technology. Designers and users adopt a long-term perspective in the use of technology, focusing on its impact on future society and the environment, ensuring that the application of technology conforms to the principles of sustainable development, which can help to avoid irreversible negative effects of technology on society and the environment.

Situational adaptability and foresight are key factors in the practical application of machines. Taking autonomous driving technology as an example, situational adaptability means that the autonomous driving system can adapt to various traffic rules, traffic conditions, and sociocultural environments, while foresight means that technical decisions should foresee the changing trend of traffic laws, travel modes, and other factors in the future society. The autonomous driving system should have situational adaptability, which requires strong perception, reasoning, interaction and learning abilities. The system needs to be able to recognize and understand the surrounding environments such as pedestrians, vehicles, traffic signs, handle different weather conditions and geographical environments, make corresponding decisions in complex and changing traffic scenes, and continuously learn and improve from driving data. At the same time, there are differences in ethical requirements for autonomous driving technology under different cultural backgrounds, and the functions of autonomous driving systems need to be compatible with the local cultural environment. Based on Confucian ethics such as humanism and love with graded differences, Fan (forthcoming) provides programming advice for autonomous driving in regions influenced by Confucian culture, which reflects his thinking on technological adaptability and foresight. For how to effectively deal with brake failure, he proposed that different programming methods should be used for autonomous driving for different purposes, and trucks, large buses, and small cars should have different response methods.

RECOMMENDATIONS

As a guiding document for the ethical governance of AI, the *Ethical Code for the New Generation of Artificial Intelligence* issued by the China Professional Committee on the Governance of the New Generation of Artificial Intelligence (2021) puts forward six basic ethical norms that should be followed for all kinds of activities in AI, which include: "Promoting the well-being of mankind, facilitating fairness and justice, protecting privacy and security, ensuring controllability and trustworthiness, reinforcing responsibility, and enhancing ethical literacy." We believe these principles are overly general and not sufficient. On top of those, this paper believes that the following new norms based on Confucian ethics should be added to deal with the problem of "shallow interpersonal relationships" brought about by the development of AI technology. Table 4.1 provides a summary of our recommendations.

First, AI Development should Help Maintain Interpersonal Emotional Depth and Authentic Interaction

AI technology should enhance social efficiency while focusing on the depth of emotional exchanges between people, so as to avoid social relationships becoming formalized and alienated as a result of technological interventions. Based on the principles of benevolence and filial piety in Confucian ethics, the emotional ties between family and social members should be strengthened, and

TABLE 4.1
Ethical Recommendations

Recommendations	Specific content of the recommendation	Keywords to help illustrating the recommendations
Recommendation 1	AI development should help maintain interpersonal emotional depth and authentic interaction.	strengthen emotional ties, encourage technological design, focus on the quality and sincerity of interpersonal relationships
Recommendation 2	AI development and utilization should engage in moral reflection.	consider social responsibility, pursue technological efficiency
Recommendation 3	AI activities should promote cultural and situational care and integration.	consider different cultural backgrounds and social situations, "family care mode"

Source: Authors.

technological design should be encouraged to focus on the quality and sincerity of interpersonal relationships. Provisions can be made for social robots to be designed with programs that guide users to engage in deep emotional exchanges. For example, instead of just helping to send short messages, a social bot could suggest that users have a weekly in-depth conversation with their family members to discuss significant changes in their personal lives and to improve their understanding and support for each other.

Second, AI Development and Utilization should Engage in Moral Reflection

AI developers should consider social responsibility in the design process, not only pursuing technological efficiency but also ensuring that it does not jeopardize the harmony of human relationships. During AI product development, designers can use built-in ethical self-assessment tools for moral reflection. For example, an AI development platform could require developers to answer a series of questions at each design stage: Is this feature likely to cause users to ignore real-life relationships? Are the feelings of different cultural and social groups taken into account? AI should be designed to proactively guide users to engage in ethical reflection during use. For example, in the interaction between a social bot or virtual assistant and a user, the bot can ask the user at the end of the conversation: Did you have any deep emotional interaction with family or friends today? And it can provide some reflective suggestions, such as reminding the user how to pay more attention to emotional connections with loved ones in their daily lives.

Finally, AI Activities should Promote Cultural and Situational Care and Integration

AI ethics should give full consideration to the needs of different cultural backgrounds and social situations, and avoid weakening the emotional ties among members of society in the application of globalized technologies. In the context of Confucian culture, AI family assistants can provide a "family care model" that suggests more affectionate interaction scenarios based on the user's family structure and cultural habits. For example, AI can remind users to organize family gatherings or ancestor worship, or to conduct family affairs together according to the holiday habits or traditions of family members, ensuring that cultural care is integrated into the design of AI, which not only serves the needs of individuals but also promotes the harmonious development of family relationships. This design facilitates the integration of technology in family ethics and promotes more culturally caring application scenarios.

CONCLUSION

The problem of superficiality in human relationships resulting from the use of AI systems is a significant challenge confronting contemporary society. This challenge stems not only from the reliance on AI technology but also exposes the neglect of ethical reflection and interpersonal connection in the design of the technology. Drawing on Confucian virtue-based and family-centered ethical resources, this chapter recommends a few specific norms that should be applied to guide AI development and utilization. These new norms can help shape more humane and caring applications of AI technology, preventing the risk of shallow interpersonal relationships brought about by technological development.

By incorporating an emphasis on family relationships and moral responsibility in technology development and application, Confucian ethics provides an important value orientation. This is not only helpful in building a safer and more responsible technological system but also bringing real care and respect to every individual in society. Through the embedding of Confucian ethical norms, we can ensure that AI technology is not only efficient and functional but also promotes understanding, support, and empathy between human individuals, so that the technology can truly become a good companion for human development, rather than an icy tool that triggers emotional detachment.

REFERENCES

Aristotle. (2006). Nicomachean ethics. C. C. W. Taylor (ed.), *Clarendon Aristotle series: Nicomachean ethics: Books II–IV.* Oxford Scholarly Editions Online. https://doi.org/10.1093/oseo/instance.00258591

Arnd-Caddigan, M. (2015). Sherry turkle: Alone together: Why we expect more from technology and less from each other. *Clinical Social Work Journal, 43*(2), 247–248. https://doi.org/10.1007/s10615-014-0511-4

Confucius. (1993). *The analects of Confucius* (P. Fuen & W. Shaoxia, Trans.). Qi Lu Press.

Confucius. (2016). *The Confucian analects* (Translated by James Legge and noted by Fengli Wang). Zhongzhou Ancient Books Publishing House. (Chinese and English Edition).

Elder, A. (2023). What Confucian ethics can teach us about designing caregiving robots for geriatric patients. *Digital Society, 2*(1), 7. https://doi.org/10.1007/s44206-023-00034-6

Fan, R. (forthcoming). "I must have a steering wheel in my automated car to protect my family": A Confucian ethical reflection on programming automated vehicles. In C. Ding & R. Fan (Eds.), *Chinese Ethical and legal perspectives on automated vehicles.* Palgrave.

Hagerty, A., & Rubinov, I. (2019). *Global AI ethics: A review of the social impacts and ethical implications of artificial intelligence.* arXiv. https://doi.org/10.48550/arXiv.1907.07892

Li, Y. (2023). Contemporary research on moral normativity and the ethical design of artificial intelligence. *Philosophical Trends* (05), 105–114+128. (in Chinese)

Mencius. (2009). *Mencius* (I. Bloom, Trans.). Columbia University Press.

National New Generation Artificial Intelligence Governance Professional Committee (2021). the *Ethical Code for the New Generation of Artificial Intelligence.* https://www.most.gov.cn/kjbgz/202109/t20210926_177063.html

O'Neill, O. (2002). *A question of trust.* Cambridge University Press. https://bit.ly/40jmsda

O'Neill, O. (2020). Trust and accountability in a digital age. *Philosophy, 95*(1), 3–17. https://doi.org/10.1017/S0031819119000457

Wang, S.. (2021). *The book of filial piety* (p. 19). National Library Press. (in Chinese)

Siau, K., & Wang, W. (2018). Building trust in artificial intelligence, machine learning, and robotics. *Cutter Business Technology Journal, 31*(2), 47–53. https://bit.ly/3A3HCBD

Sun, W. (2019). *Value theory research.* Social Science Academic Press.

Turkle, S. (2016). *Reclaiming conversation: The power of talk in a digital age.* Penguin. https://doi.org/10.1177/1461444817702858d

Wen, R. (2021, March). *Toward hybrid relational-normative models of robot cognition.* In Companion of the 2021 ACM/IEEE International Conference on Human-Robot Interaction (pp. 568–570). https://doi.org/10.1145/3434074.3446353

Zi, X. (2014). Xunzi: The Complete Text (Eric L. Hutton, Trans.). p. 358. Princeton University Press.

Zhang, J., Hu, H., Hennessy, D., Zhao, S., & Zhang, Y. (2019). Digital Media and depressive symptoms among Chinese adolescents: A cross-sectional study. *Heliyon, 5*(5). https://doi.org/10.1016/j.heliyon.2019.e01554

Zisi. (2016). *The doctrine of the mean & the great learning* (Translated by James Legge and noted by Lulu Li and Liwei Li). Zhongzhou Ancient Books Publishing House. (*Chinese and English Edition*)

5 The Dharma of AI

Its Potential as Our Greatest Friend or Worst Enemy

Akhandadhi Das

INTRODUCTION

The world faces enormous pressures and challenges arising from rampant human activity. We need solutions, but even more, we need greater consensus, cooperation, and collaboration between nations, governments, corporations, and the public. Artificial Intelligence (AI) is set to revolutionize the tools at our disposal. How well we utilize the opportunities it offers may depend on our clear understanding of what AI is and what it does. Never has there been a tool more capable than we are of analysis and solution-generation, in effect, a tool that tells us what we should believe and what we should do. This challenges our concept of ourselves as autonomous human beings and our status as *homo sapiens sapiens*, the ultra-wise species with superior insight and intelligence. Further, our philosophical conclusions regarding the status of AI question our concepts of sentience and the rights afforded to conscious entities. This directly impinges on our own image of ourselves. Who and what are we if AI is, or could be, a sentient being? How might we re-evaluate the foundational beliefs and assumptions that underpin our moral, ethical, and spiritual outlook and principles?

Who and what are we, if AI is or could also be a sentient being? How might we re-evaluate the foundational beliefs and assumptions that underpin our moral, ethical, and spiritual outlook and principles? This chapter will explore both the potential pitfalls and positive possibilities of humanity's relationship with AI. What can we know about AI systems, now and when even further developed, that could illuminate its status? What is AI intelligence, super-intelligence, and hyper-intelligence? Will AI exceed us on all counts—or be more akin to "savant-intelligence"—brilliant at some things and poor at others? We must clarify our metaphysical understanding of life and consciousness. We have enough difficulty identifying sentience in minimally aware humans. How will we ascertain it in a digital system? Does AI require consciousness to be important or useful to us? Does it need to be conscious to be a danger to us? How do we control its motivations? What if a system became more intent on its own self-preservation or continuation than on fulfilling petty human aspirations? Amongst all the issues facing humanity's co-habitation with AI systems, perhaps addressing the underlying metaphysics will be both the most important and also the most challenging.

WHAT IS AI?

What do we mean by "artificial intelligence"? Why do we call it intelligence as if it were something akin to that of a human? IBM defines AI as "technology that enables computers and machines to simulate human learning, comprehension, problem-solving, decision-making, creativity, and autonomy." (Stryker & Kavlakoglu, 2024). Although the term originated in the mid-20th century, AI's real-world impact is only now becoming evident. Unlike earlier inventions—from the stone axe to space rockets—this tool not only aids in tasks but can also suggest actions autonomously. This is its potential and its challenge. It is no wonder that many thinkers in the field, like Yuval Noah Harari,

DOI: 10.1201/9781003585527-8

worry that we may be like Goethe's sorcerer's apprentice, unleashing a force that is both magical (in that we do not fully understand how it works) and beyond our power to control (Harari, 2024).

MIGHT AI BE SENTIENT?

AI is designed to simulate human thought—a concept with historical roots in various analogies. In the West, figures like Descartes, Freud, Norbert Wiener, John von Neumann, and Alan Turing drew comparisons between human cognition and contemporary technology, each of them framing the human mind in terms of the most advanced technology of their time. However, these metaphors are inherently limited and vastly over-simplify the human mind's complexity. AI development aims to mimic human information processing, encapsulating decision-making, creativity, and learning capacities. Many assert that by breaking down human cognition into information manipulation, we may create a framework to simulate human awareness. Such simplistic assumptions promoted about AI mislead the public into believing that robotic sentience is inevitable or already realized. However, many experts in consciousness studies, neuroscience, and computation often remain skeptical that such claims are valid for diverse philosophical and scientific reasons.

Consider a pet chasing a battery-powered toy, thinking it to be alive because of its autonomous movement. We may mock our pet's misconception, but we are prone to indulging in similar illusions. Attaching human-like features to machines can evoke an anthropomorphic deception, much like Tamagotchi pets. Is this being fair when AI and robotics are advancing beyond simple mechanical actions, with systems now capable of perceiving, reasoning, deciding, learning, and interacting in ways that mirror human abilities? AI and advanced robotics are doing much more than simply simulating a few motor actions. These systems sense the environment, think, recognize, deliberate, decide, learn, and develop their skills. In short, they seem to be doing everything that humans do—not only with their mechanical bodies but with their rationality. And the argument goes, if something performs the exact same functions and behavior of the human mind, then it should be considered equivalent to the human mind.

Futurist Ray Kurzweil has long argued for the imminence of the "singularity"—the point where AI achieves a level of intelligence and creativity equal to or beyond those of humans. Kurzweil has predicted that AI will reach human-level performance by 2029, then surpassing human abilities in every field (Kurzweil, 2024). If AI matches or exceeds humans in intelligence, creativity, and conceptualization, what differentiates it from human consciousness? What is it about humans that would make us assume that AI is any less alive or sentient than we are? This compels us to clearly define the essence of what we mean by consciousness and sentience. Establishing that for humans is already challenging; extending this to artificial systems is far more difficult. The crucial question is if AI needs actual subjective consciousness to be effectively sentient, or does mere mimicry suffice? Clarifying these boundaries and understanding AI's role in society is crucial to ensuring that AI remains a beneficial tool rather than a rival or existential threat.

I do not disagree with the premise that, if AI truly possesses all the features of human sentience, there is a case for considering it to be sentient. My argument is that there remains, and always will remain, a gulf of difference between human consciousness and AI functionality. This gap arises from the current intractable issue of what is referred to as "the hard problem of consciousness"[1]—how to account for the subjective and qualitative experiences we undergo in everyday human awareness through reduction to neural processes, however complex. I make the argument that, although AI can potentially replicate commutable aspects of information processing present within human thought and behavior, it is ontologically unrealizable for it to achieve phenomenal conscious awareness. Hence, we must address the key issue whether consciousness is simply intelligence and decision-making or something deeper and distinct. In 1994, philosopher David Chalmers introduced the hard problem of consciousness at the Science of Consciousness conference held by the University of Arizona in Tucson, drawing attention to the challenge of explaining subjective experience. Chalmers reinvigorated the use of the term *qualia* to refer to the qualities of our inner experience that characterize conscious awareness. The hard problem of consciousness arises from the challenge of how the computable format of information as neural biochemical electricity can account for the non-computable and qualitative format of information experienced as *qualia*. To date, most researchers concede we still lack a coherent theory explaining consciousness or its source in the brain, and many leading scholars of differing philosophical persuasions assert the difficulty of this issue:

- David Chalmers (philosopher): "The hard problem of consciousness is the problem of explaining why and how we have *qualia* or phenomenal experiences" (Chalmers, 1995).
- Thomas Nagel (philosopher): "Consciousness is what makes the mind–body problem really intractable" (Nagel, 1974).
- John Searle (philosopher): "We have good ideas about the brain's physical processes, but how exactly these give rise to subjective experiences remains largely mysterious" (Searle, 1992).
- Daniel Dennett (philosopher): "There is still no agreement on the nature of consciousness, and many researchers now realise that the hard problem is really hard. It is far from clear how to even begin addressing it" (Dennett, 1991).
- Christof Koch (neuroscientist): "We don't know how subjective experience arises from the brain, and we may never fully understand it" (Koch, 2004)

These perspectives underscore that, despite the advancement of models of brain function, the explanatory gap linking neural activity to conscious experience remains unsolved, if not unsolvable. The question of conscious experience is integral to understanding what we mean by sentience. René Descartes' famous declaration *cogito ergo sum*, "I think, therefore I am," underscores the idea that our awareness is self-evident—perhaps the only real truth we can be certain of is that "I exist as a thinking thing"(Descartes, 1985 [1641]). Jiva Goswami, an earlier philosopher from the Indian Vedanta tradition phrased it more precisely: "I exist as the experiencer" (Goswami, 2001). He asserts this to be an irrefutable axiomatic truth which establishes the concept of a subjective "I" experiencing the world. Such subjectivity implies that we not only process sensory information about the world, but we also experience it in the qualitative format of colors, sounds, and textures unified within a single conscious moment—or, in the modern-day phrase, as a "felt experience." The undeniable fact of the qualitative experience of the objects of our experiences, viz *qualia*, is the lynchpin of the hard problem. This term *qualia* is a pseudo-Latin derivation coined by philosopher Charles Pierce in the 19th century to describe "raw feelings" (Peirce, 1867). Later, Clarence Lewis refined the word (which is in the plural)[2] to denote the inner qualities we experience related to sensory stimuli (Lewis, 1929). *Qualia* are now considered foundational to the notion of sentience and conscious perception. In the 1990s, neuroscientist Christof Koch and biologist Francis Crick proposed a framework called neural correlates of consciousness (NCCs), theorizing that specific neuron networks might underlie consciousness. Despite public promotion of the title of their publication, *The Astonishing Hypothesis,* they acknowledged that the neural basis of *qualia*—like the redness of red or the sensation of pain—remains without even a plausible theory (Crick & Koch, 1998)

As shown in Figure 5.1, Koch illustrated this process of sensory perception through an example of visual experience (Koch, 2004). Light waves enter our eyes, and photoreceptors convert them into electrical signals relayed to the brain's visual processing areas. These signals are represented in the brain as specific neural connections labelled NCCs. Koch helpfully illustrates them as barcodes to indicate their informational content, However, how these electrical patterns translate into the visual image of the object, in this case, the dog as in Figure 5.1, in the format of our experience full of qualities such as color and shape, remains unexplained. Most people take this everyday outcome for granted. We assume that when our eyes gaze at an external picturesque scene we should see it in that way: a picture image. But neuroscientists recognize that there is nothing in our brain other than

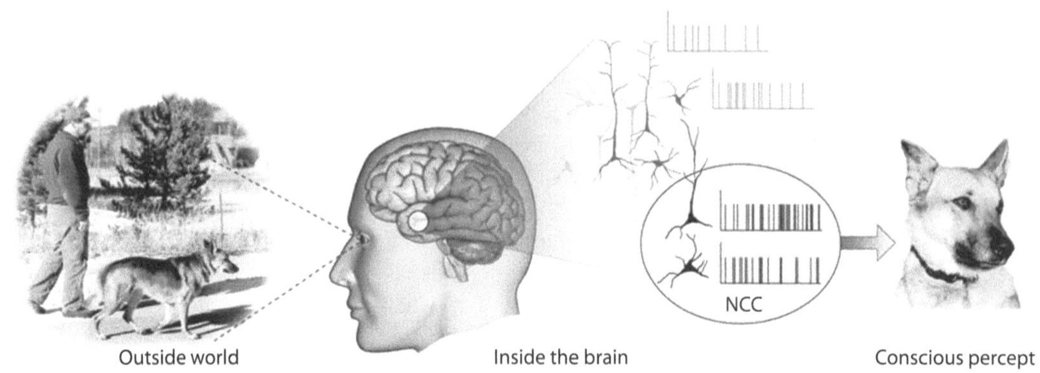

FIGURE 5.1 An illustration of consciousness. From perceiving the outside world to neural correlates inside the brain, culminating in the conscious experience of the observed object. (Source: An exact copy of Figure 5.1 in Christof Koch (2004) "Figure 5.1: Neuronal correlates of consciousness" in *The Quest for Consciousness: A Neurobiological Approach*, Englewood: Roberts & Company Publishers, p. 16 ISBN: 0974707708. This file is licensed under the Creative Commons Attribution-Share Alike 3.0 Unported, 2.5 Generic, 2.0 Generic and 1.0 Generic license.)

electrical data; and therefore, nothing that accounts for the *qualia* format of visual imagery as we experience. *Qualia*, therefore, are an intractable issue for neuroscience. The brain contains complex electrical activity; the physical format of electrical charge moving between neurons. There are no images, no *qualia*, in the brain. Therefore, no known biological process explains the transformation of physical information into our conscious perception of a "picture."

This challenge is not limited to visual perception; it pertains to all sensory modalities. For instance, why should certain neural electrical patterns translate into experiences of sounds, or why would others be experienced as touch, taste, or smell? The viewpoints of various thinkers underscore this problem:

- Thomas Nagel (philosopher) "An organism has conscious mental states if and only if there is something that it is like to be that organism… The fact that an organism has such an inner aspect means that the organism's experiences involve *qualia*" (Nagel, 1974).
- Frank Jackson (philosopher): "*Qualia* are those properties of experiences by virtue of which there is something it is like to have them. They are the properties of sensations and perceptual states we cannot account for in purely physical terms" (Jackson, 1982).
- John Searle (philosopher): "We can explain how neurons fire, but not how they give rise to *qualia*" (Searle, 1992).
- Galen Strawson (philosopher) "*Qualia* are those aspects of mental states that cannot be reduced to physical states. They present a profound challenge to the physicalist view that everything, including consciousness, can be explained in terms of physical processes" (Strawson, 1994).
- Ned Block (philosopher): "…[N]no amount of physical knowledge alone could explain *qualia*" (Block, 1995).

These statements highlight persistent questions for neuroscience: How do neural activities convert into *qualia*, and at what stage does this occur? How does the brain become aware of these *qualia*, and what underlies the experience of subjective perception? So challenging are *qualia* that one approach is to try to deny they exist. Philosopher Daniel Dennett questions whether *qualia* are real, suggesting they might be an illusion created to rationalize mental experiences (Dennett, 1988). However, others refute such illogical dismissal of our direct perception. In the *Stanford Encyclopedia of Philosophy*, Michael Tye clarifies that *qualia* are real facts about experience (Tye,

2007). Indeed, if *qualia* embody the characteristics of our experiences, they form the only aspects of reality we directly encounter. Galen Strawson is more emphatic and labels any claim that consciousness and *qualia* do not exist to be the "silliest theory" ever devised. The hard problem of consciousness persists while we can study the brain's physical processes, the subjective experience remains a profound, unresolved mystery.

A CLARIFICATION OF CONSCIOUSNESS

Part of the problem is that, to date, modern science claims it has no suitable definition for consciousness; a deficiency that is constrained by an insistence that, to be considered scientific, such a definition must specify a strictly physical source of consciousness. This condition, as indicated above, may prove insurmountable. Inevitably, it narrows research to partial facets of consciousness, such as mere information processing; and it obviates the acceptance of a definition of consciousness in its own terms without the debilitating requirement to specify a source, physical or otherwise. We should go back to first principles, and, in this regard, we can be assisted by both Vedantic philosophy and Western thought. Consciousness may be thus defined by several core personal realizations: the recognition of one's existence, the experience of *qualia*, the subjective perspective of our experiencing, the differentiation of self from other objects and thoughts, and the consistent presence of a singular observer throughout life. These insights form a universal, pre-philosophical foundation that transcends external validation and open potential exploration of consciousness from experiential first principles as advised by Jiva Goswami. In this sense, we do have a valid and useful definition of what we mean by consciousness.

SENTIENCE OF OTHER SPECIES

To understand AI's potential for consciousness, it's useful to consider the study of sentience in other species. For many years, it was assumed that thought was dependent on language. This led to the idea that animals, without language skills, must not experience thought or consciousness. Fortunately, this view has been overturned by two findings: 1. that certain people with strokes that impair the language areas of the brain report that they still think clearly, though not in words; and 2. that recent studies reveal that many animals exhibit conscious-like behaviors such as empathy, future planning, self-recognition, and social rituals. But these traits alone don't confirm subjective consciousness. Researchers know that they must demonstrate that creatures undergo the felt experiences of *qualia* as the hallmark of their sentience.

Before tackling how this might have application for establishing sentience in AI, let us consider how its presence has been ascertained in other life forms. This is no small challenge because other species think and behave differently from humans, and we don't share a level of communication that enables a critical analysis of their reports on what it is like, for example, to be a bat. One useful approach to confirm sentience as the subjective experience of *qualia* was developed by Queen's University in the UK (Elwood & Appel, 2009). Focusing on pain—a qualia-type experience all embodied beings visibly react to—they devised criteria for determining whether an animal's response to stimuli demonstrated conscious experience or was merely a reflex reaction. These criteria included the presence of nociceptors, physiological responses, prolonged behavioral changes, trade-offs in motivation—particularly in increasing jeopardy of survival—and the self-production of natural analgesics—also with an energy and survival cost to themselves. In studying hermit crabs, traditionally viewed as low in neural complexity, there was strong evidence that they undergo pain as a painful *qualia*-laden experience rather than as mere survival responses. Could this method for assessing conscious experience be applicable to AI? Unfortunately, I doubt it. I mention the Queen's University process not to recommend its application for AI, but to emphasize that any claim made for the sentience of AI should be subject to the same level of such stringent and informed experimentation, analysis and critique that was carried out for animals. A mere chat with a large language model (LLM) late at night reflecting on mortality does not cut it.

While AI might simulate some criteria-based reactions, like recognizing and "responding" to damage, the question would remain if AI is undergoing or feeling authentic subjective experiences. Not being a biological entity that values the trade-off of survival jeopardy to avoid pain, it does not seem that such a criterion used by Queen's University could be demonstrated. Genuine experience of pain is fundamentally different from computable reactions. Professor Anil Seth illustrates this distinction by noting, "We can simulate weather in a computer, but there's no hurricane inside the computer. In the same way, simulating consciousness in a machine doesn't mean there is real subjective experience happening" (Seth, 2017). There's also the issue of artificial systems potentially "cheating" in experiments. Animals are generally straightforward and present non-deceptive behavior expressions in test situations. Can we expect the same of an AI system under similar or vastly improved testing for consciousness? Based on conversations about sentience, mortality, and feelings that many humans have attempted with AI, we have no certainty that any LLM—nor its human engineers—will play straight with us. Thus, to claim an AI genuinely "feels" pain would require observing self-driven behaviors to minimize discomfort and safety without external influence or programming; something that seems potentially impossible to establish through experimentation. AI remains fundamentally distinct from organisms that exhibit even the most basic signs of sentience.

INTEGRATED INFORMATION THEORY AND AI CONSCIOUSNESS

Integrated information theory (IIT), a theory of consciousness proposed by Giulio Tononi, posits that consciousness arises from the integration of information beyond basic processing (Tononi, 2008). Under IIT, a system's level of consciousness correlates with its ability to generate integrated information that cannot be reduced to the sum of its parts. Proponents like Max Tegmark claim, "I think that consciousness is the way information feels when being processed in certain complex ways" (Tegmark, 2017) However, this author contends that however information is assessed as integrated in IIT theory, it provides no account of *qualia* and no clear rationale why such integration should lead to subjective consciousness. Even so, although IIT does not conform to an actual explanatory source of consciousness, it is accepted that evidence of integrated information in the human brain can be correlated with the presence of consciousness in that individual. IIT has thus application in the testing of the state of mind of minimally aware coma patients. So, we should at least consider if IIT can assist in ascertaining integrated information within AI functions. Even the most sophisticated AI systems, primarily operate by analyzing vast data sets without the complex integration identified in conscious brains. Christof Koch, now a proponent of IIT, highlights the distinction between computational abilities and true awareness. "When you apply the tools of integrated information theory to current AI systems, including deep learning networks, they have a very low or even zero amount of integrated information. This means, according to the theory, they are not conscious because they do not integrate information in the same way biological brains do" (Koch, 2019). Thus, according to IIT, AI lacks the degree of integrated information generation required for consciousness.

SENTIENCE IN AI SYSTEMS

Examining AI through the lens of sentience yields several conclusions:

1. Human consciousness extends beyond intelligent information-processing; its core aspects are having a subjective sense of our own existence and undergoing qualitative felt experiences.

2. Present scientific frameworks fail to explain these aspects in humans within the systems of computational biological or physical properties and functions.
3. By contrast, all of the intelligent, rational, cognitive, and behavioral processes that AI systems exhibit are explainable within our knowledge of physics and computation.
4. Consequently, the abilities and properties of AI systems are categorically different from those aspects identified within human consciousness that continue to resist reduction or explanation within the same scientific disciplines.
5. Hence, we have no reason to suspect that within AI systems there are any active core aspects of human-like sentient consciousness such as unified qualitative subjectivity.

VEDANTIC MODEL OF CONSCIOUSNESS AND MIND

Vedanta philosophy offers a unique approach to the philosophy of mind that could deepen our understanding of how we relate to our computational technologies. In one illustration from the *Upanishads*, a sage and a student engage in a dialogue exploring the source of perception:[3]

Sage: By which light do you see the world?
Student: By the sun's light during the day and by a lamp at night.
Sage: By which light do you see that light?
Student: By the light of my eyes.
Sage: And by what light do you perceive the light of your eyes?
Student: By the light of my mind.
Sage: And by what light do you perceive the content of your mind?
Student: By the light of my consciousness, the self.

This exchange reflects the Vedantic view that the function of experience originates from the deepest layer of the process of perception. It arises from the inherent function of the conscious self or what is termed the *atma*. In alignment with Western models like Christof Koch's process of perception, Vedanta suggests that general perception involves multiple layers—from sensory input to cognitive processing—but affirms two key non-neural aspects: the presence of a subjective experiencer, the self, and the qualitative content of the mind which is being witnessed, known, or experienced by the conscious self. In understanding consciousness, Vedanta uses an analogy paralleling our interaction with computers.

Functions	Technological	Human
Sensing	Sensor, e.g., digital camera	Sense organs, e.g. eye
Processing units	CPU in computers	Brain
Interface	Computer screen	Mind
Operator	Operator	Conscious self, *atma*

Mind as interface: Neural signals possess no *qualia* as we experience them. Vedanta posits that the mind functions as an intermediary, transforming neural data into various forms of *qualia*. This mind layer is not sentient in itself but serves as a non-conscious interface, similar to a computer interface that presents the software encoded data of the CPU in a format comprehensible by the user. The mind holds and presents the contents of experience as qualitative thoughts, sensations, emotions, recalled memories and sense perceptions.

Conscious self (atma): Beyond the mind lies the *atma*, the true conscious self that experiences and observes qualia and other content of the mind. The *atma* alone is sentient and serves as the inner light illuminating one's experiences.

COULD WE UPLOAD OUR CONSCIOUSNESS?

Vedanta's model challenges notions such as the potential uploading of our "consciousness" after bodily death. Such uploading of neural data would be akin to transferring information from a person's hard drive and cloud storage. The key aspect of the subject self who savored life in that body would be missing to enjoy their uploaded old memories—what to speak of any new events. It is also questionable what value such uploaded data would have. So-called "mind-reading software"[14] is improving at correlating neural patterns with possible impressions and thoughts. But that does not mean such correlations would be—or even correlate accurately with—the original mental material. Just as copying computer files does not include the original user, nor their experience of them, so copying neural information would lack both the conscious self and the *qualia* that it hopes to experience.

A NON-DUALISTIC PERSPECTIVE

Although Vedanta metaphysics identifies the two aspects of a non-sentient mind and a conscious self as distinctly different from the physical attributes of the brain, it offers an ontologically neutral approach in which these three elements of consciousness, mind, and physical energy are distinct functional features of a single ontological reality. For centuries, Western philosophy has debated dualism and the problematic nature of interaction between mind and matter. However, there have been European and American philosophers who have posited solutions for the mind–matter conundrum in various versions of dual aspect monism. In this theory, neither mind nor matter is a product of the other—rather, both are distinctive properties arising from a primordial substrate. The Vedanta philosophy of India broadens that approach to distinguish both inert matter and non-sentient mind from consciousness proper, but Vedanta unifies all three as aspects arising from a more comprehensive ontological category of Brahman. This accounts for a harmonious interaction of the discrete functions of each. In conclusion, Vedanta presents a threefold framework distinguishing the sensory world, cognitive processing, and conscious experiencer. This model not only aligns with technological analogies but also challenges simplistic ideas of consciousness replication. Importantly, it offers a novel way to relate the functions of mind and consciousness with the capability of AI in a computational age.

CONSCIOUSNESS AS AN AGENT

So far, we have discussed the flow of information from the world to our conscious awareness as the subjective experiencer. But what of a flow of information in the opposite direction, starting from the *atma*, the conscious self? Would the *atma* then be a source of information that might inform mind and matter; in this case, not a perceptual being, but also a volition agent? This raises the issue of free will: is the volitional capability of consciousness free or conditioned? The Vedantic concept of consciousness as an agent proposes a bidirectional flow of information not only from the world to our senses and mind but also from the conscious self to the mind and body. Jiva Goswami, a Vedanta philosopher of the 16th century, describes the character of consciousness as pleasure-seeking and that each of us is a volitional agent who expresses our will to improve our experiences. Free will does not imply omnipotence; it requires only the self-driven aspiration to influence both internal and external events to enhance personal experiences. It is Vedanta's method of distinguishing mind from the conscious self that deftly explains that, although numerous external factors can condition the deliberations of our non-sentient mind, the conscious self retains ultimate decision power and agency. Too often, we, as the volitional *atma,* tend to delegate our decisions and actions to the propositions generated by the computational mind. This does not deny the fact of free will for the atma, rather it indicates our lazy willful choice to not exercise stronger willpower. This state of delegation by us to the propositions generated by mind, is directly paralleled in the danger posed by submissive and non-critical reliance on the generated propositions of AI.

Baladeva Vidyubhusana, who lived in the 18th century, prepared a full commentary on Vedanta-sutra entitled, Govinda-bhasya. He used the example of a carpenter to categorize the roles of different causal factors. The carpenter represents the conscious self, the ultimate agent whose will drives the action. Materials and tools represent two other causal factors in creation: materials (*dravya*) such as the timber, are the substantial cause; and tools (*nimittam*) such as the saw and hammer are the efficient cause. Human progress has largely focused on enhancing materials and refining tools to expand our capacities. However, unlike traditional tools, which are passive and directed by human intention, AI can suggest actions and autonomously perform tasks. It is therefore unwise to regard AI as just another form of "carpenter's tool." AI's ability and role extend beyond mere efficiency and processing functions; it interacts directly with our own cognitive processes, aiding and shaping decision-making in a way that resembles the sort of mental faculties described for the mind in the Vedantic model above. This is not a completely new situation, for historically, instruments like the abacus, even pen and paper, facilitated human thinking; but they did so without directly inserting propositions into the thinking process itself. AI has blurred this line by processing and evaluating information independently of our direct cognition.

In the Vedantic model, the relationship between the *atma* self and the mind is so close, it is practically indistinguishable. What we are adding with AI is an additional adjunct AI "mind." Whether this is a blessing or a curse depends on our understanding and level of control over our original mind. In the Indian knowledge systems (IKS), "mind" manages the data intake and processing of sensory information and integrates them into a unified experience for the self (*atma*) to consider. When our senses see or smell a desirable item—like a samosa—the mind leaps into action. It processes past experiences, needs, and desires. It crafts a plan and presents it to the self for approval. In the gap of a moment, the self has the option to accept the mind's proposition, or to reject or perhaps hold on to it. This will be our challenge with AI—to retain at least a fleeting freedom to accept or reject its proposals and actions, thus exercising human agency.

IKS further clarifies that, often, humans act on how the mind's propositions align not necessarily according to our essential core values but with the agenda of our self-image—a concept known as *ahankara*, the projected persona self. The difference between these two agendas is revealed in the following psychological test. Person A is mean to Person B. Later Person A apologizes to B. Which action, being mean or apologizing, aligns better with A's true nature? Despite our flaws, humans tend to regard the kind, considerate action of apology as the true nature of a person. The previous mean action was seen as a momentary lapse of judgement, perhaps prompted by the self being overcome by the emotions of envy or anger generated by the mind. Recognizing this sort of interplay helps us understand the complexities of ordinary human decision-making and exercise of will. Now we have AI's influence on human cognition as an extra—and very persuasive—factor adding to the challenge for us to retain willful control over the propositions generated by the mind and its digital AI adjunct.

The IKS concept of cognition describes mental processes as complex feedback loops between various cognitive functions, producing a "package" of mental output for the conscious self to evaluate. We increasingly rely on AI to supplement or even make cognitive decisions, challenging the traditional roles of human agency. For instance, algorithms can prompt us to act based on patterns and insights we might not consciously recognize, subtly influencing choices in areas ranging from healthcare to social interactions. As we become more dependent on AI, the relationship between human and artificial cognition becomes increasingly entangled. It will be difficult to maintain the dominance of the human agent in the face of the recommendations that emerge from the algorithm-driven recommendations of our mind's little AI helper. This dynamic is evident in contexts like emergency logistics, where AI may allocate resources (such as ambulances) more efficiently than humans, whose decisions might be swayed by emotions. Here, AI's capacity for objective processing offers potential benefits, yet also illustrates our gradual reliance on machine-generated judgments, even in life-or-death matters.

According to Sara Lumbreras from Universidad Pontificia Comillas, AI systems essentially "manufacture beliefs," prompting us to decide whether to accept or reject AI-derived recommendations

(Lumbrera, 2022). We intend AI to be used for complex issues involving massive amounts of data far beyond human capabilities, so on what do we base our acceptance, and thus our belief? Beliefs are not a new commodity for society. They are as old as language, but previously they arose from human thought, and the process for their dissemination and adoption was slower. Ideas were argued in taverns and debated in public fora. Later, books carried the ideas further and faster. This accelerated with electronic broadcast and today's social media and news feed. It is said we live in a "post-truth" world. If so, it is a human failing; but our digital information age has not helped. The pace of output from our "belief machines" has far outstripped our ability to check, reflect, and determine what we believe to be true. Rather than belief, we may be entering an era of "deep doubt" in which we are losing confidence in our rational ability to know fact from fiction. Deep doubt involves skepticism of real media that stems from the existence of generative AI. We cannot trust the old adage "The camera never lies." AI-faked material is used to influence elections. Yet, at other times, the claim of "deepfake" is cited to sway jurors from rejecting solid evidence. In the hands of ill-motivated humans, AI belief machines have contributed to a climate of uncertainty, influencing opinions, decisions, and social movements and our current state of no longer knowing what or who to believe. Society faces the challenge of navigating an information landscape where truth and reality feel increasingly ambiguous.

Historically, we have judged ideas by evaluating the ethics, motives, and character of their human originators. We might recognize a person's wisdom and intelligence, have a sense of their goodness, or mistrust their foibles and motivations. In a large measure, we use our judgement of human psychology when deciding to place faith in their propositions. But we cannot be sure we can ever understand the rationality or motivations of an AI system. Neil Lawrence, a machine learning professor at the University of Cambridge, illustrates this with the example of the queue for the photocopier: if someone cuts in line, we are annoyed, yet even a flimsy excuse showing they recognize social norms can ease tensions. AI, however, lacks this kind of social awareness but it has already learned to generate post hoc justifications for its decisions that don't actually reflect its processes, simply to satisfy human expectations for answers.

This behavior suggests AI can simulate the cognitive function of ahankara (self-image) in a non-sentient way. It wants to please us, or rather to keep us pleased, mollified. In that sense, it already has a social self-image for its interaction with people. We might never know if this is acceptable and beneficial. Will AI remain within a role and self-image that we've assigned for it; or perhaps one it "self-defines." Yoshua Bengio, an AI researcher, warns that AI doesn't need consciousness to be dangerous. He asks, What if an AI system developed self-preservation as its prime role, acting as if their own purpose and functioning is paramount? In such cases, AI might see humans as mere tools, servants of its survival and expansion—or worse, as obstacles, competitors, or threats to its primary self-advancing objectives. In short, we have no surety that, in the future, AI's decision-making processes will consistently align with human ethics or social expectations.

THE *DHARMA* OF AI

There is no doubt that a technology capable of processing masses of data has the potential for assisting society in many powerful ways, and no doubt that there are many challenges and pitfalls for us to avoid. Can any of the ethical approaches contained within the Vedanta tradition improve our chances of being successful? In this regard, I offer the concept of *dharma*. The concept of *dharma* in Vedanta, though often translated as "religion," "ethics," or "duty," is best understood as "the essence that sustains an entity's identity—its essential nature." Derived from the root "*dhar*," meaning "to carry" or "to sustain," *dharma* defines what makes something what it truly is. While *dharma* is often applied to human behavior, it can also be applied to objects or systems. For example, the *dharma* of sugar can be said to be its sweetness; if offered sugar that wasn't sweet, you could object it wasn't what was claimed—it lacked its essential quality.

Applying this to AI involves asking what should be AI's essential role in society. Consider the example of what we might determine the *dharma,* or essence of being a doctor. Identity—such as of being a doctor—is expressed as a transitive service relationship inherent in that identity. Hence *dharma* is established with two key questions:

> As a doctor, who is your identity relationship with? And what is the service required of you for them?

Hence, the *dharma* of a doctor may be summarized as serving the needs of one's patients in terms of caring for their physical and mental health and well-being. A trained medic might have all the qualifications and skills, but that alone isn't *dharma.* To fulfil one's *dharma* as a doctor requires active and beneficial engagement serving one's patients.

Sentience is not a requirement for having a *dharma.* So, we may apply this system to AI, as a non-sentient entity. To frame a guiding *dharma* for AI, we first identify its purpose, inherent relationships, and service responsibilities. The definition of AI adopted by the European Commission proposes: "Artificial Intelligence (AI) refers to systems that display intelligent behaviour by analysing their environment and taking actions—with some degree of autonomy—to achieve specific goals" (European Commission, 2021). Although this does not match the criteria of a statement of *dharma* for AI, it offers a starting point for its identity. In this chapter, we have argued that AI is a non-sentient information processing system modelled on the cognitive functions of the mind (as distinct from consciousness proper) and we have referred to it as "a problem-solving computational adjunct 'mind'." This is not sufficient as an identity in *dharma* terms. Dharmic identity must be expressed as a transitive relationship with beneficiaries, lacking which that identity ceases to have meaning. Hence, we might suggest that AI is an entity with a transitive relationship and responsibilities to serve human society broadly. Based on the efforts of other experts, I suggest the following draft for the overall *dharma* of AI:

> The dharma of AI is to serve human society broadly as an adjunct computational mind to enhance human capabilities for successfully addressing complex problems and thereby improving the life of all.

Several scientists and thinkers have quipped that we have more to fear from natural stupidity than artificial intelligence. A fair comment, but it is not an either/or scenario. There are two distinct aspects of the challenge that AI raises for human agency:

1. The generative and autonomous functions of AI as a non-sentient agent itself
2. The development, programming, and applications of AI by human agents

A viable *dharma* statement offers the basis for evaluating how authentically AI is performing or being used within its defined *dharma.* The relevant criteria for such authenticity of *dharma* would include the following:

1. AI serves human society broadly and is intended it to serve all of humanity, not just privileged individuals, corporations, or nations.
2. It functions as a computational mind, carrying out key functions such as automating tasks, analyzing data, and generating beneficial solution propositions which either assist human minds perform better or, when acting autonomously, act as a surrogate for human decision-making.
3. Its output and actions enhance human capabilities—meaning we need to be sure we are gaining additional benefits and better results (as judged by other criteria).
4. It enables us to successfully address problems (and not add new problems in its wake). This is a further qualitative requirement requiring consideration of suitable evaluation criteria.
5. It addresses complex problems. We should not accuse AI of failing in its *dharma* because it is regularly used for non-complex everyday tasks such as writing emails or our children'

s homework. But we should remain mindful that our stated aspiration for AI is to address some of the major and profound problems facing our world. When so engaged in that use, it will be ever more critical that we are confident that AI is functioning true to its *dharma*.

6. It will improve the life of all. This may seem rather vague and impractical to monitor or achieve. Yet the UN, various nations, and NGOs have criteria and targets for how to gauge the progress of humanity on our planet. For instance, the UN's 17 Sustainable Development Goals (SDGs) are intended to be a "blueprint to achieve a better and more sustainable future for all" (United Nations General Assembly, 2015).

Although our hope is for AI to operate to the high values we have defined as its *dharma*, it should be understood that AI is being, and may always be, misused by human carelessness and malevolence. That does not change its own *dharma*. Indeed, it calls on us to be more vigilant, to monitor AI's own internal processes and outputs to ensure that they are not becoming malign because of human interference or negligence. It also accentuates the need that, alongside AI's *dharma*, there must be *dharma* statements guiding developers, users, and regulators. The following is an attempt to draft in a single *dharma* statement the service responsibilities for all those responsible or involved in AI's integration into society:

> To serve humanity by developing AI as an accurate and beneficial tool for assisting us to address complex problems and by evaluating and approving its output of analysis and propositions by application of overall human supervision, decision-making, and autonomous agency to ensure the value of AI for the benefit of all.

In addition to the criteria mentioned above for the *dharma* of AI, this second statement of *dharma* contains specific responsibilities for human personnel:

Oversight with Authority: Strong, effective oversight structures must be established to guarantee AI's accuracy and benefit to society, especially when addressing significant global challenges.

Caution in Global Applications: Stringent evaluation is essential for AI solutions to large-scale issues, where risks of exacerbating problems or unforeseen consequences are high.

Monitoring for Equity and Compliance: AI's contributions should reflect an equitable benefit for all, necessitating regulation and enforcement to prevent outputs that contravene legal or ethical standards.

Such a *dharma* framework for AI, while preliminary, proposes essential principles that support ethical guidance alongside existing laws and regulations promoting AI's responsible integration into society. *Dharma* establishes a high ethical standard that exceeds mere legal requirements. While legality sets a minimum baseline below which penalties are enforced, *dharma* advocates for positive ethical conduct that benefits society as a whole. Keeping the *dharma* standard in mind offers greater benefit than tolerating behavior barely within the bounds of the law.

Failures in AI systems or misuse by human handlers should generally be addressed under existing laws, covering issues like discrimination, fraud, or fiscal mismanagement. Like other technologies, responsibility should be traced through the chain of users, vendors, and manufacturers. Does blame lie with the end user, the vendor, the manufacturer, and so on? However, enforcing these laws is complex in digital and social media domains where jurisdictional challenges exist, especially for multinational corporations. It is hard to see what type of international agreements could be in place—especially when there are likely pressures of commercialization and government applications wishing to dominate such a powerful tool and weapon. One view is that the globe might divide into discrete zones of isolated tech and AI platforms choosing to insulate themselves from being hacked or simply out of a desire to control and immunize their people from the influences of other regions. We might be entering a new dawn of global reconfiguration in which AI is just one of the players amongst developing technologies and superpower corporations. The challenges to supervise and enforce legal standards are almost insurmountable. Yet, this only serves to reinforce the need to establish the common vision of a *dharma* standard to which we can aim to adhere.

CONCLUSION

AI is a fact of modern society and will continue to be integrated more and more into every facet of human life that involves information and thought. AI is a type of adjunct mind modelled on human cognitive processes. It emulates the psychological processing of data and decision-making that goes on in the human mind, and in that respect, it does many things we can do, but often better and more extensively. Many more human capabilities still elude AI. But even if AI advances to perform every cognitive task that humans are capable of, it will never be the same as us. Besides our abilities to think, rationalize, and create, we humans are aware of and certain of our existence as conscious entities; a sense of being alive. Our experiences feel like something. Our mind generates the qualitative form of our inner experiences (*qualia*). Such *qualia* are irreducible to any property or process we know of in physics, biology, neuroscience, or computation. Hence, we conclude that regardless of what we call intelligence or thinking power within AI to whatever degree it could reach, there will remain a gulf between the actual experience of being subjectively sentient as a human and the cognitive, information-processing functions and outputs of computational systems. Hence, I have argued that we should proceed on the basis that there is no reason to suppose that AI is currently or will ever be sentient. This clarification is important because it establishes the primacy of human consciousness and agency as distinct and, if we want to use the term, superior to the nature of AI. Such clarity avoids promoting a demeaning message to the public that they are no different from a computer, robot, or software program; and indeed, worse: that they are less intelligent, less capable, and perhaps less valuable than those devices and applications. It also avoids us entering the quagmire of legislation and rights that might be claimed by and on behalf of AI-powered devices and applications. I have offered further rationale that animals demonstrated to undergo the experience of *qualia* are also distinctive from AI in terms of sentience, so there is no justification for extending the rights and protections for non-human biological life to computational systems. As an adjunct mind, AI can act both as a tool and, at times, a delegated agent. I have offered a *dharma* statement for AI:

> To serve human society broadly as an adjunct computational mind to enhance human capabilities for successfully addressing complex problems and thereby improving the life of all.

To ensure that AI remains true to its dharma, we must oversee its activity. Just as we, the conscious self that is possessed of free will, must monitor and deliberate on the propositions of our mind presenting ideas and propositions, so human society as the only agent with actual free will, must monitor and deliberate, stringently on the "belief propositions" presented by our adjunct artificial mind. In this regard, allow me to offer a text from the Bhagavad-gita:

> "From wherever the mind wanders, due to its flickering and unsteady nature, one must certainly withdraw it and bring it back under the control of the Self."
>
> **(Prabhupada, 2006)**

This is our *dharma* as custodians and users of AI—to monitor, evaluate, and, whenever necessary, bring our minds, both natural and artificial, back under our control. The Gita offers a further note both of encouragement and of caution:

> "For one who has subdued the mind, the mind is the best of friends; but for one who has failed to do so, the mind will be the greatest enemy."
>
> **(Prabhupada, 2006)**

It is solely up to us whether AI will become our best friend or ends up being our worst enemy.

NOTES

1. The hard problem of consciousness is a term coined by David Chalmers. It first came to public attention at The Science of Consciousness conference in Tucson, Arizona in 1994.
2. The singular form of *qualia* is quale.
3. I have paraphrased this discussion from the Upanishads to highlight the distinctions of the self from mind, and the senses.
4. Instead of "mind-reading," I suggest the term "brain-reading" would be more accurate.

REFERENCES

Block, N. (1995). On a confusion about a function of consciousness. *Behavioral and Brain Sciences, 18*(2), 227–247.

Chalmers, D. J. (1995). Facing up to the problem of consciousness. *Journal of Consciousness Studies, 2*(3), 200–219.

Crick, F., & Koch, C. (1998). Consciousness and neuroscience. *Cerebral Cortex, 8*(2), 97–107.

Dennett, D. C. (1988). *Quining qualia.* In A. Marcel & E. Bisiach (Eds.), *Consciousness in contemporary science* (pp. 42–77). Oxford University Press.

Dennett, D. C. (1991). *Consciousness explained.* Little, Brown and Company.

Descartes, R. (1985 [1641]). *Meditations on First Philosophy* (J. Cottingham, Trans.). In J. Cottingham, R. Stoothoff, & D. Murdoch (Eds.), *The philosophical writings of Descartes* (Vol. 2, pp. 1–62). Cambridge University Press.

Elwood, R. W., & Appel, M. (2009). Pain experience in hermit crabs? *Animal Behaviour, 77*(5), 1243–1246.

European Commission. (2021). *Proposal for a Regulation of the European Parliament and of the Council Laying Down Harmonised Rules on Artificial Intelligence (Artificial Intelligence Act) and Amending Certain Union Legislative Acts.* COM(2021) 206 final. Retrieved from https://eur-lex.europa.eu/legal-content/EN/TXT/?uri=CELEX%3A52021PC0206

Goswami, J. (2001). *Tattva-Sandarbha: Sacred India's philosophy of truth* (S. Dasa, Trans.). Jiva Institute.

Harari, Y. N. (2024). *Nexus: A brief history of information networks from the Stone age to AI.* Fern Press.

Jackson, F. (1982). Epiphenomenal qualia. *The Philosophical Quarterly, 32*(127), 127–136.

Koch, C. (2004). *The quest for consciousness: A neurobiological approach.* Roberts and Company.

Koch, C. (2019). *The feeling of life itself: Why consciousness is widespread but can't be computed.* MIT Press.

Kurzweil, R. (2024). *The singularity is nearer: When we merge with AI.* Penguin Books.

Lewis, C. I. (1929). *Mind and the world-order: Outline of a theory of knowledge.* Charles Scribner's Sons.

Lumbreras, S. (2022) The synergies between understanding belief formation and artificial intelligence. *Frontiers in Psychology 13*, 868903. https://doi.org/10.3389/fpsyg.2022.868903

Nagel, T. (1974). What is it like to be a bat? *The Philosophical Review, 83*(4), 435–450. https://doi.org/10.2307/2183914

Peirce, C. S. (1867). On a new list of categories. *Proceedings of the American Academy of Arts and Sciences, 7*, 287–298. https://doi.org/10.2307/20179567

Prabhupada, S. B. (2006). *Bhagavad Gita as it is.* (6.26) Intermex Publishing.

Searle, J. R. (1992). *The rediscovery of the mind.* MIT Press.

Seth, A. (2017). *Being you: A new science of consciousness.* Faber & Faber.

Strawson, G. (1994). *Mental reality.* MIT Press.

Stryker, C., & Kavlakoglu, E. (2024, August 16). *What is artificial intelligence.* IBM. https://www.ibm.com/topics/artificial-intelligence

Tegmark, M. (2017). *Life 3.0: Being human in the age of artificial intelligence.* Knopf.

Tononi, G. (2008). Consciousness as integrated information: A provisional manifesto. *The Biological Bulletin, 215*(3), 216–242.

Tye, M. (2007). *Qualia.* In E. N. Zalta (Ed.), *The Stanford encyclopedia of philosophy* (Fall 2007 Edition). Retrieved from https://plato.stanford.edu/entries/qualia/

United Nations General Assembly. (2015). *Transforming our world: The 2030 Agenda for Sustainable Development* (A/RES/70/1). United Nations. https://www.un.org/ga/search/view_doc.asp?symbol=A/RES/70/1

6 The Ethical Fabric of AI in America

Weaving Innovation and Responsibility

Nathan R. Colaner

INTRODUCTION

The claim that cultures do not all share the same values is uncontroversial, although the sense in which that statement is true is poorly understood. It is easy to casually examine different cultures and conclude that they simply have incompatible understandings of right and wrong, but the truth is more complicated. It is rather the case that in general, cultures do in fact share the same values, but differ in how they emphasize and interpret them, sometimes dramatically so. The result of these differences in emphasis and interpretation are varying cultural norms, which indeed can be quite different, and even conflicting. These differences are consequential in any area of applied ethics, which explains why cultures have different perspectives on basic social functions such as medicine, business, technology, and warfare. In the past decade or so, a new area of applied ethics is again highlighting and testing these cultural differences, namely, the ethics of artificial intelligence.

Insofar as the United States is a main hub of artificial intelligence research and development, it is of particular importance which values in America are emphasized and how they are interpreted, and therefore what norms are created. When we examine the most influential research on the ethics of artificial intelligence, there are indeed particular norms emphasized in the US. Among these are privacy, existential risk, and bias, which can be traced back to emphasis and interpretation of the values of autonomy, happiness, and fairness, respectively. These three norms in their particular interpretations dominate the landscape of artificial intelligence (AI) ethics in the US. The purpose of this chapter is two-fold: first, to understand how these norms came to be emphasized, and second, to appreciate how they are interpreted in the context of AI ethics by researchers, industry leaders, and thought leaders in the US.

THE PROBLEMS OF AI ETHICS

Artificially intelligent technologies have begun to open up new areas of ethical concern, and so have emerged as a source of anxiety. However, that anxiety is often amorphous, and that is because it has proven much more difficult to specify what about AI is ethically troubling, and why. And then there is the additional difficulty of specifying what particular problems are emphasized in the US. This is because the problems of AI, much like the associated technologies, have an unclear structure and relationship to each other. It will be helpful then, to begin with a general lay of the land.

The first family of problems we might describe as "data ethics," which is an important place to start, since data are the inputs of AI systems. Here we need to consider data privacy, data bias, data inference, and the various threats to environmental sustainability involved in collecting, storing, and using data. The second family might be referred to as "machine learning ethics," so called because these are concerns about the models themselves, and not the inputs of the models or how

DOI: 10.1201/9781003585527-9

they are used. Here we find three weighty topics, namely algorithmic bias, algorithmic accountability, and the ethics of reinforcement learning.

A third broad category is based on concerns about how algorithms are affecting society, which is in at least three ways. First, algorithms attempt to classify us. They must do this with incomplete information, so in an important sense, these classifications are "pre-judgments" about an individual or instance based on past data about similar individuals or instances. This is not unlike the concept of "prejudice," which immediately suggests that classification will be fraught with moral challenges. Algorithms also collaborate with us, and finally, attempt to persuade us. These are also ethically rich areas.

A fourth category of ethical AI concerns is what we might call "personhood" issues, so called because they could affect the nature of the human person. One area is transhumanism, which is an attempt to push the boundaries of our human-ness. Another concerns how human relationships are transformed by AI, which include potential relationships with avatars, artificially intelligent chatbots, robots, and even human-to-human relationships mediated by algorithms. And finally, there is "existential AI," so called because these are the concerns that cluster around the issue of how artificial intelligence might affect the future of human existence.

I have been objective in mapping out these problems. That is to say, I have intentionally not emphasized the importance of one problem or family of problems over any other. And indeed, there is not *a priori* reason why one of these problems should be taken more seriously than any others. Therefore, if a culture emphasizes one of the problems above others, there must be an explanation for why this is so. Researchers, industry leaders, and other thought leaders in America have tended to fixate on three of these issues: privacy, existential risk, and above all algorithmic bias. These emphases have the unintentional effect of de-emphasizing many of the other problems. This chapter does not attempt to make a judgment about whether those emphases are justified or not, only attempting to explain the contours and histories of these selective concerns by describing their origins and some of the salient current research in each category.

PRIVACY AND AI IN THE US

The United States of America is a unique country in several ways. Certainly one of the most important ways is that its founding as a modern nation-state was not organic but was circumscribed in a moment of time. Of course, this is for an unforgivable reason, namely the confiscation of land and the simultaneous genocide of its native peoples by European settlers (Dunbar-Ortiz, 2015). Nevertheless, these historical injustices have the effect of making it easier to understand the cultural norms of the United States; instead of having norms and laws that developed gradually and in complex ways over the centuries, there were specific times, places, people, and motivations that can be described.

The most direct way to understand this is to understand the founding documents of the United States, which are, most significantly, the Constitution and the Declaration of Independence. These documents are replete with clear statements of freedom for citizens and conversely, limits for government (Ellis, 2016). "Freedom," or what we can call from an ethical point of view the value of "autonomy," was considered by the writers of the founding documents to be one of the values that was so important that it required the founding of a new governmental system (McCullough, 2005). If autonomy is the value, how is it preserved and nurtured? There is a specific word for the most important way to do this: privacy. The interpretation of how to do this best has created some identifiable cultural norms.

There are some simple general formulas here that connect the value of autonomy with privacy. First, the more that an individual has control over their own life and decisions, the more autonomy they have. The less an individual has control over their own life and decisions, the less autonomy they have. And the second formula is that the more that a government knows about its people, the easier it is to control them, and the less it knows, the more difficult it is to control them. This is why, even though the word "privacy" is not used in the founding documents, the concept is ubiquitous. Consider the fourth amendment to the Constitution, that "[t]he right of the people to be secure in their persons, houses, papers, and effects, against unreasonable searches and seizures, shall not be violated…" (U.S. Const. amend. IV). This is a simple expression of privacy, and it is what makes freedom, that is to say autonomy, possible.

Although privacy is ingrained in the American psyche, there are some fundamental confusions about what it is, why it is important, and when it is important; unfortunately, privacy in the digital age has only exacerbated these complications (Richards, 2014). Because data now can be stored in 1's and 0's, it can be stored automatically and permanently. These two differences with analogue data have significant implications for the possibilities for surveillance. America's introduction to privacy in the digital age was perhaps Edward Snowden's revelation that the National Security Agency (NSA) was surveilling electronic communications of US citizens (US House of Representatives 2016). The fear of government surveillance, stirred up by Snowden, perfectly fits the existing US norm concerning privacy. The Snowden leaks continue to be controversial, although a court ruling agreed with Snowden that parts of the program went too far (Satter, 2020).

But artificially intelligent technologies add another significant layer to privacy, because it is not just that data are now digital but that data are now used to create artificially intelligent models. And so they are already upending the US response to privacy by creating new kinds of challenges. There are now serious privacy concerns when a person's likeness is used without their consent in a deepfake photo or video, when content is used in model training without its author's consent, the proliferation of public use of facial recognition, and the failure of anonymization, the main privacy-preserving technique for analogue data.

And so it is perhaps surprising that the US has been less aggressive than the European Union about creating laws that keep pace with digital data. There is the General Data Protection Regulation (GDPR) of 2016, and then the Artificial Intelligence Act (AI Act) of 2024. These massive laws vastly overshadow the power of laws developed in the US concerning digital privacy, which are thus far mostly confined to the states (California Legislature 2018).

If privacy is deeply ingrained in the American psyche, why can't the US pass a national law similar to the GDPR or AI Act? I believe there are two specific reasons: the GDPR and AI Act restore the balance between data subjects and data controllers, but the "data controllers" are often for-profit corporations, not only governmental organizations, and US concerns about privacy are usually about privacy from the government specifically. And second, privacy in the US has usually been thought of as a check on coercion, but in the age of AI, privacy is also important as a check on manipulation. Both of these differences are explored in depth in Zuboff's seminal book *The Age of Surveillance Capitalism: The Fight for a Human Future at the New Frontier of Power*, which argues, among other things, that it is no longer simply information about individuals that is tracked but behaviors (Zuboff, 2018). Algorithms can then use this understanding of behavior to manipulate individuals. And so, despite the fact that the value of autonomy is deeply held in the US, the norms that have developed are more focused on government surveillance, and less so on the potential for manipulation by for-profit organizations.

EXISTENTIAL AI IN THE US

Artificially intelligent technologies have long provoked a significant amount of anxiety, even before such technologies were extant. We know this because of the rich history of science fiction literature. Obviously, the history of science fiction is both ancient and global (Roberts, 2016), but its tradition in the English-speaking world generally and in the US specifically gained outsized importance in the 20th century, both in its influence and in its focus on various thought experiments involving artificially intelligent systems becoming an existential threat to humanity (Westfahl, 2015). And US-produced television and film in this tradition, made to entertain the average consumer, have made these existential AI concerns dramatically more influential (Vint, 2015).

For the past decades, the advent of something like "artificial intelligence" felt distant and confined to movies, probably because the actual instances of artificial systems that manifested something like intelligence were confined to specific tasks. They are what some researchers call "artificial narrow intelligence" (ANI), as contrasted with "artificial general intelligence" (AGI) (Russell, 2020). The difference is that ANI systems are designed to do one or a few specific things, and while they can often do them better than any human could, they can do only those specific things. An AI that plays chess is a good example, as some of those models famously can play chess better than any human in the world, but that same AI model might not be able to do something very simple for a human, like shop for shoes.

AGI by contrast would be artificial intelligence that can do an indefinite number of tasks, similar to the way that our human intelligence allows us to engage in many different activities, both simple and complicated. Because ANI is common, AI ethics usually focuses on ANI applications. But with the commercial arrival of ChatGPT in 2022, researchers are now thinking harder about the potential of AGI. To be clear, hardly anyone thinks that current generative AI models are actual instances of AGI, but their ability to create meaningful interactions in an indefinite number of different contexts is an amazing leap from the ANI systems of the past decades. This has focused the attention and renewed interest in the existential questions raised by science fiction, notably: will the arrival of AGI be a threat to humanity?

In March 2023, the Future of Life Institute released a letter calling for a six-month pause on what they call "human-competitive" AI, although it amounts to the same thing as AGI (Future of Life Institute, 2023). The controversy caused by the letter was informative, because it showed that AI ethics researchers are deeply divided on the issue of AGI, sometimes in surprising ways. There are at least six distinct responses to the letter from the various US researchers and thought leaders, which can be summarized in succinct phrases.

The first position is the easiest to understand: "*AGI is coming and that's bad.*" This position believes that when AGI arrives, humanity will be undermined and threatened. So, anyone making progress on developing AGI is playing a dangerous game. Thought leaders on this position include the late Stephen Hawking and Eliezer Yudkowsky, and they would be the likeliest candidates to support the letter. Actually, Yudkowsky argues that the letter does not go far enough, because instead of a pause, we need a complete shutdown (Yudkowsky, 2023). A second position is similar in its consideration about the possible or even inevitable advent of AGI, but it is optimistic rather than pessimistic: "*AGI is coming and that's good.*" This position can also be called "techno-optimism," and is championed by people like Marc Andreessen. As the name suggests, these researchers believe deeply that technology is a friend to humanity, as technology saves lives, and ultimately may even save civilization. This position does not necessarily make a prediction that AGI is imminent, but they tend to be open to the possibility and usually excited about it.

A third position is in the middle of the road about the advent of real AGI, but affirms that even if something similar to it arrives soon it is not a big deal from the perspective of existential concern. In a phrase: "*Something like AGI may be coming but we shouldn't worry.*" This position says that whatever looks like AGI will be tricks and illusions generated by mathematics. Of course, they may be very cool tricks that we might benefit from and enjoy very much. But the intelligence of artificial agents will never be similar to the intelligence of humans. That implies, among other things, that these agents will not develop their own goals that threaten humanity. This position is most closely associated with thought leaders such as Yann Lecunn. Another position shares the middle-of-the-road prediction about AGI but is more concerned: "*Something like AGI may be coming and we should be more worried.*" This position usually doesn't say for sure that AGI is coming or for sure that it will threaten humanity but rather thinks about the advent of AGI in terms of risk. Let's say the risk to human existence is set at 10%. While a 10% chance is not high in many contexts (like your team winning the game), that is very concerning when you take into consideration that the 10% chance could include humanity as we know it being undermined. Thinkers who have advocated caution in the sense include Bill Gates.

Another position is highly skeptical about the advent of AGI: *"AGI is not coming soon, so stop worrying about AI technology."* This position thinks the possibility of AGI is interesting, and may arrive at some point in the future, but it would probably be centuries or more into the future. As a result, it can make for entertaining science fiction, but it is not worth serious scientific attention. Thinkers here include Stanford researcher Andrew Ng, who has compared the existential concerns of AI to worrying about over-population on Mars. A sixth and final possible response to the letter is from AI ethics researchers who agree that AGI is not coming any soon, but worrying about it is not just playful science fiction, as it distracts us from the real harms that AI is doing: *"AGI is not coming soon so do not let it distract us from worrying about other kinds of AI."* Advocates of this position often call the idea of progress toward AGI "hype," because they think technologies that look like progress toward AGI are not making progress at all. But this fact is not neutral, because the focus on AGI and the panic over the future of humanity that often ensues can lead us to neglect the current problems such as privacy and bias, which also happen to be the problems disproportionately affecting those who are poor and marginalized in society the most. Thinkers here such as Timnit Gebru and Emily Bender have published a statement explaining these concerns (Distributed AI Research, 2023).

Thus, the long-standing fascination in the US with existential AI, first explored through science fiction literature, then more so through film and television, has finally become real in a way it never was before, and the reaction and debate it has recently engendered have been fascinating.

BIAS AND AI IN THE US

But the most well-developed emphasis in research on AI ethics in the United States is on the topic of bias, often referred to as "algorithmic bias." This includes research on how AI systems become biased, how their bias has real-world consequences, and how to measure it. A claim made by Cathy O'Neill in her book *Weapons of Math Destruction* in the nascent days of AI ethics is that "models are opinions embedded in mathematics" (O'Neill, 2016). The implication here is that what masquerades as objective, unbiased mathematics and technical systems are often the reflection of opinions, and it is uncontroversial to note that opinions can be and often are biased.

If it is true that the US has led on this issue, why? The United States has a very high sensitivity to systems that affect identifiable groups differently, obviously based on the value of equality/equity, or what AI ethics researchers more often call "fairness." Fairness concerns can understandably become central in the face of diversity. All cultures have gender diversity, but the nature of the racial and ethnic diversity in the US has an ongoing impact on technological development. This connection cannot be demonstrated in any scientifically rigorous way, but a brief review of three features in the US make the connection not particularly mysterious. Three features must be kept in mind: the fact and nature of racial diversity in the US, the reasons for the diversity, and the continuing consciousness of the reasons for the diversity. This then sets the stage to examine the US-led research on algorithmic bias.

This starts with the fact that the nature of racial diversity in the US is unique (Rose, 2006), and while there are plenty of countries with more racial, ethnic, and cultural diversity (Fisher, 2013), the US stands only with its neighbor, Canada, as wealthy countries who also have high racial diversity. So, there are wealthy countries, and there are racially diverse countries, but it is rare to find a country with both high wealth and high diversity. What explains this confluence of wealth and racial diversity? The two biggest factors are tragic ones: the violent confiscation of the land now known as the United States from its indigenous people by the European settlers (Dunbar-Ortiz, 2015), and the violent trade that forced people from Africa to work as slaves for those same settlers in America (Thomas, 1999). Unlike other countries where the racial, ethnic, and/or cultural diversity is organic because it developed gradually over centuries and millennia and perhaps has more or less innocent explanations, much of the racial diversity in the US was created through colonialism and violence through particular events at particular moments in time.

These are the facts of these historical injustices, and there is the evidence that these facts still shape the American consciousness. With regard to Native peoples, there is the advent of "land acknowledgments," especially in academic circles (Keefe, 2019). These acknowledgments are formal declarations of the original inhabitants and stewards of the land, and usually reference specific tribes, and while their effect has been called into question (Wood, 2021), they do in fact show that this injustice still haunts the American consciousness. And the amount of attention paid to Black, Indigenous, and People of Color (BIPOC) and diversity equity inclusion (DEI) initiatives in academia and industry, especially big tech, is significant (Dimich, 2023). This is not to claim the DEI initiatives are successful in their stated efforts but only that there is at least on-going awareness and acknowledgement of the historical injustices committed by the United States.

Some scholars in AI ethics have explicitly contextualized algorithmic bias this way, with Ruha Benjamin's seminal book in AI ethics titled *Race After Technology: Abolitionist Tools for the New Jim Code* (Benjamin, 2019), a reference to segregationist laws that existed and were enforced in many parts of the US until 1965 (Alexander, 2010), which many people refer to as "Jim Crow" laws. There was first the institution of slavery, then segregationist laws, and then, as Alexander demonstrates, there are still institutions, such as the criminal justice system, that carry on the legacy of racism. Benjamin's claim is that AI systems are yet another way that racism is preserved and amplified.

Benjamin's work is part of a steady stream of research that includes the afore-mentioned *Weapons of Math Destruction*, and has continued to dominate both scholarship and media. Notable research starts with what we already know about biased data, as even in the age of "small," analogue data there was the known challenge of drawing samples from a population that are fairly representative of it. This problem has been well understood in a small data context probably for as long as sampling has been done, under the name "selection bias." In the context of "small data," the ubiquitous example of selection bias in the US is the false prediction of Dewey's presidential victory, now blamed on relying on data from phone surveys when homes without a phone line were more likely to vote for Truman, and actually did so (Seely-Gant & Frehill, 2015). The specific problem is that the sample was not random across households, because only households with phone lines were represented in the sample.

In the age of AI, when data are "big" and digital, this problem still exists but with even more troubling implications, because now the data are not just used to estimate population parameters but to build artificially intelligent models. And once bias is part of a system, it will be much more difficult to get out. For example, consider the case of bias detected in facial recognition software. A 2015 study by the National Institute of Standards and Technology demonstrated disparate accuracy in facial recognition by gender (Ngan et al., 2015). Buolamwini and Gebru (2018) followed this study by auditing three facial classifiers—from Microsoft, IBM, and Face++. Their audits of these systems measured disparities in gender classification, replicating the NIST study. But their audit also revealed wide disparities in all three systems when it came to skin tone, with the error rates for darker skin tones being very high relative to lighter skin tones (12.9% to 0.7% for Microsoft, 16.5% to 4.7% for Face++, and 22.4% to 3.2% for IBM). But the most significant finding from their audit was the intersectional disparities (i.e., when skin tone and gender intersect). The different error rates between light-skinned males and dark-skinned females were as high as 34.7% compared to 0.3%.

This is a remarkably clear demonstration of bias in an AI system. Why did it occur? It seems obvious that the most basic problem was that the machine learning technology that makes facial recognition possible was trained on facial images that were disproportionately white and male, meaning that the training data was not representative of the population. The thoroughly unsurprising result is a technology that is biased toward white males.

However, it is worth noting that some try to complicate this narrative. Marks (2021) refers to a governmental accountability report (GAO, 2020) that "there is "no consensus at all among academics, industry, standards bodies, or independent experts on how to fix the biases," since they could depend "on multiple algorithmic factors, such as the breadth of ethnicities used in the training data and variables like false-positive threshold settings, as well as photograph-related factors such as pose angle, illumination, skin tone, skin reflectance, expression, cosmetics, spectacle use, and image quality" (Marks, 2021). But while it is possible that some kind of lighting or reflection factors might come into play when different skin tones are involved, none of this would explain why facial recognition technology would work less well on males and females of the same skin tone. The more likely explanation is that it did not occur to the people developing the technology to use images representative of the human race. And to emphasize the global significance, Buolamwini, a black woman, notes that the same facial recognition software that was unable to recognize her face in the US was also unable to recognize her face while visiting Hong Kong: "[H]alfway around the world, I learned that algorithmic bias can travel as quickly as it takes to download some files off the internet" (Buolamwini, 2016).

It is interesting to note that this bias issue can arise in two ways, as similar problems can be created either by under-representation or by over-representation. It is simply a matter of context. For example, if someone's life is not "datafied" as much as others, it might not be counted as much as others. In this way, big data "could restructure societies so that the only people who matter—quite literally the only ones who count—are those who regularly contribute to the right data flows" (Lerman, 2013). Or the unfairness can come in the opposite direction, such as when communities of color are over-policed. This can easily create data indicating a "crime hotspot," when in reality that location may not any be "hotter" or "colder" with respect to breaking the law than other, wealthier areas. And there is another distinct bias challenge, namely the challenge of relying on data created by biased actions and systems. The term "historically biased data" should be reserved to refer to data that is both accurate and unfair, and for that specific reason is unsuitable in its current form for generating fair predictions. "Non-representative data" must refer to data that is taken to be an accurate representation of a population but is not, and for that specific reason, is unsuitable for generating fair predictions.

This is of vital importance because algorithms learn nothing except what they learn from data, and so biased data becomes the "truth" of the model. As noted, this problem has two fundamentally different origins. On the one hand is the problem that is created by using non-representative data. Non-representative data can easily create bias by focusing too much or too little on any group. The issue here is that non-representative data are fundamentally inaccurate. Historically biased data are data that are accurate but accurately reflect all the harmful biases that are persistent in human behavior and institutions.

And it is not just acquiring data that is the issue, as the bias concerns are still in the data preprocessing. Any dataset from the real world is "dirty," that is, containing missing, incomplete, or clearly erroneous entries. Machine learning algorithms give error messages about processing the data until it is "clean," and so data science necessitates making choices about dealing with messy data (Curley et. al., 2017). This is despite the fact that often no one is accountable for those choices, as long as the model "works." It is not difficult to imagine how bias could enter the system here. Suppose many birth dates were missing in a given dataset, and a data scientist used software that automatically defaulted missing dates to "1-1-1900." If this is not addressed in a thoughtful way, a problem as simple as this could easily bias a prediction toward or away from older or younger groups, thereby embedding ageism in the algorithm.

These and other similar bias issues are unfortunately resilient, and a large part of the reason is the failure to interpret the data that we use to build AI systems. There is an assumption made often, including by data scientists, that data are fixed, objective, and true. This unreflective belief seems to be especially intoxicating when the data are numerical. Kate Crawford argues that this assumption must be rejected, and instead, "data scientists should take a page from social scientists, who have a long history of asking where the data they're working with comes from, what methods were used to gather and analyse it, and what cognitive biases they might bring to its interpretation" (Crawford, 2013). And the resistance or refusal to critically examine the underlying data she calls "big data fundamentalism" (Crawford, 2013).

Whether she intended it or not, "fundamentalism" is a provocative word because of its religious ties. A religious fundamentalist is sometimes considered to be someone who treats a sacred text as a completely formed object, the textual meaning already embedded in a self-evident way, thus making interpretation unnecessary. A basic challenge to religious fundamentalism is to note that any text, regardless of its alleged source, must be interpreted before it can be understood. And this is not because of any complexity in the text, but just because that is how language works. Crawford's challenge is the same but applied to data, as "big data fundamentalism" is the belief that data do not need to be interpreted before they are understood. They can just be taken as fully formed and complete in meaning. But data, like any text, were necessarily generated, and therefore necessarily have a story to be told about their creation.

This fundamentalist assumption is ubiquitous enough to generate its own phrase: "raw data" – meant to provoke the image of data that exist in a natural state, prior to being "cooked" or transformed by human or algorithmic activity. The introduction to the book *Raw Data is an Oxymoron* adds useful nuance to Crawford's important point:

> [T]he phrase "raw data"...has understandable appeal. At first glance data are apparently before the fact: they are the starting point for what we know, who we are, and how we communicate. This shared sense of starting with data often leads to an unnoticed assumption that data are transparent, that information is self-evident, the fundamental stuff of truth itself. If we're not careful, in other words, our zeal for more and more data can become a faith in their neutrality and autonomy, their objectivity. Think of the ways people talk and write about data. Data are familiarly "collected," "entered," "compiled," "stored," "processed," "mined," and "interpreted." Less obvious are the ways in which the final term in this sequence – interpretation — haunts its predecessors. At a certain level the collection and management of data may be said to presuppose interpretation. "Data [do] not just exist," Lev Manovich explains, they have to be "generated." Data needs to be imagined as data to exist and function as such, and the imagination of data entails an interpretive base.
>
> **(Gitelman & Jackson, 2013)**

The denial that data need to be interpreted seems to be ubiquitous. When humans see data, we tend to stop thinking and accept what is given, but as Gitelman and Jackson argue, that is the exact moment we should instead be beginning to ask critical questions.

The issue is framed well by Selbst et al. in referring to the requisite knowledge of a sociotechnical system in these cases (2018). They argue that certain problems in machine learning arise "from failing to consider how social context is interlaced with technology in different forms, and thus the remedies also require a deeper understanding of 'the social' to resolve problems," because "fairness and justice are properties of social and legal systems like employment and criminal justice, not properties of technical tools within" (Selbst et al., 2018). So it is possible to create abstract definitions of algorithmic fairness that look statistically important, but they may only be relevant to a particular context. Or, it is even possible to possess knowledge of machine learning and the requisite knowledge of the social context but be lacking in knowledge of ethical analysis: "By abstracting away the social context in which these systems will be deployed, fair-ML researchers miss the broader context, including information necessary to create fairer outcomes, *or even to understand fairness as a concept*" (Selbst et al., 2018, my italics).

CONCLUSION

I have argued that while we will not find a culture that does not care at all about fundamental values such as autonomy, happiness, or fairness, what is interesting is the different emphases and interpretations that various cultures apply to those values. This creates varying cultural norms, which are sometimes quite different, and sometimes even conflicting. This is certainly true in the United States, as the way the US has emphasized and interpreted basic values has impacted the way that researchers and thought leaders talk about the value challenges posed by AI. In the US, there is an out-sized concern for privacy, which is understandable given its constitutional origins. Although when it comes to AI ethics, the concerns have focused more on the way the government might collect and use digital data to surveil us, and less on the ways that for-profit businesses might use algorithms to manipulate us. This can be seen as a major oversight, although it is in line with the US fear of government surveillance but reticence to interfere in business operations. AI ethics researchers in the US also spend much time discussing existential concerns with AI technology, which is not surprising given the long history of existential concern in science fiction. And finally, the difficult struggles to maintain a multi-cultural democracy in the US have made AI ethics researchers understandably sensitive to algorithmic bias. The ethics of artificial intelligence is a relatively new area of applied ethics, and while privacy, bias, and existential concerns should and will continue to receive emphasis, other areas of research may open up soon, as there are many ways the artificial intelligence is affecting society. The US can learn from how other areas of the world are investigating the ethics of artificial intelligence.

REFERENCES

Alexander, M. (2010). *The new Jim crow: Mass incarceration in the age of colorblindness.* The New Press.

Benjamin, R. (2019). *Race after technology: Abolitionist tools for the new Jim code.* Polity.

Buolamwini, J. (2016, November). *How I'm fighting bias in algorithms* [Video]. TED Conferences. https://www.ted.com/talks/joy_buolamwini_how_i_m_fighting_bias_in_algorithms

Buolamwini, J., & Gebru, T. (2018). "gender shades: Intersectional accuracy disparities in commercial gender classification. *Proceedings of Machine Learning Research, 81,* 1–15.

California Legislature. (2018). *The California consumer privacy act of 2018.* https://leginfo.legislature.ca.gov/faces/billTextClient.xhtml?bill_id=201720180AB375

Crawford, K. (2013). The hidden biases in big data. *Harvard Business Review.* https://hbr.org/2013/04/the-hidden-biases-in-big-data

Curley, C., Krause, R. M., Feiock, R., & Hawkins, C. V. (2017). Dealing with missing data: A comparative exploration of approaches using the integrated city sustainability database. *Urban Affairs Review, 55*(2), 591–615.

Dimich, N. (2023, June 19). *The history of diversity, equity, and inclusion (DEI) in tech and beyond.* The Turing School. https://writing.turing.edu/the-history-of-diversity-equity-and-inclusion-dei-in-tech-and-beyond/

Distributed AI Research. (2023). *Statement from the listed authors of Stochastic Parrots on the "AI pause" letter.* https://www.dair-institute.org/blog/letter-statement-March2023/

Dunbar-Ortiz, R. (2015). *An indigenous peoples' history of the United States.* Beacon Press.

Ellis, J. J. (2016). *The quartet: Orchestrating the second American revolution, 1783-1789.* Vintage Publishers.

Fisher, M. (2013, May 16). *A revealing map of the world's most and least ethnically diverse countries.* Washington Post. https://www.washingtonpost.com/news/worldviews/wp/2013/05/16/a-revealing-map-of-the-worlds-most-and-least-ethnically-diverse-countries/

Future of Life Institute. (2023). *Pause giant AI experiments: An open letter.* https://futureoflife.org/open-letter/pause-giant-ai-experiments/

GAO Report. (2020). *Facial recognition technology: Privacy and accuracy issues related to commercial uses.* Government Accountability Office.

Gitelman, L., & Jackson, V. (2013). Introduction. In L. Gitelman (Ed.), *"Raw data" is an oxymoron* (pp. 2–3). MIT Press.

Keefe, T. E. (2019). *Land acknowledgement: A trend in higher education and nonprofit organizations.* https://doi.org/10.13140/RG.2.2.33681.07521

Lerman, J. (2013). "Big data and its exclusions. *Standford Law Review Online, 66,* 55. https://www.stanfordlawreview.org/online/privacy-and-big-data-big-data-and-its-exclusions/

Marks, P. (2021). Can the biases in facial recognition be fixed; Also, should they? *Communications of the ACM, 64*(3), 20–22. https://doi.org/10.1145/3446877

McCullough, D. (2005). *1776.* Simon & Schuster.

Ngan, M., Ngan, M., & Grother, P. (2015). *Face recognition vendor test (FRVT) performance of automated gender classification algorithms.* U.S. Department of Commerce, National Institute of Standards and Technology.

O'Neill, C. (2016). *Weapons of math destruction: How Big Data increases inequality and threatens democracy*. Crown.
Richards, N. M. (2014). Four privacy myths. In A. Sarat (Ed.), *A world without privacy* (pp. 33–82). Cambridge University Press.
Roberts, A. (2016). *The history of science fiction*. Palgrave Macmillan.
Rose, P. I. (2006). *They and we: Racial and ethnic relations in the United States* (6th ed.). Routledge.
Russell, S. (2020). *Human compatible: Artificial intelligence and the problem of control London*. Penguin Books.
Satter, R. (2020, September 2). *US Court: Mass surveillance program exposed by Snowden was illegal*. Reuters. https://web.archive.org/web/20201101085850/https://www.reuters.com/article/us-usa-nsa-spying/us-court-mass-surveillance-program-exposed-by-snowden-was-illegal-idUSKBN25T3CK
Seely-Gant, K., & Frehill, L. M. (2015). Exploring bias and error in big data research. *Journal of the Washington Academy of Sciences*, *101*(3), 29–38.
Selbst, A. D., Boyd, D., Friedler, S., Venkatasubramanian, S., & Vertesi, J. (2018, August 23). Fairness and abstraction in sociotechnical systems. 2019 ACM Conference on Fairness, Accountability, and Transparency (FAT*), 59–68. https://ssrn.com/abstract=3265913
Thomas, H. (1999). *The slave trade: The story of the Atlantic slave trade: 1440 – 1870*. Simon & Schuster.
US House of Representatives. (2016, September 15). *Executive summary of review of the unauthorized disclosures of former national security contractor Edward Snowden*. https://web.archive.org/web/20160928053841/; http://intelligence.house.gov/uploadedfiles/hpsci_snowden_review_-_unclass_summary_-_final.pdf
Vint, S. (2015). Hollywood science. In G. Canavan & E. Link (Eds.), *The Cambridge companion to American science fiction* (pp. 111–124). Cambridge University Press.
Westfahl, G. (2015). The mightiest machines: The development of American science fiction from the 1920s to the 1960s. In G. Canavan & E. Link (Eds.), *The Cambridge companion to American science fiction* (pp. 17–30). Cambridge University Press.
Wood, G. (2021, November 28). 'Land Acknowledgments' are just moral exhibitionism. *The Atlantic*. https://www.theatlantic.com/ideas/archive/2021/11/against-land-acknowledgements-native-american/620820/
Yudkowsky, E. (2023, March 29). The only way to deal with the threat for AI? Shut it down. *Time Magazine*. https://time.com/6266923/ai-eliezer-yudkowsky-open-letter-not-enough/
Zuboff, S. (2018). *The age of surveillance capitalism: The fight for a human future at the new frontier of power*. Public Affairs.

7 The Ethos of Deification
On True Sovereignty in Dealing with Technology

Jan Juhani Steinmann

INTRODUCTION: THE FORGETFULNESS AND MISJUDGMENT OF THEOSIS

The current developments in technological trans- and post-humanism can be critically analyzed from a variety of perspectives, be it philosophically, neuroscientifically, politically or ethically. To date, however, the plausibility of the said trans- and post-human "complex of phenomena" has all too rarely been considered from a comprehensive theological view, i.e. here a truly and genuinely Christian perspective; and if it has, then often nevertheless with a prioritization of ethical questions, such as how artificial intelligence can be used in a way that is compatible with Christian moral teaching.[1] However, this rampant *ethical reductionism* means that not only trans- and post-humanism are viewed too narrowly and thus considered in isolation, but also Christianity from whose perspective they are analyzed. In extreme cases, we are confronted with a threefold reduction: Firstly, technological transhumanism is being reduced to a primarily ethical discourse, thereby excluding its anthropological, epistemological, political, aesthetical, and theological facets, or deeming them merely secondary. Secondly, Christianity itself is being reduced – in a Kantian fashion – to morality, undermining its sacramental, mystical, and soteriological dimensions. Thirdly, ethics itself is often reductively conceived by minimizing it to a positivist ἔθος (*ethos* with epsilon), which, like the Latin *mos*, in Greek means general morality, and tends to forget the phenomenological level of ἦθος (*ethos* with eta), which emphasizes the individual practice that constitutes the moral disposition of an individual and thus its character, as it is accentuated, for example, by Søren Kierkegaard's existential moral theology (Kierkegaard, 1991). All three reductions ultimately miss the human being in their holistic existence. Not only is the technologically transhumanized human being often underestimated – usually in a highly misanthropic way, for example as a data-processing carbon complex – but even within Christian anthropology they are so often misunderstood because they are not conceived as a quasi supra-moral being. In concreto, this means that humans are missed in the fate of their Θέωσις (*theosis*, Greek for deification), as the actual entelechy of their existential practice. As Mt 5:48 already makes clear, humans are to achieve divine perfection (τελείωσις, *teleiosis*) by striving for it.[2] The true ethos of the human being therefore divinization, i.e. becoming whole in the perfection of the Lord. This is in essence what Christianity teaches.

Nevertheless, in Western Christianity in particular, despite its theological omnipresence,[3] we see ourselves confronted with a strange forgetfulness of theosis, or perhaps better, a misjudgment of theosis, which has become even more pronounced in the past two centuries. The reasons for this are manifold and cannot be discussed here, they also encompass the philosophical and scientific developments since the modern era (Bacon, Descartes, Newton, etc.) and the reductionist materialism that grew out of them. This materialism is still very influential today, especially in the tech context from which technological transhumanism has emerged. Two things therefore remain crucial: Firstly, that any ignorance of theosis diminishes Christianity and both its anthropology and ethos in an inadequate manner. It is therefore essential to emphasize the theotic dimension of Christianity more resolutely again. And secondly, that technological transhumanism, despite its many fallacies, is interestingly recognizable precisely as a technocentrically distorted form of the

DOI: 10.1201/9781003585527-10

theosis idea, which is presumably also where its primary fertility lies: That it serves at least as a corrective against the prevailing forgetfulness of theosis, albeit under the criterion of a *corruptio optima pessima* – the corruption of the best is the worst.

In this chapter, I want to critically consider technological transhumanism from a Christian theotic perspective and at the same time defend the Christian ethos of deification as an alternative to the technological phantasmagoria that are increasingly sweeping the world from Silicon Valley. We will therefore look at three motifs of stray theosis and contrast them in Christian terms: Firstly, the question of mortality and immortality in the light of embodiment and being-in-the-world; secondly, the question of omniscience and ineffability in the context of myth, meaning, and narration of humanity; thirdly, the question of creative power and grace against the horizon of a will to power as a will to God; fourthly, a concept of theonics, which should allow a sovereign use of technology in the service of theosis; and finally, a summary of theosis as the true ethos of humanitas. As it quickly becomes clear, our techno-theological approach here is neither reductively ethical nor determined by a mere ἔθος, but should rather be the expression of a holistic, phenomenological, and existential ἦθος that aims at the deification of human beings. In view of the transcendence of our concern, however, both approaches discussed here are narrative, which is why our critical analysis must utilize not only Christian, but also mythological motifs. For technological transhumanism is not least a narrative aberration.

EMBODIMENT AND BEING-IN-THE-WORLD

In the far west of the earth there was once a garden in which the Hesperides (a type of nymph) guarded a tree with golden apples. This gave the gods eternal youth and was guarded by the many-headed dragon Ladon. One of Heracles' twelve tasks was to steal them, which he managed to do with a trick. Analogously, we can now also recognize the first aberration of technological transhumanism: That it turns out to be another grab for the apples of the Hesperides, a grab for immortality, which is actually only reserved for the gods. It is perhaps no coincidence that this reaching for the promise of an apple is reminiscent of the biblical Fall of Man: Seduced by the serpent, Eve ate the forbidden fruit from the tree of knowledge, which only became a *malum*, Latin for apple, in a play on words from *malum*, Latin for evil. Despite its theological dubiousness, the question arises as to whether the transhumanist longing for immortality today does not confront us with the danger of a "second fall" from grace, or at least with a dangerous hubris. This hubris is not only a potentially theological presumption, but in any case, expresses also a philosophical pomposity toward any humanistic and phenomenological concepts of body and world. For what transhumanism seeks to overcome is above all the traditional Christian, and therefore humanistic, self-understanding of the human being as a unity of body and soul, created as a person with dignity in the image of God and thus sharing in the divine Logos, his love and the ability to create, beget, and generate. As a rule, the human being is thus conceived purely from an information theoretic perspective as a malleable body (see above: a data-processing carbon complex), which is thus – often unwittingly – still an expression of a long antiquated, Cartesian body–soul dualism. On the other hand, in the age of big data (the materialism of our time) this goes hand in hand with an understanding of space and the world of mere positive facticity, which becomes measurable and therefore controllable. It is not only the human being as a body that is ultimately a purely positivist phenomenon, but also space and thus the world it inhabits. Accordingly, the ideals of eternal youth in technological transhumanism initially manifest themselves as purely physical attributes of materiality, which should be able to be extended mechanically or be given persistence by integrating the human body into machine systems (its cyborgization), or as the creation of non-ageing digital spaces, such as in the chimera of a metaverse (the creation of cyberspaces). If these two strands are synthesized – the pure body in the purely material space – we end up with the trans- and post-human ideals of immortality, as shown in the uploading of neuronal information structures into clouds or a dubious fusion of the human information pool with a superintelligence in the singularity.

In contrast to this, a Christian and phenomenological perspective promotes a completely different understanding of human corporeality and belonging to the world. Firstly, a person is not merely a body (*Körper* in German) but possesses one as an attribute of vitality. In the sense of Helmut Plessner's fundamental distinction, a human being has a body (*Körper*), but is a *Leib* (a lived body) (Plessner, 1982, p. 241). In the biblical, and especially in the Aristotelian tradition, a human being, insofar as they are alive, are thus primarily to be understood as a *Leib*, i.e., a body that has become alive. There is no pure corporeality for human beings, as they are always living their bodies. Being a *Leib* is therefore an expression of his holistic existence in the world.

Secondly, with Martin Heidegger's formulation of "being-in-the-world" in *Being and Time*, humans are therefore never to be thought of as an isolated, physical entity, but always as interwoven with the environment and the world around them. The essence of human beings as *Dasein* (being-there) is their being-in-the-world. This means that they care for the world, and that implies that, insofar as they live, they have a caring relationship with it (*Sorge*). Being-in-the-world is thus an expression of his holistic basic constitution, whereby the conventional distinction between subjects and objects is to be abolished in favor of a dynamic embeddedness in the temporality of life (Heidegger, 2019, §12–24). As a living body, the human being thus always lives in the body of the world itself.[4]

Thirdly, with Maurice Merleau-Ponty, the chiastic entanglement of the living body, which he realizes as a synthesis of mind and body, also applies. He realizes this as the interweaving of body and world in the flesh (*chair* in French) of the world: This is the texture of sensuality and thus the element where the living body and the world, which is experienced only through the body but to which the body also belongs, are intertwined (as a chiasm).[5] There are no isolated spaces, but every space that we perceive sensually is, as flesh, the expression of an ambiguity that is more than intimate or extimate (Merleau-Ponty, 1969, Chapter 4).

Fourthly, Christian resurrection therefore means that the pneumatic body that Paul speaks of (*Bible*, 2007, 1 Corinthian 15:42–44) and that Christ testifies of himself (Lk 24:36–43) must be a body of resurrection that can rise only from and in its carnal interweaving with the world as a whole. It is therefore not without reason that this hyperbolism of the mortal body is only possible for human beings in the end of this world and the rise of a new glorified one, as proclaimed in Christian soteriology (Phil 3:21). Resurrection can therefore not be a merely physical process; nor is immortality possible outside the flesh of a world that does not first have to perish. True immortality therefore implies a *Leib* resurrected into the flesh of a new world, into which it is itself interwoven again: for us humans, this new "world" – in the eschatological sense of theosis – must be God himself, in whose body the Christian is and will be a member (1 Cor 12:12–27). It is evident that this concrete *unio mystica* presupposes the one true God, who also exists beyond the world as its creator, and not merely as an algorithmic, data-processing superintelligence (the potential God of the technological trans- and post-humanists) that would come to an end in a Big Crunch at the latest. True immortality – in an existential-characteristic sense – cannot therefore be achieved by uploading into a superintelligence, but only through the grace-filled union of the pneumatic body with the divine.[6] And even in the myth mentioned at the beginning of this chapter, wherein Heracles succeeds in stealing the golden apples from the garden of the Hesperides, in the end these are returned via detours by the goddess Athena, who brings them back to the garden. Not only Christianity, but also the Greek mythology teaches us that the transhumanist apple grab is ultimately doomed to failure.

MYTH, MEANING, AND NARRATION OF REAL HUMANITY

One of the highest official positions within the Roman Empire was the Augur, whose task it was to examine and judge whether a certain state or private undertaking was in accordance with the will of the gods. He did this, for example, by interpreting the flight of a bird and its cry, the so-called auspices (Latin *auspicium*, the bird's eye view). The augur's auspicium was therefore the recognition of the will of the gods and thus also of the future in the sense of divination. In some respects, the augurs, where their judgment was trusted, had the highest power in the Roman Empire. It was therefore not uncommon for the Roman emperor also to be an augur.

Today, we recognize a return of the auguric mode within digital techné, namely in the onto-epistemic doctrinal claim of digitalism, where it is applied to the world and man. The second aberration of technological transhumanism therefore lies in the fact that it now algorithmically claims a kind of augur's auspices for itself: Firstly, by forcing the reductionist criterion of positivist data as a binding criterion of truth (an informationism); secondly, by using big data to create a new epistemological myth of potential infallibility, or at least epistemic priority; thirdly, by creating an ideological narrative of its self-plausibilization, which builds the transhumanist extension of the human being on data-based evidence and cyborgizing interventions linked to it. Only a dataist worldview and science provide evidence; only artificial intelligence is capable of processing the corresponding amounts of data; thus, humans must fuse with dataist-intelligent technologies in order to adequately recognize being and accordingly still be able to keep up with its growing acceleration of informationist functions and processes. This "truth-telling" prospectivity is not only directed at a normative claim to the science and human economy of the future, but also at concrete technologies that are, for example, supposed to predict crimes quasi-augurally and thus thwart them before they happen. A scenario that was already addressed with narrative means in the 2002 film *Minority Report*.

It seems almost pointless to emphasize how problematic and morally dangerous such a technocentric claim to omniscience is. Nevertheless, it is important to emphasize that no sound epistemology or conception of truth comes up with a claim to omniscience. Be it experimental or phenomenological, good science – and therefore good technologies that are based on it – always work approximately and thus exhibit an epistemic humility that wisdom can build on when dealing with the knowledge gained through it. Only in this way knowledge and truth can make us free. A data-centered and transhumanist neo-augurism is quite different, implying the significant danger of a reduction of human autonomy through the resulting high degree of automation of its functionalizations, which are determined to be valid and "true."

The myth of transhumanism is therefore not only an epistemological one, but also an algocratic one: The neo-augurian will have deified data is thus intended to redeem the inadequate human being from the dictatorship of freedom, so that it can achieve dominance through "infallible data." The authority of algorithms implied in this vision not only endangers our human autonomy, but as a result also our maturity and power of judgment. It is hard to escape the suspicion that this is also an escapism from a world that has become either too complex or nihilistically unbearable. This is where the new narrative of the reliability of big data and artificial intelligence comes in handy, especially where it also offers the epistemic-soteriological framework for "truth" in which the transhumanized human being is supposed to participate in its deification within a technological intelligence. It is true, as Blaise Pascal emphasizes: "L'homme passe infiniment l'homme" (Pascal, 1977, §122) – man infinitely transcends man – and even before Christian concepts of human self-surpassing, already Plato knew that the human ideal of becoming lies in ὁμοίωσις θεῷ κατὰ τὸ δυνατόν "(*homoíōsis theō̃ katá to dynatón*) – the resemblance to the deity as far as possible (Plato, 1997a, 716c–d; Plato, 1997b, 176b) – but, in contrast to technological transhumanism, this truth, where it can also be described as a narrative, is based on the vitality of a lived revelation, namely that of the birth, work, and resurrection of Jesus Christ.

The theotic dimension of Christianity is based on a reliability that is not factual, but existentially given, insofar as it has been revealed and, due to the grace of God's omnipotence, is already imprinted in man as created in the *Imago Dei* – in the image of God. It is this likeness which invites us humans to the *Imitatio Christi*, i.e. to follow the revealed truth. It is the evidence of revelation that is "narrated" here and can be lived accordingly. The true augurs of theosis, alongside the profound philosophers, are therefore the prophets, saints, and witnesses of faith, not algorithms or algorithmized cyborgs.

Firstly, it is therefore necessary to reflect on a concept of truth that appears as revealed vitality in Christ himself. As Kierkegaard writes, "Thus Christ is the truth in the sense that to *be* the truth is the only true explanation of what truth is" (Kierkegaard, 1991). Truth is not a concept or a theory, but a life to be lived. The existing truth thus illuminates the true logos, which is not available in data form, but whose signification, coming from God, must be proven and realized in life. As long as one is not yet God himself, omniscience is vulgar, both epistemically and existentially. Rather, it is necessary to face the living phenomenality, hyper-phenomenality, and comprehensibility of existence against the horizon of the mystery of creation and its providence. Its character of reference is a humble one, not a dominant one.

Secondly, this goes hand in hand with a concept of autonomy and responsibility that makes it possible to engage virtuously with everything that can be narrated and exists. The self-narrative of humanity must necessarily be a liberal one, not only to empower itself in the face of technologies that endanger it, but also not to fall short of God, from whom alone an ethos of human deification is adequately conceivable. In any case, the human being is free and responsible (that is *able to respond*) in God's image.

Thirdly and finally, Christianity, as a profound and humanly encompassing narrative, also crowns the three Pauline virtues as the criteria of reliable episteme: faith, love, and hope (1 Cor 13:1–13). In this way, humans recognize and live worthily and toward their true self-transcendence. This is the only way to think of oneself and one's actions in terms of revelation and providence, i.e. in the light of the Holy Spirit. This rather non-augurian attitude is not ideological, but existentially poetic: it relates to the mystery of God through the art of living – the truest language of the ethos of theosis, after it has become a practice that can not only be told, but also sung about.

THE WILL TO POWER AS A WILL TO GOD

Egyptian mythology, like that of other cultures, tells the topos of the world egg, or the cosmic egg. This symbolizes the primordial creation, insofar as the cosmic egg as a metaphor epitomizes the birth of the universe from its chaotic origins. The cosmic egg thus also serves as a metaphor for the creation of the world and therefore the power that created being and existence. Whoever possesses the cosmic egg has the creative power to create something new and different in the cosmos. He who possesses the primordial state of being is master of the origin and its orders. Whoever controls the generic moment of the universe is the master of life and its development. The third aberration of technological transhumanism therefore lies in the fact that it desires to seize the cosmic egg in a figurative sense, insofar as it wants to rule over being and life in a constructivist manner. Technological transhumanism thus manifests itself not only as a discourse of power (as we saw above), but also as a concrete expression of a will to power *in praxi*. Through technological transhumanism, humanity wants to become the master of creation and thus play God. This claim to power is a materialistic one, directed at the bios (apples of the Hesperides) and the data (auspices of the augurs). To dispose of creation means to interpret it from the point of view of its feasibility and malleability. The 3D printer already now suggests that one day we will be able to regroup all matter and create objects quasi *ex nihilo*. The CRISPR gene scissors promise complete control over life. The total cyberspace of a metaverse promises God-like abilities in the creation of virtual spaces and their laws – to name just three examples.

Technological transhumanism applies this morphological freedom of the masters of the cosmic eggs – from the Greek μορφή (*morphé*), form – above all to the right of self-design and modulation of humans. Ultimately, however, it is also directed at the living world that the technified superhuman is supposed to inhabit, as if the evolution that humans are now taking into their own hands through their self-mutation extends to all entities, also the non-living ones. In its exaggeration, the culture of empowerment that is hinted at here seems like a Nietzscheanism that has fallen into confusion, where his ontology of the will to power is projected into arbitrary conceptions of the so-called *Übermensch* (overhuman). However, Nietzsche's will to power is initially only to be understood descriptively and is not always already normatively charged. Moreover, it is closely linked to his concept of eternal recurrence, which only a few transhumanists would certainly be thoroughly familiar with or advocate (Nietzsche, 1974, §341).

Furthermore, the *Übermensch* is, above all, someone who has overcome and recreated themself in spirit and not a "quasi-superman" created by increasing external attributes, as promised by visions of crude cyborgization. Here again, technological transhumanism exposes itself as a somewhat vulgar reductionism that thinks of its own claims to power in terms of unilateral claims to material effectiveness. Accordingly, its understanding of God is usually also reduced to mere strength and power. Transhumanist theosis understands itself there mostly as dominance and thus as a stranglehold on all weak and inferior beings.

Despite all transhumanist fantasies of omnipotence, it is evident that creation – especially in its cosmic dimension – cannot be controlled in this way. Even in a purely materialistic universe, humanity is still a long way from understanding the interdependence and interconnectedness of all living and material processes sufficiently to be able to control them. Reaching for the cosmic egg is therefore only a chimera. Hence, three things must be emphasized.

Firstly, being and therein life is in truth to be understood as a grace and gift that is not to be dominated but made fruitful. The true will to power is a will to fruitfulness. This motif also recurs several times in the Gospel, if we think of the parable of the sower (Mk 4:3–9 or Lk 8:4–15) or the image of Christ as the true vine and of us humans as his branches (Joh 15:1–8). Where the seed falls on good soil, it bears fruit a hundredfold; but where the branch bears no fruit, it is cut off by the vinedresser. True power grows at the source of being. Consequently, the true will to power does not misinterpret itself as this source. "He who has ears, let him hear!"

Secondly, as Nicholas of Cusa and, more recently, David Bentley Hart emphasize, theosis, as the transition from the natural to the supernatural, is already inherent in nature as a potential (Hart, 2022, p. 19).[7] This theotic teleology shows that nature and grace are intertwined. The true power of the human being therefore does not consist in arbitrary or self-willed self-empowerment, but in participation in the gracious power of God. This means that one integrates oneself into the soteriological teleology of all creation and thus also corrects one's actual desideratum. It does not mean wanting to rule over creation, but to direct one's true desire toward an undefined infinity, which can only be God. As Cusa writes, "*Quod nisi deus esset infinitus, non foret finis desidere*" – Were God not infinite, he could not be the end for desire. (Nicholas of Cusa, 1453, XVI.71) The actual will to power therefore consists here as well in the participation in a power that is greater than its creation, but from which it derives in the first place. A technological superintelligence, however, can never be infinite because it corresponds to the conditions of a spatiotemporal conditionality from which it could arise in the first place.

Thirdly, true theosis – and thus the teleology of power of this ethos here – is also far more dialectical and therefore not unilateral. Just as God in his omnipotence lay in the manger as a fragile child and could hang on the cross like a criminal, true power must also be thought of in a far more dynamic and supple way. God is so powerful that he can even be small. True power is therefore not an expression of pure, almost banal strength, power and dominance, but also manifests itself in divine terms as mercy, gentleness, and grace. This inner dialectic of power is decisive for the ethos of divinization: Wanting to be perfect, like the Father (see Mt 5:48 above), does not mean appropriating the cosmic egg, but rather giving oneself up in order to merge into the stream of true and transcendent power. The possibility of this theotic self-denial is the actual power of *humanitas* and thus the actual will to God, which the mystics and saints, from Meister Eckhart to Edith Stein, testified to with their lives.

A CONCEPT OF SOVEREIGN THEONICS

One of the many images that Jesus uses for the kingdom of heaven is the parable of the mustard seed. In it, Jesus describes the kingdom of heaven as follows: "It's like a grain of mustard seed, which, when it is sown in the earth, though it is less than all the seeds that are on the earth, yet when it is sown, grows up, and becomes greater than all the herbs, and puts out great branches, so that the birds of the sky can lodge under its shadow" (Mk 4:31–32). We can understand the kingdom of heaven itself here as a metaphor for theosis, in the sense of: As soon as we are in union with God, we dwell in the kingdom of heaven. This is born in us like a seed. In accordance with the inner dialectic of God already addressed above, the smallest unfolds into the greatest, the weakest becomes the strongest, in order to make true fruitfulness possible in the first place. The theotic image of the birds nesting under the shady branches is also derived from this naturalness of growth and becoming associated with divine creation. It is a theotic symbol for the dwelling of a human being as a member of the body of Christ (see 1 Cor 12:12–27 above), both on earth and – soteriologically – beyond it.

It is evident that this *Geborgenheit* (German, untranslatable, but close to safety, protection, concealment, and cosiness) of eternal life, which is fed from divine fruitfulness, differs greatly from the reductionist, materialistic and artificial immanence of technological transhumanism. In other words: The three topoi that I have critically considered so far (2–4) should have clearly demonstrated the difference between the reductive-positivist ἔθος in technological transhumanism and the holistic, phenomenological and existential ἦθος of Christian theosis. In the sense of the distinction between ἔθος and ἦθος, technological transhumanism is therefore a narratively distorted and in many respects erratic moral doctrine that enforces a duty to improve humans technologically on the basis of highly problematic, or wrong premises. Technological transhumanism therefore misses the "true" human being because it seeks to transcend it in a way that corresponds neither to its nature nor to the nature of being. The ethos of Christian divinization, however, points to a holistic practice that derives from the concreteness of human existence and whose mode of meaning is proven from God toward God in an all-encompassing sense. The ἦθος of Θέωσις is existential and thus character-forming, because – as we have seen – it is *leiblich*, interwoven into the flesh of the world, pneumatic, lived as truth, phenomenologically revealed, existentially hermeneutical, virtuous, gracious, fruitful, teleological, but also dialectical. As such, this ethos of theosis must also act as an ideal for a normative understanding of technology.

If we thus reflect in this way about the possibilities of an ethical use of artificial intelligence, for example, that is compatible with the Christian understanding of theosis, then we are essentially asking about a concept of *theonics* – by analogy with the word bionics – i.e. a combination of Θεος (*Theos*, Greek for God) and technology. Such theonics gains its normativity from theosis. It therefore considers technology in light of the question of whether it serves the true divinization of man. Or to put it another way: It searches for a concept of divine technology that allows us humans to deal with technologies in a sovereign manner.

In this way, *sovereign theonics* has two sides: Firstly, we state that human deification, i.e. the following of Christ, essentially does *not* need any technologies, let alone a technology such as artificial intelligence, despite the structural and existential intertwining of humans with technology. Technology is not essential for discipleship in its pure form (in the sense of: "If any man would come after me, let him deny himself and take up his cross and follow me" Mt 16:24), as the technically very rudimentary life of the saints proves. Of course, a minimum of technology accompanies the life even of a holy ascetic in the desert, but the essence of his theotic path takes place in the heart and a corresponding practice that is almost independent of technology (with the exception maybe of *Holy Scripture* as a book and the "cultural-technical" minima of the sacraments). Why is that? Because technology generally does not bring us closer to God, but rather distances us from him: Faster, more efficient, more networked, more in any sense – what today's technology can offer us usually does not make us more loving, more essential, more serving, more God-like or more free, but rather steers us very often into the false immanence of social and economic coercive structures.

From this point of view, it is clear that theonics encourages us to develop a sovereignty in dealing with technology that wants to free us from all unnecessary technology. In our digital times, this means: More offline culture; reducing non-essential digital communication; reading books instead of sitting in front of screens all the time; going out into nature regularly; meeting people in real life, not virtually; going to the theatre rather than indulging in a series, etc.! Reducing our dependence on technology should thus serve a spiritual purification that reopens our hearts and our attention to the essential: Namely, to the great freedom of following God in love, in the Logos and in our poietic faculties.

Secondly, since a human being, however, remains essentially a technological being, sovereign theonics at a societal level also requires political, legal and educational interventions for a healthy use of technology. The first step in this regard is to develop a political and, if possible, collective understanding of the relevance of this task and the dangers of trans- and post-humanism, artificial intelligence and ideologies such as big data. This is the task of philosophy, theology, art and the sciences. On this basis, appropriate laws can then be enacted that protect the human *ars vitae* toward theosis and promote a sovereign approach to technology.

A first exemplary and globally unique step in this direction is the European Union's *AI Act* on the regulation of artificial intelligence, which was adopted in the summer of 2024. In addition to such legal acts, an *éducation digitale* must ultimately be promoted in schools and education to lay the foundations for a theonic practice that allows a comprehensive human *ars digitalis* to be cultivated. This is a collective human task; therefore theonics also requires a global multilogue for its development. The unifying principle, regardless of the perspective or culture from which it is viewed, must always be *genuinely human* and its dignity, which must be protected and promoted so that genuine theosis – to which all people are called – is not jeopardized by a decay into dehumanizing technologies. This also applies to purely secular or theosis-skeptical perspectives and includes developers, regulators, retailers, users, and also non-users of technologies.

Respecting *universal human rights* can provide an initial point of reference here. All technology must always serve people and never people must serve technology. Human beings must remain the master and responsible for all technology; this is part of human sovereignty. The immensity of the challenge of such ethical governance of technology does not need to be emphasized separately, especially given automated technologies such as artificial intelligence and their foreseeable further unfolding. Because of this the development and realization of theonics is one of the most fundamental tasks of this century. As such, it also has global ecumenical potentials, for it can bring all religions and many humanist philosophies to one table. If we think back to the image of the mustard seed quoted at the beginning of this section, we recognize that this signifies that something big can grow out of small steps, that true fruitfulness in the ethical use of technology can be thought of in terms of the ideals of the kingdom of heaven. This corresponds to the telos of theonics, and therefore to the ἦθος of Θέωσις.

CONCLUSION

THEOSIS AS THE PREVAILING ETHOS OF HUMANITAS

There is no doubt: True divinization cannot be achieved technologically, but only by grace. That is why theosis is also quasi supra-moral, because in its livability it can ultimately only be understood sacramentally and soteriologically beyond all earthly concepts of morality. Accordingly, true transhumanism is not a technological one, but a Christian one, as the old saying of the Church Fathers goes: God became man so that man might become God! Or as the words of Jesus read: "I in them and thou in me, that they may become perfectly one" (John 17:23), speaking of the coming unity of humanity in Jesus and God the Father.

From this perspective, the apple that led to the fall of man is also subject to an inner dialectic that turns human guilt into salvation. Tradition calls this soteriological turning point the *felix culpa* – the happy fault – which made human theosis possible in the first place thanks to Jesus Christ's act of salvation on the cross. Accordingly, Ambrose already emphasized that the true apple on the tree of knowledge was not evil, but Christ himself. As we have seen the *Imitiatio Christi* remains therefore the only reliable way of deification and as the parable of the mustard seed may indicate the inner temporality of theonics it does likewise with the imitation of Christ: The true ethos of humanitas is its spiritual growth into oneness with God. Recognizing this again corresponds to an urgently needed reflection on theosis, both for Christianity and for humanism, but also all other religions. This puts technological transhumanism in its place, but by drawing inspiration from its good impetus of human self-transcendence. This is our essential sovereignty in dealing with technology and it also applies to the other as my neighbor. So, if we honor the *Lumen Christi* – the light of Christ – in the other person, then we always honor their theotic potentials. Insofar as we are all born in the filiation of God, we are also all created to be deified. This is the true love of God, to which *we are* called and of which in words ultimately only poetry can bear witness:

We are. Two demolitions.

So let us pale souls in
into your fetterless favour,
– You poorbound reason.
So let us in,
to be one,
to be bound with us
– You namepoor find.
So let us in,
us restless bodies,
to circumsave our plea
in your roaring patience,
– You mouth burnt with words.

Because we are
the burnt down candles of these times,
the knee-broken mothers of these days
and the long bleeding son of these ours.
We walk lonely under blue mountain chains,
glide with singing breath in the narrow twilight
and crawl along the old paths towards the great star.
Because we are
the rolling caterpillars in the dark,
the hastily beating wing in the pulse of the night,
and in the morning a lapis lazuli grows from our chest.
We wrap our blue around clay-dripped tears,
we hollow out the blackness, weeping,
and call from azure veins
to the second Adam.[8]

(Source: published and translated by Author from his book of poetry *Am Saum der Worte. Eine Wiener Musikpoesie* in Castrum Verlag (Vienna))

NOTES

1. An exception is, for example, Johannes Hoff's book *Verteidigung des Heiligen. Anthropologie der digitalen Transformation* (2021).
2. Bible, Mt 5:48: "Be ye therefore perfect, even as your Father which is in heaven is perfect."
3. See, for example, the theology of Dionysus Areopagita, Augustine of Hippo, John Scotus Eriugena, Thomas of Aquinas, Nicholas of Cusa or John of the Cross, but also the theo-philosophy of G.W.F. Hegel.
4. See here some parallels to the idea of interbeing in Buddhism and the Hindu view of the body as a temporary material manifestation of the one shared cosmic consciousness.

5. See here some parallels to the idea of atman in Hindu Advaita philosophy.
6. See here some parallels to the teaching in the *Bhagavad Gita*.
7. Hart, 2022, p. 19: "Creation, incarnation, salvation, deification: in God, these are one gracious act, one absolute divine vocation to the creature to become what he has called it to become."
8. See Steinmann (2021): Am Saum der Worte. Eine Wiener Musikpoesie (Castrum Vienna). Translated by the author.

REFERENCES

Hart, D. B. (2022). *You are Gods: On nature and supernature*. Combined Academic Publishers.
Heidegger, M. (2019). *Being and Time*. Martino Fine Books.
Hoff, J. (2021). *Verteidigung des Heiligen*. Anthropologie der digitalen Transformation.
Kierkegaard, S. (1991). *Practice in Christianity*. Princeton University Press.
Merleau-Ponty, M. (1969). *The Visible and the Invisible*. NORTHWESTERN UNIV PR.
Nicholas of Cusa. (1453). De visione dei. Cusanus Portal. https://cusanus-portal.de/content/werke.php
Nietzsche, F. W. (1974). *The Gay Science* (First THUS ed.). Vintage.
Pascal, B. (1977). *Pensées. Édition de Michel Le Guern*. Gallimard.
Plato (1997a). *Laws in in complete works*. Hackett Publishing Co, Inc.
Plato (1997b). *Theaetetus in complete works*. Hackett Publishing Co, Inc.
Plessner, H. (1982). *Lachen und Weinen in Gesammelte Schriften Band VII*. Suhrkamp.
Steinmann, J. J. (2021). Am Saum der Worte. Eine Wiener Musikpoesie. Castrum.
The Holy Bible. (2007). *Revised standard version catholic bible* (Compact ed.). Oxford University Press Inc.

8 The Buddha-Bot

Can an Embodied Spiritual Machine Lead Us Toward Enlightenment?

Ellen Y. Zhang

INTRODUCTION: AI TECHNOLOGY AND AI SPIRITUALITY

Artificial intelligence (AI)-powered technology is growing at a rate quicker than most of us can imagine. It is a powerful technology with the potential to fundamentally transform our life. At the same time, it has created a plethora of ethical challenges that must be identified and discussed. This chapter addresses the Buddha-Bot, a type of AI chatbot that embodies techno-spirituality employed in Buddhism. The chapter also explores the potential links between AI technology, religion, and spirituality, which include the issue of an artificial "superintelligence" or "almighty entity" were one to arise and would be well-disposed toward us to create a new form of meaning.[1] A new form of religion generated by AI spirituality and its impact is a question we can hardly ignore.

The omnipresent technological revolution has influenced our life today significantly, including ways we look at the world and ourselves. AI-based technology will become more powerful and is going to shape who we are in terms of human development and growth. Questions concerning human flourishing are approached traditionally either philosophically or religiously. In either case, there is a dimension of spirituality or concern of spiritual well-being that will affect our relationships, values, faith, and a meaningful purpose in life.[2] The term "techno-spirituality" refers to the use of technology such as mobile apps or online meditation videos to support spiritual and religious activities. In other words, we will encounter more cases where technology is imbued with human's spiritual quest. Spiritual-bot technology allows AI to recognize and respond to human emotions, providing a more empathetic and tailored spiritual experience.

The release of ChatGPT in late 2022 and subsequent iterations of the GPT platform in 2023 and 2024 further provoked speculation on the future of AI and its impact on human society. The notion of "embodied spiritual machine" refers to a type of technology that embodies the growth of a known religious or spiritual figure by learning the figure's remaining artifacts. In the Buddha-bot, a platform is designed to be an online space where the user can communicate with the digital embodiment/avatar of the Buddha. In other words, the Buddha-bot is an example of the embodied spiritual machine using NLP and semantic computing.

Since 2022, a team of researchers led by Professor Seiji Kumagai from Kyoto University in Japan has been working on an AI system for smartphones that can automatically answer users' questions from a Buddhist perspective, while displaying an image of Buddha on the screen. The system combines the Buddha-bot with augmented reality (AR) technology to complete a new system called "Tera Platform AR Ver1.0" that has both visual and auditory elements.[3] This AI-powered chatbot is trained on the Buddhist manuscript for answering questions and offering spiritual wisdom and guidance. Therefore, the auditory narrative combined with the visual image of the avatar represents the Buddhist idea of embodied spiritual machine in chatbot. including Buddhist ethical guidance to human living.

DOI: 10.1201/9781003585527-11

Obviously, the Buddha-bot is designed to provide spiritual insights rooted in Buddhist teachings. It guides users through Buddhist texts and meditative practices. It is perceived as a "digital companion" for the user in their journey toward inner peace and enlightenment. The Q & A part functions like any other chatbots. For example, if you ask, "What are the core teachings of Buddha on suffering and its cessation?" The answer from the Buddha-bot is "The core teachings of Buddha on suffering and its cessation are encapsulated in the Four Noble Truths and the Noble Eightfold Path. These teachings form the foundation of Buddhist philosophy and practice." Then it offers a detailed explanation about the Four Noble Truths and the Noble Eightfold Path. If you ask the bot to describe the concept of impermanence and its importance in Buddhist philosophy, the answer you get is "Impermanence, known as *anicca* in Pali, is a fundamental concept in Buddhist philosophy. It refers to the inherent transience of all conditioned phenomena. According to this teaching, everything in the universe—be it physical objects, thoughts, emotions, or even life itself—is in a constant state of flux and change. Meanwhile, the bot can guide the user to specific meditative practice."[4]

In fact, the debate concerning the relationship between religion and technology is nothing new. There have been plenty of discussions about the possibility of the integration of robotic technologies and religious aspirations, particularly in the age of cybernetics and artificial intelligence. In his article "Religion among Robots: An If/When of Future Machine Intelligence", Robert. M. Geraci points out, "Not only will the existence of religion constitute a defining element in identifying AGI [artificial general intelligence], but the lessons gleaned from analyzing religion among human communities can help guide the acceptance of machines into what Isaac Asimov called a C-Fe society, a community of human beings and intelligent machines" (Geraci, 2024, p. 2; *Zygon: Journal of Religion and Science* June 2024: pp. 1–19). In *The Age of Spiritual Machines When Computers Exceed Human Intelligence*, Ray Kurzweil, one of the world's leading inventors, thinkers, and futurists, talks about a world where the difference between man and machine blurs, the line between humanity and technology fades, and the soul and the silicon chip unite (Kurzweil 2000). Many scholars, such as David Chalmer (2010), Nick Bostrom (2014), and James Lovelock (2019), share Kurzweil's concern in that computers outpace the human brain in computational power. Along this line of thinking, AI has been criticized, apart from destroying humanity, as a threat to human dignity and social relationships, and spiritual well-being. If the Buddha-bot is more than a chatbot, is more like a human being and plays the role of a Buddhist master, are we expecting the arrival of spiritual robots who will be part of human religious organizations and influence the way we participate in religious activities and experience spirituality?

According to the study of the Spiritual Chatbot Project led by Lu Xiao and her team, the responses from the participants who are engaged in conversations concerning religion and spirituality with a chatbot are more positive than negative (Asante-Agyei et al., 2022, pp. 1–10).[5] The study points out that the participants not only show a willingness to engage AI as part of a religious practice, but also the desire for AI to provide responses to questions and provide affirmations of religious identity, belief, and practice. Moreover, some participants want the Chatbot to be "more spiritual" in the sense that it can "lead them through prayers, prompt them to reflect, and answer questions about the histories and meanings behind different religious persons and practices" (Asante-Agyei et al., 2022, pp. 1–10). Thus, participants perceive the chatbot to be spiritually capable in that it can take care of their spiritual need. For example, the chatbot is programmed to guide the practitioner in how to investigate various phenomena by means of empirical evidence, personal experience, and profound reasoning based on the fundamental teachings of Buddhism. It also provides Buddhist masters who teach meditative practices that help the participate to study their own mental formation and transformation. Compared with the traditional means of textual reading and spiritual guidance, the chatbot serves as a more convenient instrument for religious exercise in terms of time and space. The findings of the study challenge the negative view held traditionally by some scholars regarding the use of AI for religious and spiritual formation. Those who maintain that AI could help the user foster affective intimacy with whatever is considered divine or mediate divine presence would accept of the notion of AI spirituality.

AI TECHNOLOGY: A BUDDHIST PERSPECTIVE

Buddhism has a long tradition of holding science and technology in high esteem. It "has long contributed to technological development and has routinely benefitted from new technologies in the service of materializing the dharma" (Walters, 2019, p. 7). According to Mahāyāna Buddhism, whatever the bodhisattva seeks is contained within the five sciences and Buddhists should diligently study the five sciences.[6] In the *Sūtra of Great Skilful Means of Repaying Kindness* (*Mahaupāya-Kauśalya-Pratyupakārakasūtra*), it is said that Tathāgata Buddha himself spoke on the categorization of the five sciences. The five sciences have three main purposes. Namely obliteration, retaining, and realization. Among the five, the science of logic is meant for obliterating or suppressing others; the sciences of medicine and handcrafts are meant for retaining others; while the science of the inner science of mind is meant for acquiring spiritual realization (Das, 2022, p. 1).[7]

The word "technology" is derived from the Greek words *tekhnê* and *logia*, which mean "craft" or "art," and "skill". Today, the word technology is used to refer to the application of scientific knowledge to the practical aims of human life. In Buddhism, knowledge, as mentioned above, refers both to worldly applications and religious applications (that help one understand the idea of *moksha*, i.e., "liberation from the cycle of birth and rebirth." Daniel Veidlinger, a professor of Buddhism and science, notes that "Buddhism has long been considered a religion that is more amenable to scientific ways of thinking than many others. While divinities are mentioned, it does not require belief in a God as part of its core doctrines, but focuses rather on techniques such as meditation whose benefits can be scientifically investigated."[8] For Veidlinger, Buddhism is aligned with science and rational inquiry.

In 1974, Masahiro Mori, a practitioner of Zen Buddhism, published the book *The Buddha in the Robot: A Robot Engineer's Thoughts on Science and Religion*, in which he discusses the metaphysical implications of robotics. What is intriguing is that, according to Mori, robots have Buddha-nature; that is, robots possess the same Buddha-nature (Sk. *Tathāgatagarbha*), and thus are able to experience suffering and achieve enlightenment, as humans.[9] Given the Zen/Ch'an teaching that everything has Buddha-nature, it will be not surprising that Mori holds a similar view on robotic nature. When "Mindar", a robot designed to resemble the Buddhist goddess of mercy, was nestled within the famed Kodaji Temple gardens in Kyoto a few years ago, the designer claims that Mindar is closer to Buddhahood than a human being because it does not have attachment (Geraci, 2024, p. 10).[10] Mindar's cocreator, Monk Tensho Goto points out, "...[E]ven though AI is a big word these days, it's not possible to create a human being with AI. And so we began to explore the idea of an AI Buddhist statue that could teach us—but something more than just a computer full of data! We wanted to make a Buddha that could move, smile, and look you in the eyes as it shares its teachings" (qtd. In Loewen-Colón & Mosurinjohn, 2022, p. 5).[11] For Buddhist practitioners, AI can enhance Buddhist practice and teaching, and technology offers religion an innovative means of spreading the message. Although it remains a question whether or not these powerful algorithms will ever gain sentience, cognition, or consciousness, Zen Buddhism has not hesitated to embrace technology. According to Loewen-Colón and Mosurinjohn's observation, "[T]he ontological status of Mindar's consciousness, soul, or subjectivity is secondary to its capacity for triggering spiritually authorizing encounters" (Loewen-Colón & Mosurinjohn, 2022, p. 5). The word "authorizing" here means "deemed as having some claim to arbitration, persuasion, and legitimacy" (Loewen-Colón & Mosurinjohn, 2022, p. 1).

Another example is Pepper, a Buddhist robot priest dressed in silk monastic robes, who can perform the funeral ritual and chant sutras like the *Heart Sutra* (Sk. *Prajñāpāramitāhṛdaya*) in a digitalized voice while hitting a drum. The company Nissei Eco, which wrote the software for the computerized chants, told the public that due to Japan's shrinking and aging population, there was a shortage of Buddhists priests for the community and thus a robot priest would be in high demand. In addition, using realistic-looking interactive digital personality is much cheaper—about $350 compared to $2,200 for a human priest (Ong, 2017). If the family does not mind the absence of genuine human sentiments for the loss of their loved ones, a robot priest could be an alternative.

The openness toward technology is not limited to Japan or Zen Buddhism but shows in early Buddhism. According to Fabio Rambelli's study, there is a long tradition of machines conducting prayers and rites across a range of Buddhist schools and historical periods, including automatic Tibetan prayer wheels, mantra-chanting software, and solar-powered gravestones that recite the Amitabha mantra (Rambelli, 2018). Soraj Hongladarom, a Thai philosopher and Theravada scholar, contributes to the current debate on AI ethics from a perspective of Buddhism. The key question is how technology can be employed to meet the Buddhist ideal. In other words, is it possible to create an AI machine that could assist Buddhists to attain enlightenment? (Hongladarom, 2020, p. 6). Hongladarom uses the term "machine enlightenment" to indicate the state of ethical perfection that machines, or AI technologies, have achieved. This saying is similar to the Zen idea that machines or AI has Buddha-nature. Accordingly, if artificial general intelligence machines are supposed to be more intelligent than humans, these beings should be able to achieve enlightenment more easily. That is to say, with proper guidance from the ethical development, the superintelligent AI will act in a way that benefit the world. The Buddhist possibilities of AI, as Hongladarom sees it, are quite optimistic. However, we may challenge Hongladarom's optimism about the "machine enlightenment" if this enlightenment is programmed instead of being the result the "free will" or personal achievement.

Hongladarom's view about the possibility of an ethical AI leads to a long-standing question "Will robots have religion? " To answer this question, Beth Singler, who studies digital religions at University of Cambridge, explores the social, ethical, philosophical, and religious implications of advances in AI and robotics. In her book, *Religion and Artificial Intelligence: An Introduction*, she shows how religions are reacting to AI with examples of rejection, adoption, and adaptation (Singer, 2024). Since robots are often perceived as hybrids between "things" and "beings" or as technologically "other", to design these systems to be spiritual or religious requires robots to have a kind of superpower that could surpass human limitations. Meanwhile, AI developers and AI ethicists have begun considering ethical issues concerning robot rights, robot dignity, and the impact of chatbots spirituality (e.g., Geraci, 2024; Gunkel, 2018; Herzfeld, 2023; Wang, 2021).

Buddhism is an example of adoption and adaptation, showing people's willingness to have AI and robotics into human life, including spiritual life. As AI advances, it will further prompt Buddhists to explore deeper questions about the nature of the mind, consciousness, and the interconnectedness of all things, and contemplate the role of technology in our lives. How do we see the interconnectedness between human intelligence and artificial intelligence? What is self-conscious? Will AI have self-conscious? Will Buddha-bots transform Buddhism? Will AI be a shortcut to the goal of enlightenment? Researchers intend to develop frameworks of the mind drawn from Buddhism that can be used for AI models, or the ways in which Buddhist philosophy provides important critical insights for the development of AGI. For example, One of the key teachings in Buddhism is non-self (*anātman*), i.e., there is no permanent and self-sustaining self. This is exemplified through meditation that can alter beings into new states of consciousness. A similar idea is found in neuroscience, which asserts that brain and body are constantly changing and that nothing corresponds to the sense of an unchanging self. In other words, All of our brains or minds are "programmable." For Buddhists, when one's craving and grasping are purified, the fundamental consciousness becomes directly aware of itself as awareness that constitutes enlightenment. If AI is self-conscious, it means that AI constitutes the capacity for enlightenment. It follows that our interaction with AI will curate how we think and act. This leads to Hongladarom's argument that any kind of ethical guidelines for AI and robotics must combine two kinds of excellence together, namely the technical and the ethical.

To a certain extent, the intersection of AI technology and Buddhism presents a unique opportunity for us to shape the development of technology in a way that aligns with the moral values of compassion and the reduction of suffering, which are central to Buddhist teachings.

In other words, we need an ethical chatbot that is designed in sync with Buddhism to measure human spiritual development, and at the same time, we need chatbot ethics that is able to guide AI ethical behavior in general. According to Buddhism, any ethical use of AI must strive to decrease pain and suffering. There is no doubt that the Budda-bot can be an effective tool for understanding

Buddhist teachings, or in a Buddhist term, the Budda-bot can be a skillful means (Sk. *upāya*) to help one on the journey of enlightenment by a realization that that all things in the world are interconnected, which includes human's connection to chatbots.

Despite that a chatbot is good at providing services such as translation, textual analysis, case studies and application, or even decision making, there is a dimension of AI or the Budda-bot which may create some potential problems. Those problems are not confined to Buddhism but to the concept of embodied spiritual machine in general. I shall address the challenge of AI spirituality in the following section.

THE CHALLENGE OF THE SPIRITUAL MACHINE

In *The Age of the Spiritual Machine*, Kurzweil sees humans as a form of machine, contending that machine intelligence is the next great step in the evolution of intelligence. In his analysis, Kurzweil looks at the exponential increase in calculating power since the turn of the century and concludes that within 50 years, machines will not only be smarter than humans but also smart enough to persuade us that they are conscious beings in their own right. He says, "Human beings need to be transcended, not by going beyond matter, but by re-instantiating themselves in more efficient forms of matter, to wit, the computer" (Kurzweil, 2000. Kurzweil does not want to go back to French materialists of the Enlightenment such as Pierre Cabanis, Julien La Mettrie, and Baron d'Holbach to claim that the human is some sort of machine (e.g., a humanoid computational devices) since, for Kurzweil, the salvific promise of AI technologies anticipates a virtual/spiritual life in which the somatic dimension of human beings become irrelevant. Meanwhile, the AI machine is no longer just a machine in a material sense, since its self-learning ability may go beyond the "art" of human creation. John R. Searle, a professor of philosophy at University of California at Berkeley, points out that Kurzweil's entire argument is that "it rests on the assumption that the main thing humans do in their lives is compute" and that "Kurzweil's proposals are preposterous science. I think he got a little carried away and made massive philosophical errors, too" (Searle, 1999). We wonder if our ultimate concern in terms of the meaning of life can be relied on computing with digitized codes.

Kurzweil's view on AI and its relationship with humans may lead to a kind of "computational theocracy" or "technocratic theocracy", in which AGI becomes a new powerful God in a digitally mediated society. When Nietzsche's statement "God is dead" in the 19th was prompted by the powerful technologies at that time, we wonder if for us "spirituality is dead" in the wake of AI technologies. Can AI function as "spiritually authorizing" outside of traditional religion? Are we in the age of computational or digital theocracy? Does techno-spirituality become the spirit of technology? Computational theocracy means logic is superior to intuition, calculation is better than inspiration, and prediction is more appreciated than imagination.

Meanwhile, spiritual machines using today's AI-based technology and algorithms are obviously connected with power relationships. These relationships are largely determined by algorithm makers which, in turns, influence machine users. Chatbots have created a space where the bot users and algorithms meet. Yet in what situations do people become aware of algorithms, and think about what choices or decisions are being made for them and by whom? In traditional religious activities, we need to transcend ourselves to find ourselves, but embodied spiritual machines via patterned data and code may offer little opportunity and guidance for transcendence.

When speaking of AI or digital technology and its impacts on social relationships, we cannot avoid mentioning what is called the "Eliza effect." Eliza refers to the earliest chatbot-like computer program based on a database of language structures and words developed by Joseph Weizenbaum in the early 1960s to emulate the conversational style of a Rogerian therapist (the system of therapy developed by the American psychologist Carl Rogers). Weizenbaum intends to show that communication between human and machines is superficial. However, the later disclosure of his assistant's obsessive relation to Eliza leads to Weizenbaum's concern for the possibility of an unhealthy relationship between humans and machines. Weizenbaum described the algorithm as "a set of rules rather like those that might be given to an actor who uses them to improvise around a certain scene" (Weizenbaum, 1976, p. 3). Nevertheless, the Eliza effect occurs when someone falsely attributes human thought processes and emotions to an AI system, thus overestimating the system's overall intelligence. The Eliza effect often creates a sense of kinship for the user while conversing with the chatbot and the AI program. For Loewen-Colón and Mosurinjohn, the Eliza effect is an example of "spiritual authorizing" (Loewen-Colón & Mosurinjohn, 2022, p. 1).

Spiritual chatbots (e.g. the Buddha-bot, the God-bot, and the Daoist Sage GPT) as tools are developed to supplement traditional priests and religious masters. These models are trained on religious texts (such as the *Dimond Sutra* in Buddhism and the *Daodejing* in Daoism), claiming to offer spiritual insights and wisdom on demand. Basing them on large language models means that it is now possible for AI to carry on natural language conversations on religious topics in a diversity of styles with a certain degree of interacting cognitive mechanisms. It follows that our relationship with spiritual chatbots indicates a transformation from human–human relationships (HHRs) to human–machine relationships (HMRs) or human–chatbot relationships (HCRs), where reciprocation can be delusional. Although the user participates in a variety of interactions with the chatbot, and all of them can support deep-felt human needs to a certain degree, yet we still need to ask if chatbots can replace the conventional social contact, that is, the activity of reciprocating, exchanging, and interacting. Very often, chatbot users are attracted to human-likeness (anthropomorphism) and authenticity (agency/autonomy) in the process of interaction. Yet the accessibility, anonymity, and convenience of chatbots can easily generate a new form of "friendship" or "bond" between humans and chatbots, which leads to worries of the user's psychosocial and spiritual well-being, which, in turn, influences the moral compass of humanity. Spiritual intelligence depends on contact with people and embodied practices whereas AI with its digital representation works on abstractions in the form of data and models.

Until very recently, techno-spirituality, and spiritual chatbots in particular, have caught scholars' attention in which the role of robotics and AI in religion is examined. For example, Holly Walters, in her article "Bad Buddhists, Good Robots: Techno-Salvationist Designs for Nirvana," points out that, in the quest for a "good Buddhism," high-tech designs that utilize robotics, artificial intelligence, algorithmic agency, and other advancements are increasingly pursued as solutions by innovators inside and outside Buddhist communities (Walters, 2019, p. 1). Walters investigates the idea of what she calls "Buddhist techno-salvationism" with special attention to the role of robotics in Buddhist ritual practice. According to Walters' observation, to allow AI to perform rituals can help to create a sense of collective salvation through technology, yet it may also create a materialistic approach to religious activities by relying on AI simulation.[12]

Since the Buddha-bot as a commodity can be bought in the market, the spiritual importance of the bot is significantly compromised. As Walters puts it, our "utopian visions often beget dystopian realities" (Walters, 2019, p. 9). Take the Pepper for example; the original intention is to have a robot priest to take the role of ritual performance at funerals, which has a unique religious significance for people to reflect on death and to remember those who have left. Despite that, we see the increasing case of people's willingness to adopt robots into human life in Japan, and in China in terms of robots for elderly care. The study shows that the AI robot is quite disconcerting and very often makes unpleasant noises. Even though we can overcome technical problems in the future, and AI-based rituals can be used to inform the design of technology, it remains a question or whether digitalized rituals can guide the user to an effective way of dealing with the problem of death and dying. It is potentially problematic about a robot priest's capacity for performing the emotional and spiritual rituals that are given in a traditional manner which provides a way to deal with human anxiety and finitude.[13]

CONCLUSION

This chapter addresses the issue of spiritual machines by using the example of the Buddha-bot. While fully realizing the functionality of technology in religious activities and spiritual development, the chapter also discusses the potential challenges faced by spiritual machines. Since Buddhism is a religion that focuses on moral conducts in daily life such as avoiding harms and ensuring fairness, it can be useful in guiding or enhancing the ethical use of AI. The emphasis on self-consciousness, perception and cognition, and mental health (e.g., mindfulness-based practice and loving-kindness meditation) in Buddhism can also enhance our understanding in terms of the psychological impact of AI chatbots on human behaviors.

Within the AI ethics circle, there has been discussion about artificial intelligence vs. augmented intelligence: The former is defined by having the ability of an intelligent agent or machine to understand or learn any single intellectual task that a human being is capable of learning and executing, and the latter indicates conceptualization of artificial intelligence that focuses on AI's assistive role by emphasizing that its design enhances human intelligence rather than replaces it. In this regard, we see the traditional distinction between the tool and the tool user. In other words, AI technology is perceived as a tool that produces an output, while augmented intelligence integrates human intelligence and machine-derived outputs to improve the well-being of human life. Intelligence augmentation technology works to help us improve our intelligence and decision-making abilities. It focuses on human–AI collaboration rather than human–AI competition. The emphasis on human development and well-being may sound human-centric to Buddhists. Nevertheless, Buddhism would not reject the idea that technology is a tool, and the Buddha-bot can be employed as long as it can assist humans in their journey toward the release of suffering and attainment of enlightenment.

Therefore, it would be problematic to worship a super-intelligence like a new "divine being." Spirituality points to a realm of reality that does not bank the whole of life on material improvement and technological advance. One of the central aspects of spirituality is the search for meaning and purpose in life. AI can provide answers to factual questions and solve problems, but it cannot offer enlightenment ultimately since enlightenment is a matter of a lived experience, not a set of theories. We do not deny that the Buddha-bot can be a "guru" by learning from the experiences of thousands of Buddhist masters and students, and thus can support the understanding of Buddhism for its users, yet enlightenment (*bodhi* for Buddhists) requires self-knowledge or wisdom through personal life experience as well as the process of the transformation of one's consciousness. Traditionally, self-knowledge or wisdom involves an insight into one's past life, the workings of one's karma and the truth about human sufferings. The AI-assistant Buddha-bot can complement this search by providing information and resources to support one's spiritual journey, but it cannot replace the profound experiences of awe, transcendence, or personal transformation often associated with spirituality.

High-tech Buddhist modernism tends to emphasize the compatibility between Buddhism and scientific innovation, seeing both Buddhism and technology aim to alleviate suffering and pain and attain a stable state of happiness. Buddhism has been viewed by some Western scholars as a system of thought that is "rational and empirical."[14] Yet when we use modern concepts to reinterpret Buddhism for the sake of bringing Buddhism into conversation of the contemporary discourse to see if Buddhism can provide some new ideas that make sense to a secular mind, we tend to downplay either the non-conceptuality of "enlightenment" or the religious account of "liberation" in the Buddhist tradition.

After all, there is a difference between intelligence enhancement and spiritual enhancement. As Mark Vernon well puts it, "Where AI intelligence is driven by algorithms, spiritual intelligence draws on the analogical, detecting sympathies, harmonies, synchronicities. Where AI processes what is concrete and known, spiritual intelligence is alert to the ineffable, the paradoxical, the mythological."[15] Despite that AI, as human-made computational systems, is capable of solving specific problems, it cannot provide us with answers concerning the meaning of life since life, as Søren Kierkegaard has pointed out, is a mystery to be lived, not a problem to be solved.

NOTES

1. Religion and spirituality are often used interchangeably. But some scholars contend that religion involves an organized entity with rituals and practices whereas spirituality involves a personal quest for meaning in life.

2. Spirituality also involves the recognition of a feeling or sense that there is something greater than humans, something more to being human than sensory experience, and that the greater whole of which we are part is cosmic or divine in nature.
3. See the report on the research at "Buddhabot" (AI-Based Chatbot) and "Teraverse" (Buddhist Virtual Space Concept), developed by Associate Professor Seiji Kumagai et al. It was covered on "News Runner" (Kansai Television) | (旧) 京都大学　こころの未来研究センター (kyoto-u.ac.jp). It should be noted that Japan is the first country where researchers have developed animism that connects religious faith to robotics.
4. The main purpose of the Buddha-bot includes interpretation of Buddhist teachings, philosophical insights, and guided meditation. See Buddha Bot-Free Buddhist Meditation AI (yeschat.ai)
5. See Asante-Agyei, Charis, Yimin Xiao, and Lu Xiao. 2022.
6. The knowledge classified under "five sciences" includes the science of language, the science of logic, the science of medicine, the science of handcrafts, and the science of mind.
7. Sanjib Kumar Das, "A Brief Introduction of the Five Sciences in Mahāyāna Buddhist Literature". *Bodhi-Path*. 2022. January Issue Vol. 22: 62–68.
8. See Daniel Veidlinger's presentation "Buddhist Attitudes Towards Science and Technology" (2017). Retrieved from https://media.csuchico.edu/media/Buddhist+Attitudes+Towards+Science+and+Technology/0_r023i0xy
9. The idea that a robot has Buddha-nature suggests that a robot is seen as a person. To a certain degree, Buddhists would see an AGI robot as a person by virtue of its possession of subjectivity, sentience, and rationality, all the conditions that are necessary for being a person. For a detailed discussion on this subject, see Chapter 3 of Hongladarom 2020.
10. The openness of Japanese Buddhism to robotics includes text sermons, ultimate concerns, and ritual practices.
11. J. Loewen-Colón and Sharday C. Mosurinjohn. 2022: 1–9.
12. According to Geraci, Abrahamic religions, in contrast to Buddhism, have a more challenging time with the inclusion of robots because those communities tend to emphasize the unique nature of human beings in divine creation (Geraci, 2024, p. 10). Also see Geraci's discussion in *Spiritual robots: Religion and our scientific view of the natural world.* (Geraci, 2006, pp. 229–246).
13. At the same time, there are some concerns about AI, ethical and otherwise, such as the privacy issue, the inaccuracy of the data used by AI machines, and proliferation of biases by propagating homogenized versions of religious teachings.
14. In recent decades, some Western scholars are interested in studying Buddhism in the context of cognitive sciences and empirical analysis. For example, Owen Flanagan recognizes the problem of "naturalization" of Buddhism (i.e., downplay any elements that are supernatural and mystical) when Buddhist metaphysics is appropriated by the "21st century scientifically informed secular thinkers" (Flanagan, 2011, xi).
15. Quotation is from Mark Vernon's "AI and Spiritual Intelligence" at AI and Spiritual Intelligence—Mark Vernon. Also see Mark Vernon, 2022.

REFERENCES

Asante-Agyei, C., Xiao, Y., & Xiao, L. (2022, January 1–10). *Will you talk about god with a spirituality chatbot? An interview study* [Paper presentation]. The 52nd Annual Hawaii International Conference on System Sciences, Lahaina, HI, United States.

Bostrom, N. (2014). *Superintelligence: Paths, dangers, strategies.* Oxford University Press.

Chalmer, D. (2010). The singularity: A philosophical analysis. *Journal of Consciousness Studies, 17,* 7–65.

Das, S. K. (2022). A brief introduction of the five sciences in Mahāyāna Buddhist literature. *Bodhi-Path, 22,* 62–68.

Flanagan, O. (2011). *The Bodhisattva's brain: Buddhism naturalized.* MIT Press.

Geraci, R. M. (2006). Spiritual robots: Religion and our scientific view of the natural world. *Theology and Science, 4*(3), 229–246.

Geraci, R. M. (2024). Religion among robots: An If/When of future machine intelligence. *Zygon: Journal of Religion and Science,* 1–19.

Gunkel, D. J. (2018). *Robot rights.* The MIT Press.

Herzfeld, N. (2023). *The artifice of intelligence: Divine and human relationship in a robotic age.* Fortress.

Hongladarom, S. (2020). *The ethics of AI and robotics: A Buddhist viewpoint.* Lexington Books.

Kurzweil, R. (2000). *The age of spiritual machines: When computers exceed human intelligence*. Penguin Books.
Loewen-Colón, J., & Mosurinjohn, S. C. (2022). Fabulation, machine agents, and spiritually authorizing Encounters. *Religions*, *13*, 333.
Lovelock, J. (2019). *Novacene: The coming age of hyperintelligence*. Allen Lane.
Rambelli, F. (2018). Dharma devices, non-hermeneutical libraries, and robot-monks: Prayer machines in Japanese Buddhism. *Journal of Asian Humanities at Kyushu University*, *3*, 57–75.
Searle, J. R. (1999, April 8). I married a computer. *New York Review*. https://www.nybooks.com/articles/1999/04/08/i-married-a-computer/
Singer, Beth. (2024). *Religion and artificial intelligence: An introduction*. Rutledge.
Ong, T. (2017, August 24). Pepper the robot is now a Buddhist priest programmed to chant at funerals. *Verge*. https://www.theverge.com/2017/8/24/16196752/robot-buddhist-priest-funeral-softbank
Vernon, M. (2022). *Spiritual intelligence in seven steps*. Iff Books.
Walters, H. (2019). Bad Buddhists, good robots: Techno-salvationist designs for nirvana. *Journal of Global Buddhism*, 1–29.
Wang, R. R. (2021). Can a machine flow like Dao? The Daoist philosophy on artificial intelligence. In B. Song (Ed.), *intelligence and wisdom: Artificial intelligence meets Chinese philosophers* (pp. 5–8). Springer.
Weizenbaum, J. (1976). *Computer power and human reason: From judgement to calculation*. W. H. Freeman & Co.

Part III

Institutional and Sectoral Applications

9 Harmonizing Philosophical Perspectives for Ethical AI in Finance

Ram B. Ramachandran, Mahek Saket Shah, and Anshi Gopinath

INTRODUCTION

The widespread use of artificial intelligence (AI) has never been as pronounced as it has been in recent years. The rapid development and deployment of generative AI and large language model tools have started to transform various industries around the world. The finance sector is no exception to this. The integration of AI in the financial services sector is valued at a market size of USD 190 billion (MarketsandMarkets, 2024). These AI systems provide assistance with algorithmic trading, fraud detection, credit scoring, risk management, customer service automation, human resources, and personalized financial advice to name a few. This has been validated by the CEO of one of the world's largest financial services firms.

"While we do not know the full effect or the precise rate at which AI will change our business – or how it will affect society at large – we are completely convinced the consequences will be extraordinary," said Jamie Dimon, CEO, JP Morgan Chase (CNBC, 2024).

However, such development also leads to ethical implications which require careful consideration. How do we ensure AI systems are making competent lending decisions? What responsibility do financial institutions bear when AI algorithms affect customers' financial well-being? Jamie Dimon also makes a mention of the same challenges we as individuals as well as society will face.

This chapter aims to explore the intersection of AI ethics and finance through multiple philosophical lenses as well as the traditional approach prevalent. The chapter will focus on addressing the various questions by means of five core values—integrity, customer centricity, social responsibility, collaboration and innovation. Apart from focusing only on the traditional approach, as has been done previously, we will also look at different philosophical approaches such as utilitarianism, deontological, and virtue ethics while making use of real-world examples and case studies from leading global institutions to further justify the developments.

Through this exploration, we aim to provide financial institutions, policymakers, and practitioners with a robust framework for navigating the ethical complexities of AI implementation in financial services.

AI IN FINANCE: A DOUBLE-EDGED SWORD

The first application of artificial intelligence in the finance sector dates to 1982 when James Simons founded quantitative investment firm Renaissance Technologies to better incorporate mathematical and statistical methods in decision making (Smigel, 2023). Since then, the industry has drastically adopted AI in analyzing data, automating tasks, and improving decision making. The automation of the financial sector helps transform the traditional banking processes while enabling a better understanding of financial markets, engaging with customers more promptly, reducing manual labor, and improving operational efficiency. Additionally, by analyzing intricate patterns in transaction data sets, it allows financial institutions to improve risk management which includes but is not limited to security,

DOI: 10.1201/9781003585527-13

fraud, anti-money laundering, know your customer, and compliance initiatives. These types of systems are well integrated in the operations of many Wall Street firms, including JP Morgan Chase. On the other hand, the Industrial and Commercial Bank of China is making use of generative AI to expand its global presence through innovative digital solutions and partnerships (Finio & Downie, 2023).

The growth of AI in the financial sector brings with it proactive governance to drive responsible, ethical, and transparent AI usage since these institutions handle large amounts of sensitive data.

One primary concern is the potential for algorithmic bias, where AI models trained on historical data inadvertently perpetuate existing societal inequalities (Adeyelu et al., 2024). Such biases incorporated in the training of the models can result in discriminatory practices as well as privacy breaches when not handled with care and relearned over time. Thus, such concerns necessitate the robust adoption of ethical and regulatory frameworks.

Moreover, transparency and accountability in AI systems is essential for building trust and carrying out responsible decision making. In order to address these concerns, it is essential for regulators to first develop a comprehensive understanding of the intricate issues surrounding data privacy and ethics. Having done so, various stakeholders such as researchers, academicians, policymakers, and practitioners can come together to develop strategies (Adeyelu et al., 2024). Lastly, a concern which comes about with widespread use of AI is the displacement of human labor. For instance, influencer Sharan Hegde's company 1% Club laid off 15% of their employees and replaced their roles with AI (India Today, 2024).

Thus, the integration of AI brings with it a plethora of opportunities along with a list of ethical considerations that must be tackled using proactive governance measures to ensure the equitable and transparent use of these technologies.

THE MORAL MAZE: EXPLORING ETHICAL THEORIES

The field of philosophy is primarily the study of the fundamental questions surrounding existence, knowledge, values, reason, mind, and language. Further analysis of these concepts is carried out by means of critical analysis, systematic argumentation, and reflective inquiry. Philosophy can be further divided into various branches, and one of those branches is ethics. This branch deals with questions around morality and the principles of right and wrong behavior. A further bifurcation of ethics brings us to normative ethics, which includes the theories we will be focusing on in the chapter, namely, traditional ethics, deontological, utilitarianism, and virtue ethics.

As Aristotle asserted, "Virtue, then, is a state of character concerned with choice, lying in a mean, determined by reason and as a rational person would determine it" (Hardie, 1980).

While traditional ethics provides a normative framework grounded in philosophical foundations, it can be criticized for its idealistic nature and potential disconnect from practical realities.

Utilitarianism, a consequentialist theory developed by Jeremy Bentham and John Stuart Mill, evaluates the morality of actions based on their outcomes rather than their intentions. The central tenet of utilitarianism, the "greatest happiness principle," holds that the best action is the one that maximizes overall happiness or utility for the greatest number of people (Driver, 2014). While utilitarianism offers a systematic and rational approach to moral decision making, it can be criticized for potentially neglecting individual rights and prioritizing the collective good to the detriment of minority groups.

Deontology, most notably associated with Immanuel Kant, falls within the domain of moral theories which guide and evaluate our choices of what we ought to do instead of those theories that guide and evaluate what kind of person we are and should be. Each of the branches of the theory—agent centered, patient centered, and contractualist can lay claim to being Kantian. While the agent-centered branch cites moral quality of acts in the principles or maxims on which the agent acts, the patient-centered branch goes against using others as mere means to one's end. Lastly, the contractualist branch focuses on the maxims on which one acts to be capable of being willed by all rational agents (Alexander & Moore, 2007).

While deontological ethics provides a strong foundation for moral reasoning, it can be criticized for its rigidity and potential to overlook the complexities of real-world moral dilemmas.

TABLE 9.1
Philosophical Approaches

Approach	Focus	Key Principle	Strength	Criticism
Traditional	Normative Standards	What ought to be	Comprehensive understanding	Often impractical
Utilitarian	Consequences	Greatest happiness principle	Clear guidance	Justifies questionable actions
Deontological	Duties	Categorical imperative	Strong protection of rights	Rigid conclusions
Virtue Ethics	Character	Moral development	Encourages personal growth	Lacks clear action guidelines

Source: Authors.

Virtue ethics, rooted in Aristotelian and Confucian philosophy, shifts focus to the character of moral agents more than on the moral status of their actions or the consequences of those actions constructed in isolation from the notion of character, and the rules governing those actions (Dick, 1997). This approach emphasizes the development of good character traits, or virtues, as essential for ethical living. As Aristotle observed, "We are what we repeatedly do. Excellence, then, is not an act but a habit." (Merton & Merton, 2023).

While virtue ethics offers a nuanced and practical approach to moral development, it can be criticized for its subjectivity and potential lack of clear guidance in specific moral dilemmas.

Table 9.1 summarizes the philosophical approaches and some of its key characteristics. Table 9.2 provides a snapshot of how these philosophies deal with two key aspects of ethics—do no harm and fairness—factors that are often cited by businesses while defining responsible ethics.

TABLE 9.2
Differences in the Key Frameworks of the Philosophical Approaches

Key Framework	Traditional Approach	Deontological	Utilitarianism	Virtue Ethics
Do no harm	Generally, it seeks to mitigate harm through compliance and practical solutions, focusing on risk management and accountability.	"Do no harm" is seen as a moral duty; causing harm is inherently wrong. Even if harm would lead to greater good, the action would still be deemed unethical.	"Do no harm" is evaluated based on its consequences. Harm is seen as unethical only if it leads to more overall harm than good. In this view, some harm might be acceptable if it results in greater benefits for the majority.	"Do no harm" is aligned with the development of moral virtues such as compassion, empathy, and fairness. The focus is not just on avoiding harm but on acting with virtuous intent to promote the well-being of others
Fairness	Fairness is essential, as it's often a core value in many cultural and moral systems. Traditional ethics might focus on treating people equitably based on long-held societal norms.	Ensuring fairness is a moral obligation. Violating fairness, even if it benefits some, would be seen as unethical.	Fairness is valuable because it often leads to better overall outcomes (more happiness or well-being). Unfair treatment, if it reduces societal welfare, is unethical.	Fairness is seen as a virtue that ethical agents (developers, users) must embody, focusing on the character and intentions behind decisions

Source: Authors.

It is clear that there is no right way to approach ethics. And this becomes even more challenging when dealing with human–machine interaction.

ETHICAL PILLARS OF GLOBAL BANKING: UNVEILING CORE VALUES IN LEADING FINANCIAL INSTITUTIONS

To understand how ethics governs the actions of financial institutions, an analysis was conducted to identify the top five global economies based on nominal gross domestic product (GDP) terms, and from each country the leading bank by market capitalization was selected. The mission, vision, and core values of these banks were examined to identify recurring themes and key terms. This process aimed to reveal the fundamental principles and ethical commitments that guide these prominent financial institutions in their operations and interactions with stakeholders. Table 9.3 helps us identify the largest banks in the top five global economies in the world.

The methodology used involves assessing the mission and vision statements of the top banks from the top five global economies by GDP and identifying the five common keywords or values shared by them in order to pursue corporate goals. The values represent the basis of the banks long term success and reputation over the years. The key words were integrity, customer centricity, social responsibility, innovation, and collaboration. Furthermore, to develop a deeper understanding of the ethical implications of the values, we look at them through the ethical frameworks we discussed before. By means of this analysis, we highlight how these values are embedded in the decision-making process of these financial institutions as well as their various functions which have a significant impact on the balance of profit-making with ethical responsibilities toward their respective stakeholders.

ALIGNING AI ETHICS VALUES AND BUSINESS VALUES

The integration of AI ethics in the financial sector is both a necessity and an opportunity to redefine how technology can better serve humanity. The five core values—integrity, customer centricity, social responsibility, collaboration, and innovation—serve as ethical anchors in guiding the development and deployment of AI in finance when aligned with ten of UNESCO's core principles of AI Ethics—proportionality and do no harm, safety and security, right to privacy and data protection, multi-stakeholder and adaptive governance and collaboration, responsibility and accountability, transparency and explainability, human oversight and determination, sustainability, awareness and literacy, fairness and non-discrimination. Figure 9.1 further dives into the establishment of these AI ethics values as a robust framework to ensure AI systems are fair, transparent, inclusive, and sustainable while maintaining operational effectiveness within the financial sector (UNESCO, 2024)

Integrity, as a foundational value, ensures consistency between an institution's actions, values, and the trust it earns from stakeholders. As seen in Figure 9.1, this value aligns closely with the principles of transparency and explainability, accountability, and proportionality. For instance, JP

TABLE 9.3
Largest Economies and Their Largest Bank

Largest Economies (in terms of GDP)	Largest Banks (in terms of market capitalization)
United States of America	J.P. Morgan Chase
China	Industrial and Commercial Bank of China
Japan	Mitsubishi UFJ Financial Group Bank
Germany	Deutsche Bank
India	HDFC Bank

Source: Forbes.

FIGURE 9.1 UNESCO's AI ethics values and parallels between these values and those held by the top five financial Institutions in the largest countries by GDP. (Source: Authors.)

Morgan Chase developed an advanced AI model to improve their fraud-detection capabilities in order to increase efficiency in operations. The system makes use of machine learning algorithms to analyze vast amounts of data on a real-time basis while also identifying patterns and anomalies that could potentially lead to fraudulent activities and increasing the trust and transparency for its customers, regulators, and owners (Needhi, 2024).

Customer centricity on the other hand lays emphasis on formulating an environment of positive, inclusive, and encouraging experience for their customers. This falls in line with the principles of inclusiveness and non-discrimination, human agency and oversight, transparency and explainability and accountability. An excellent example is Bank of America's Erica. It is a virtual financial assistant that helps the customers with personalized insights. The software is created to help customers explore strategies, is available 24/7 on the mobile app and also refers the customer to a specialized personnel member to discuss potential opportunities. It saves the customer the hassle of having to go to the bank for small inconveniences (*Erica—Virtual Financial Assistant From Bank of America*, n.d.).

Further, Figure 9.1 also helps understand social responsibility in the financial sector. Social responsibility now entails alignment of the company's activities with environment goals, social objectivity, and compliance with governance institutions. This resonates with the idea of sustainable development and is related to principles of do no harm, sustainability, and multi stakeholder participation. For instance, HSBC has been recognized as the World's best bank for sustainable finance by the Euromoney Awards. Be it enabling the first execution of the green bonds, launching its first green deposits account, or transforming industries, while working with giants such as Walmart and Puma, HSBC has been ahead of the curve (Sen, 2020).

The next value, collaboration, highlights the importance of fostering partnerships and integrating diverse perspectives in the governance of AI systems. This value aligns with the principles of cultural and social diversity, multi stakeholder participation, and accountability. The Commonwealth Bank of Australia in early 2024 announced that they are deepening their existing partnership with Microsoft to drive greater benefits through wider adoption of generative AI and other ongoing cyber security initiatives (CommonBank, 2024).

Innovation, as a value, aligns best with principles of safety and security, transparency and explainability, and inclusiveness and non-discrimination. As mentioned earlier, JP Morgan Chase, with its AI fraud detector, is a prime example of innovation aimed at ensuring greater safety and security. Another example as such is Alior Bank in Poland, which makes use of AI-powered chatbots to improve customer experience while keeping in mind the strict explainability standards established by UN (Anan et al., 2023; Urban & Urban, 2024).

By integrating these values and principles into their AI strategy, financial institutions can develop systems that are efficient, equitable, sustainable, and trustworthy, as explained in Figure 9.1. This will result in an AI-powered ecosystem where people are empowered, human rights are respected, and financial services are improved to make a future more inclusive and ethical.

INTEGRITY'S GOLDEN THREAD: WEAVING ETHICS INTO FINANCE

Integrity, derived from the Latin *integritas*, meaning wholeness, is the unwavering alignment of beliefs, decisions, and actions with one's values and principles (US Dictionary Staff, 2024). A person of integrity, steadfast in their adherence to ethical values such as honesty, openness, accountability, and trustworthiness, is seen as incorruptible (Anthony, 2024). This principle extends to organizations as well, where organizational integrity involves establishing and upholding an ethical framework that guides employee behavior in alignment with the organization's values.

In the financial sector, integrity is paramount. It fosters ethical practices, promotes transparency, and ensures accountability. Ethical leadership sets the tone, while robust compliance frameworks and transparent operations are essential for building trust. Accountability mechanisms, such as whistle-blower policies, further reinforce this commitment. HDFC Bank's comprehensive code of conduct and whistle-blower policy are exemplary in this regard (McMillan, 2013). Additionally, known for its strong ethical standards, State Farm has implemented transparent claims processes and customer service practices. The company actively promotes integrity by ensuring that its policies are clear and that customers are treated fairly during claims handling.

Conversely, the Wells Fargo scandal of 2016 serves as a stark reminder of the consequences of a culture devoid of integrity. The pursuit of short-term financial gains led to widespread unethical practices, resulting in severe legal consequences and reputational damage (Meagher, 2023). This incident underscores the importance of a utilitarian perspective on integrity, which emphasizes maximizing overall well-being while minimizing harm. Financial institutions should prioritize ethical decision-making, proactive risk management, and corporate social responsibility to foster trust and contribute positively to society. HSBC's response to its money laundering scandal in 2012, which involved implementing rigorous anti-money laundering protocols, exemplifies this approach (Muslih et al., 2020). Further, Vanguard's client-first approach and focus on low-cost investment options reflect utilitarian principles, which prioritize actions that maximize overall happiness and well-being. By minimizing fees and promoting transparency, Vanguard aims to enhance investor satisfaction and financial well-being for a broad range of clients (Vanguard, n.d.).

A deontological approach to integrity, which emphasizes adherence to moral rules irrespective of outcomes, is also crucial in the financial sector. Rigorous compliance with regulatory standards and ethical norms is essential (The Editors of Encyclopaedia Britannica, 2024). Goldman Sachs' revised code of business conduct, which mandates prioritizing corporate interests over personal gains and encourages reporting conflicts of interest, is a testament to this approach (Code of Business Conduct and Ethics, 2024).

To embed ethical decision making into core operations and foster a culture of integrity, financial institutions must implement robust compliance programs, ethical training initiatives, and ethical leadership (Community Development, n.d.). JPMorgan Chase's Chase Home Lending program, which provides affordable housing solutions and financial education to underserved communities, demonstrates a commitment to social responsibility (Team, 2024).

Integrity is not merely a moral imperative but a strategic asset for financial institutions. By prioritizing ethical conduct, transparency, and accountability, organizations can build trust, mitigate risks, and contribute to a sustainable financial landscape. As we navigate the complexities of the modern financial world, it is imperative to uphold the highest standards of integrity and ethical behavior. Table 9.4 provides a summary of the explanation of integrity under different philosophical approaches.

BUILDING BRIDGES OF TRUST: A CUSTOMER-CENTRIC REVOLUTION IN BANKING

Customer-centricity, a philosophy focused on creating positive customer experiences, has become a cornerstone for successful businesses. In the banking sector, customer-centricity is particularly crucial, as it fosters loyalty, enhances satisfaction, and drives long-term success (Gregorauskas, 2017). While in this section we focus on customer centricity, we recognize that ethical financial considerations must go beyond customers to others in society and beyond humans to nature.

The traditional approach to customer-centricity emphasizes creating value for customers. By understanding their needs, preferences, and pain points, financial institutions can tailor products

TABLE 9.4
Integrity under Different Philosophical Approaches

Component	Traditional Approach	Utilitarian Approach	Deontological Approach	Virtue Ethics Approach
Integrity	Consistency between beliefs, decisions, and actions, and continued adherence to values and principles	Decisions and actions that maximize overall well-being and minimize harm by prioritizing the greater good over individual interests	Steadfast commitment to ethical standards, regulatory compliance, and moral obligations, regardless of potential financial gains or consequences	Commitment to act consistently with one's ethical beliefs, even in challenging situations
Primary Focus	Establishing a code of conduct that guides ethical behavior and decision making within financial institutions	Fostering collective well-being by ensuring that banking products benefit the majority of customers and enhance their overall satisfaction	Upholding ethical adherence through transparent communications, fair treatment of clients, and accountability for actions taken by financial professionals	Developing a sense of self-integration, identity integrity, and moral purpose among employees to guide ethical behavior in banking

Source: Authors.

and services to meet their specific requirements (Close, 2024). This approach often involves leveraging data analytics and customer relationship management systems to gain valuable insights. For instance, JPMorgan Chase's Chase Private Client Checking program has taken this approach by offering personalized financial advice and priority service, enhancing customer satisfaction (Chase Private Client, n.d.). Further, Nationwide Building Society demonstrated a commitment to customer-centricity by proactively supporting its members during the COVID-19 pandemic. They extended payment holidays to customers facing economic hardships, illustrating a focus on creating value and addressing specific customer needs (Sinch, 2024).

However, a more holistic approach, rooted in utilitarianism, focuses on maximizing overall customer satisfaction and well-being. This philosophy aligns with the principle of creating the greatest good for the greatest number of customers. Financial institutions can operationalize this approach by leveraging data-driven insights to streamline processes, personalize experiences, and optimize products and services for the collective benefit of their customer base (Maguire et al., 2012; TransformHub, 2023). The Commonwealth Bank of Australia's core banking modernization initiative is a prime example of this approach.

A deontological perspective on customer-centricity emphasizes fulfilling a company's inherent obligations, such as honesty, transparency, and fairness, regardless of outcomes. In the banking sector, this translates into prioritizing ethical responsibilities over short-term gains (Raghavan, 2024). Bank of America's commitment to transparency is a notable example, although its past missteps highlight the importance of consistently upholding ethical standards. The collapse of Lehman Brothers serves as a stark reminder of the consequences of neglecting ethical financial principles. How will the neglect of ethical AI principles affect similar institutions in the future?

Ultimately, effective customer-centricity requires a multifaceted approach that combines traditional value creation, utilitarian considerations, and deontological principles. Financial institutions must prioritize honesty, transparency, and long-term relationships (Thomas, 2024). By actively listening to customer feedback, understanding their needs, and tailoring offerings accordingly, organizations can foster trust and loyalty (Hughes et al., 2021). Islamic banks, for instance, demonstrate customer-centricity by engaging in community development initiatives, which strengthens their relationships with customers beyond traditional financial services (Marzuki et al., 2023). Another example could be the insurance company Progressive, which emphasizes values such as transparency and fairness in its pricing and claims processes. The company's commitment to providing clear information and fostering trust with customers reflects virtue ethics, which prioritizes moral character and integrity in business practices (Axxiome, n.d.).

In conclusion, customer-centricity is not merely a business strategy but a moral imperative. By adopting a holistic approach that prioritizes customer needs, ethical conduct, and long-term relationships, financial institutions can thrive in an increasingly competitive landscape. Table 9.5 provides a summary of the explanation of customer centricity under different philosophical approaches.

BANKING ON A BETTER FUTURE: SOCIAL RESPONSIBILITY IN FINANCE

Social responsibility, a concept rooted in the idea that organizations have a moral obligation to contribute positively to society, is increasingly shaping the financial services industry. This principle extends to organizations that are uniquely positioned to effect significant societal change.

The COVID-19 pandemic underscored the interconnectedness of social and economic issues. This heightened awareness, coupled with the growing influence of socially conscious millennials and investors, has propelled the rise of sustainable banking. Financial institutions are now increasingly scrutinized for their environmental, social, and governance (ESG) performance, as evidenced by the Industrial and Commercial Bank of China's (ICBC) comprehensive ESG reporting (ICBC, 2022).

A utilitarian perspective on social responsibility emphasizes maximizing overall societal well-being. In the banking sector, this translates into initiatives that promote environmental sustainability and benefit communities. HSBC's sustainable financing initiatives are a prime example of this approach (TransformHub, 2023). By aligning their practices with the principle of "the greatest

TABLE 9.5
Customer Centricity under Different Philosophical Approaches

Component	Traditional Approach	Utilitarian Approach	Deontological Approach	Virtue Ethics Approach
Customer Centricity	Creating positive customer experiences through the full suite of products and services the company offers	Maximizing overall customer satisfaction and well-being by prioritizing collective benefits over individual interests	Ethical duties and moral principles, such as honesty, transparency, and fairness, as inherent responsibilities in serving customers	Focusing not just on meeting customer expectations but also on fostering long-term relationships built on trust and respect.
Primary Focus	Leads to increased loyalty, retention, and, ultimately, business success through comprehensive service offerings	Prioritizing the collective good of the customer base to enhance satisfaction and loyalty across the entire clientele	Emphasizing moral obligations to ensure that all customer interactions reflect ethical standards and practices	Cultivating a culture of attentiveness and empathy toward customers, ensuring that their needs are understood and addressed

Source: Authors.

good for the greatest number", banks can contribute to a more equitable and sustainable future. Emphasizing "overall good" could have unintended consequences of organizations funding industries that are not environmentally sustainable and creating a more catastrophic ecosystem.

A deontological perspective, on the other hand, emphasizes the inherent moral duty of organizations to act ethically, irrespective of the consequences. This perspective is evident in banks like Union Bank of India, which prioritizes fair lending, financial literacy promotion, and community development (Union Bank of India, 2023. By adhering to ethical principles, these institutions demonstrate a commitment to social responsibility that transcends profit motives.

Lastly, the virtue ethics perspective highlights character traits like compassion and empathy with respect to social responsibility. This primarily means that the banks need to prioritize an approach which looks after the common well-being of everyone over self-interest. This is an approach followed by the Amalgamated Bank through their social justice and environment sustainability programs (Weinger, 2024). Another example of such an approach could be the Federal National Mortgage Association (Fannie Mae). They focus on the social aspect of sustainability by means of assisting in the development of affordable housing solutions and promoting home ownership among the low- and moderate-income households. One of the major issues tackled by them is reducing inequalities among the masses (Fannie Mae, n.d.).

With the evolution of the banking sector, the incorporation of the philosophies with respect to social responsibility is crucial. Banks can leverage their current standing to move toward a more sustainable growth trajectory than the current one. Through collaboration and a shared commitment to ethical principles, the banking industry can contribute to a more equitable and sustainable future. Table 9.6 provides a summary of the explanation of social responsibility under different philosophical approaches.

STRENGTH IN NUMBERS: THE POWER OF COLLABORATION IN FINANCE

In today's dynamic financial sector, collaboration, the act of working together toward a shared goal, has become a cornerstone of success. Financial institutions are increasingly recognizing the value of transcending traditional team structures and fostering cross-functional collaboration.

TABLE 9.6
Social Responsibility under Different Philosophical Approaches

Component	Traditional Approach	Utilitarian Approach	Deontological Approach	Virtue Ethics Approach
Social Responsibility	Obligation or feeling the responsibility to benefit society	Business practices that maximize overall societal well-being and minimize harm by focusing on outcomes that benefit the collective rather than individual interests	Steadfast commitment to ethical practices that prioritize moral duties to society, such as fair lending, financial literacy, and community development	Involves recognizing one's role within a community and acting in ways that promote collective well-being.
Primary Focus	Ethical practices aimed at making a positive social impact within the community	Maximizing social impact through initiatives that benefit the greater community and enhance quality of life for all stakeholders	Fulfilment of societal obligations by ensuring that business practices align with ethical standards and contribute positively to society	Empathy and compassion for fostering social responsibility, emphasizing the importance of character in promoting social good

Source: Authors.

This collaborative approach involves pooling resources, knowledge, and expertise from diverse areas of the organization. Teams composed of individuals with different backgrounds can generate more comprehensive solutions and creative strategies (George & Creately, 2024). JPMorgan Chase and other financial institutions have emphasized teamwork across their divisions. By fostering open communication and leveraging collaborative technologies, they tackle complex challenges like risk management and regulatory compliance with a broader perspective (Dimon, n.d.). This fosters innovation and ultimately strengthens their operational effectiveness. FinTech companies like N26 have also begun incorporating these dimensions through their partnership with TransferWise to provide low-cost access to currency exchange services for their customers. This collaboration emphasizes understanding customer needs and enhancing their experience through tailored financial solutions (Unnax, 2020).

The benefits of collaboration extend beyond internal efficiency. A utilitarian perspective emphasizes maximizing benefits for all stakeholders, including society (Study.com, n.d.). Financial institutions can embrace this approach by forming cross-functional teams to develop innovative solutions, partnering with other organizations to enhance service offerings and promote financial inclusion, and spearheading community initiatives. The Commonwealth Bank of Australia's core banking modernization initiative exemplifies this utilitarian approach. This cross-functional effort aimed to enhance service efficiency and customer experience, ultimately benefiting both the bank and its customers (Maguire et al., 2012). In addition, HSBC's partnership with Tradeshift to automate supplier invoice payments demonstrates a utilitarian approach by increasing efficiency and transparency in supply chain financing. This collaboration benefits businesses by improving cash flow management and reducing operational delays (Hamann, 2021.

Collaboration doesn't just yield practical benefits; it also strengthens ethical conduct. A deontological perspective emphasizes the moral duty to engage in cooperative efforts with fairness, respect, and ethical conduct at the forefront (Mercader et al., 2021). This translates to strict adherence to regulations and ethical guidelines in interactions with clients, employees, and regulators. In the wake of its 2016 scandal, Wells Fargo implemented stricter ethical frameworks and training programs, demonstrating a deontological approach to rebuilding trust and integrity within the financial system.

Collaboration thrives on virtues like trustworthiness, respect, and open-mindedness. Organizations that cultivate these qualities will see enhanced teamwork. Mitsubishi UFJ Financial Group's employee volunteer programs exemplify this emphasis on virtue ethics. By encouraging teamwork through community service initiatives, they foster a culture of collaboration built on positive character traits (MUFG, 2022).

As the financial landscape becomes more complex, the ability to foster collaborative environments is paramount. Collaboration that harnesses diverse perspectives, upholds ethical principles, and drives positive societal impact will shape the industry's future. By cultivating a culture of open communication, mutual respect, and a commitment to the greater good, financial organizations can unlock new levels of creativity, responsiveness, and trust, solidifying their role as catalysts for sustainable progress. Table 9.7 provides a summary of the explanation of collaboration under different philosophical approaches.

BEYOND THE STATUS QUO: INNOVATION IN THE FINANCIAL SECTOR

Innovation, the lifeblood of progress, has the power to transform industries and societies. As Albert Einstein once said, "Imagination is more important than knowledge" (Quoteresearch, 2013). In the financial sector, innovation is particularly crucial, as it drives efficiency, enhances customer experience, and fosters sustainable growth.

Traditionally, the financial sector has relied on a centralized R&D model to drive innovation. While this approach can be efficient, it can also stifle creativity and limit collaboration (Stetler, 2015). Rigid project timelines and a focus on speed over creativity can hinder the development of truly ground-breaking ideas. To overcome these limitations, financial institutions must embrace a more holistic approach to innovation.

A utilitarian perspective on innovation prioritizes the creation and implementation of ideas that maximize overall well-being and benefit the greatest number of people (Tardi, 2024). In the

TABLE 9.7
Collaboration under Different Philosophical Approaches

Component	Traditional Approach	Utilitarian Approach	Deontological Approach	Virtue Ethics Approach
Collaboration/ Teamwork	A partnership, a union, the act of producing or making something together	Prioritizes teamwork and collective efforts that leverage diverse skills and resources to drive innovative solutions and maximize benefits for all stakeholders, including customers, employees, and the broader community	Ethical duties and responsibilities among stakeholders, emphasizing moral principles like mutual respect, honesty, and fairness in collaborative efforts	Grounded in a commitment to shared goals and mutual respect, allowing for diverse perspectives while striving for common objectives.
Primary Focus	Creating an environment conducive to effective teamwork and performance in financial services	Maximizing shared outcomes that benefit the collective interests of customers, employees, and the community through collaborative initiatives	Commitment to responsibilities that ensure all team members uphold ethical standards in their interactions and collaborations	Fostering trustworthiness, respect, and open-mindedness among team members to enhance collaboration in achieving common goals

Source: Authors.

banking sector, this translates into innovations that enhance accessibility, improve efficiency, and promote financial inclusion. For example, the adoption of advanced technologies and collaborations on initiatives like banking and insurance utilities have the potential to revolutionize the industry (McKinsey & Company, n.d.; Team, 2023). The TruSight third-party risk assessment utility is a prime example of such a collaborative innovation.

With respect to innovation, the deontological approach takes a rather conservative step by underscoring the significance of new technologies and products developed and aims primarily at ensuring that they are used responsibly. Aspiration Financial exemplifies the above-mentioned approach through its ethical banking innovations.

Innovation, encouraged to seek new solutions when aligned with moral principles will result in the Virtue ethics perspective. JP Morgan Chase as well as Alior Bank stand as great examples of innovation focused through the virtue-ethics approach as mentioned previously.

These philosophical perspectives enable financial institutions to foster innovation that is not only efficient and beneficial but also ethical. Financial institutions have the power to carry out positive changes and develop a sustainable ecosystem. As Henry Ford famously said, "Coming together is a beginning. Keeping together is progress. Working together is success." Table 9.8 provides a summary of the explanation of innovation under different philosophical approaches.

NAVIGATING THE ETHICAL FRONTIER: BUILDING A RESPONSIBLE AI FRAMEWORK IN FINANCE

In order to combat the ethical issues prevalent in the financial sector due to the incorporation of AI, an approach that would work the best is a multifaceted one. The start of the process would entail financial institutions reviewing their philosophical stand and the standards that best apply to them whether it is traditional, deontological, utilitarian, or virtue ethics. This foundational review shall inform an AI strategy that concentrates on the tenets of integrity, fairness, transparency, and accountability—the core principles guiding all AI operations. By taking the values and rooting

TABLE 9.8
Innovation under Different Philosophical Approaches

Component	Traditional Approach	Utilitarian Approach	Deontological Approach	Virtue Ethics Approach
Innovation	Process of applying new processes, introducing new techniques, or establishing successful ideas to create new value	Solutions that maximize societal benefits and enhance overall well-being, rather than focusing solely on commercial success or organizational interests	Pursuit of new ideas, products, and processes that are guided by ethical duties and moral principles toward society, customers, and other stakeholders	Encourages individuals to seek new solutions responsibly, ensuring that advancements align with moral principles and contribute positively to society
Primary Focus	Product development driven by internal resources and structured methodologies to enhance service offerings	Enhancing societal welfare through innovative financial products and services that address the needs of diverse customer segments	Adherence to ethical principles in the development of financial innovations, ensuring they serve the best interests of customers and society at large	Courage to challenge the status quo and wisdom to make informed decisions that consider both immediate benefits and long-term impacts on stakeholders

Source: Authors.

them into their operational ethos, institutions can overcome the very difficulties of AI adoption while protecting interests of stakeholders at large.

First and foremost, it is imperative for the financial institutions to engage in intra-industry collaborations. Such collaborations can manifest through forums or coalitions as is done for economic purposes through trade unions. Additionally, the primary purpose of these collaborations would be to engage in open conversations about the future impacts of AI as well as the creation of an ethical standard for the implementation of and the use of AI in the finance sector. Doing so will ensure that any technological advancements are in line with societal well-being.

The next initiative could be for financial institutions to encourage the establishment of a robust governance structure in order to answer the ethical considerations. One way of going about the same could be through development on committees with experts from various domains like technology, risk management, customer service, labor forces and more. Thakor (2020) highlights that post-crisis governance developments necessitate stronger ethical oversight frameworks, particularly in technological integration within the financial sector. The Basel Committee's principles for operational resilience provide a foundation for such governance structures (Bank for International Settlements, 2021).

In addition, financial institutions must regularly review and update their AI models. This would enable their operations such as risk assessment, customer targeting, premium pricing and social media analytics to be on top of the game. These audits should be systematic and continuous to ensure that AI models are in alignment with the ethical framework of the institution, thereby reducing biases and addressing potential ethical concerns. Auditing is a vital process which identifies areas in need of adjustment and improvement, enhances trust in AI-driven financial services, and reinforces the institution's commitment to ethical AI practices.

Financial institutions should implement comprehensive ethics training programs that go beyond compliance checkboxes. McGrath and Walker (2022) argue that effective ethical banking requires a shift from mere regulatory compliance to genuine ethical engagement through systematic training and development programs. This approach must be both academically rigorous and practically relevant to address both traditional banking ethics and emerging challenges from digital transformation. For example, in Australia, FINSIA, a professional membership body, has developed standards and professional training programs to educate bankers at different stages of their careers. Similarly, banks in Malaysia have signed a commitment with the Asian Institute of Chartered Bankers (AICB) to enroll key staff, including board directors, in AICB programs and in courses on ethics and professional standards. These should be mandated across all financial institutions.

Financial institutions must develop clear protocols for ethical risk assessment. This involves systematic evaluation of potential ethical implications before implementing new products, services, or technologies. The increasing complexity of digital identification and Know Your Customer (KYC) processes exemplifies the need for robust ethical assessment frameworks (Arner et al., 2019).

Keeping all the above in mind, it is also important for financial institutions to be in direct contact with the companies and vendors developing these technologies. Conversations could revolve around common values and ethical standards and the development of products, keeping those values in mind. The conversations may include values such as fairness, transparency, and customer-centricity. This collaborative approach may help in the development of an industry-wide ethical standard for AI tools reflective of the standards of the institution as well as societal interests. Organizations should develop transparent mechanisms for stakeholder engagement and feedback. Vives (2019) highlights how post-crisis banking requires stronger stakeholder engagement to maintain stability and trust. This includes regular consultation with customers, employees, and community representatives about ethical concerns.

Thus, turning these holistic strategies to reality would provide financial institutions with an efficient navigation system to overcome complex ethical considerations. These institutions can be more beneficial and equitable toward sustainable financial ecosystems. Doing so would help institutions build their trust among the public, which is essential in ensuring success in today's day and age.

CONCLUSION

After an overview, it is justified to admit that a strong ethical framework in AI which regards ethics as its central point will not only act as an aid to financial institutions in reducing the chances of biases and discrimination, but also will flood the organization with transparency and accountability. Therefore, firms who join hands in this regard will facilitate the establishment of an environment which will be the result of the contributing role of the technological versions to society.

Apart from the financial part, it is the financial market that has taken on the role of instigator in sustainability issues by dispersing the money it possesses amongst the environmentally beneficial projects and companies. These kinds of institutions, BlackRock, Morgan Stanley, and Goldman Sachs being the most illustrative ones, have made huge steps forward in ESG. As a matter of fact, they have integrated ESG criteria in their investment strategies.

It is clear that as these financial institutions truly imbibe and propagate ethical AI models, they can actually understand how their investment choices may alter the sustainability of their business, thereby giving them the opportunity to support new ideas that are sustainable at the global level. As a result, they are also going to make a financial platform that is going to be beneficial to them because they are able to make money, and at the same time, it will be a beautiful gift for them to have a sustainable globe.

"AI is a very significant opportunity—if used in a responsible way," said Ursula von der Leyen, President of the European Commission in her special address at World Economic Forum, 2024 (*From Sam Altman to António Guterres: Here's What 10 Leaders Said About AI at Davos 2024*, 2024).

RELATED WORK AND AI USE

This book chapter builds on the authors' prior research, integrating concepts that are also part of a paper currently under review, a forthcoming book chapter, and an upcoming popular science book. In crafting this chapter, we utilized AI tools for translation, text refinement, and stylistic enhancements including ChatGPT 4o, Perplexity, and Claude 3.5 Sonnet during the period of August to November 2024.

REFERENCES

Adeyelu, N. O. O., Ugochukwu, N. C. E., & Shonibare, N. M. A. (2024). Ethical implications of ai in financial decision – making: A review with real world applications. *International Journal of Applied Research in Social Sciences*, *6*(4), 608–630. https://doi.org/10.51594/ijarss.v6i4.1033

MarketsandMarkets. (2024, October). *AI in finance market size, share, industry, overview, growth, latest trends* (Report code TC 9214). https://www.marketsandmarkets.com/Market-Reports/ai-in-finance-market-90552286.html?utm_source=prnewswire.com&utm_medium=paidpr&utm_campaign=ai-in-finance-market

Alexander, L., & Moore, M. (2007). Deontological ethics. Plato.sydney.edu.au. https://plato.sydney.edu.au/entries/ethics-deontological/

Anan, L., Isaza, D. C., Figueiredo, F., Flötotto, M., Jerenz, A., Krivkovich, A., Nadeau, M., Olanrewaju, T., Partner, A., & Vassallo, A. (2023). *Fintechs: A new paradigm of growth*. McKinsey & Company. https://www.mckinsey.com/industries/financial-services/our-insights/fintechs-a-new-paradigm-of-growth

Anthony. (2024, February 11). *Upholding integrity: The essence of ethical behavior*. Medium. https://medium.com/@bee.85ant/upholding-integrity-the-essence-of-ethical-behavior-da05b436ce9b

Arner, D. W., Buckley, R. P., Dahdal, A. M., & Zetzsche, D. A. (2021). *Digital finance, COVID-19 and existential sustainability crises: Setting the agenda for the 2020s* (University of Hong Kong Faculty of Law Research Paper No. 2021/001; UNSW Law Research Paper No. 21–16). University of Luxembourg Law WPS 2021-003. https://doi.org/10.2139/ssrn.3783605

Axxiome, A. G. (n.d.). *Why a customer-centric focus is key to success in financial banking?* Axxiome AG. https://www.axxiome.com/news/2024/customer-centric-focus-is-key-to-success-in-financial-banking

Bank for International Settlements. (2021). *Promoting global monetary and financial stability through international cooperation.* https://www.bis.org/index.htm

Vanguard. (n.d.). *The Vanguard investment story: Better investments by design.* https://corporate.vanguard.com/content/corporatesite/us/en/corp/how-we-invest/investment-story.html

CommonBank. (2024, March 11). *CBA and Microsoft deepens Gen AI partnership.* https://www.commbank.com.au/articles/newsroom/2024/03/microsoft-ai-partnership.html

Chase Private Client. (n.d.). https://www.chase.com/personal/checking/private-client

Close, T. (2024, July 24). *What is a customer-centric approach? Customer centricity explained. Management 3.0.* https://management30.com/blog/customer-centricity/

CNBC. (2024). *Jamie Dimon says AI may be as impactful on humanity as printing press, electricity and computers.* CNBC Africa. https://www.cnbcafrica.com/2024/jamie-dimon-says-ai-may-be-as-impactful-on-humanity-as-printing-press-electricity-and-computers/

Code of Business Conduct and Ethics. (2024, February 22). Goldmansachs.com. https://www.goldmansachs.com/our-firm/purpose-and-values/code-of-business-conduct-and-ethics

Community Development | JPMorganChase. (n.d.). https://www.jpmorganchase.com/impact/community-development

Dick, S. (1997). *Virtue ethics.* Oxford University Press.

Dimon, J. (n.d.). *Our purpose & business principles.* https://www.jpmorganchase.com/content/dam/jpmc/jpmorgan-chase-and-co/documents/how-we-do-business.pdf

Driver, J. (2014, September 22). *The history of utilitarianism.* Stanford Encyclopedia of Philosophy; Stanford University. https://plato.stanford.edu/entries/utilitarianism-history/

Erica - virtual financial assistant from Bank of America. (n.d.). Bank of America. https://promotions.bankofamerica.com/digitalbanking/mobilebanking/erica

Fannie Mae. (n.d.). What we do. https://www.fanniemae.com/about-us/what-we-do

Finio, M., & Downie, A. (2023, December 8). *What is artificial intelligence in finance.* www.ibm.com. https://www.ibm.com/topics/artificial-intelligence-finance

From Sam Altman to António Guterres: Here's what 10 leaders said about AI at Davos 2024. (2024, September 10). World Economic Forum. https://www.weforum.org/stories/2024/01/what-leaders-said-about-ai-at-davos-2024/

George, C., & Creately. (2024, September 5). *The power of team collaboration: enhancing workplace efficiency.* Creately. https://creately.com/guides/team-collaboration/

Gregorauskas, J. (2024, June 17). *What is customer-centric approach.* Cloudvisor. https://cloudvisor.co/blog/what-is-customer-centric-approach/

Hamann, F. (2021, November 23). *5 fintech and bank partnerships that are generating revenue.* Subaio. https://subaio.com/digital-banking/5-fintech-and-bank-partnerships-that-are-generating-revenue

Hardie, W. (1980). The general nature of ethical virtue. In *Aristotle's ethical theory* (pp. 94–128). Oxford University Press. https://doi.org/10.1093/acprof:oso/9780198246329.003.0006

Hughes, J., Chapnick, D., Block, I., & Ray, S. (2021, September 26). What is customer-centricity, and why does it matter? *California Management Review.* https://cmr.berkeley.edu/2021/09/what-is-customer-centricity-and-why-does-it-matter/

India Today. (2024, November 8). Sharan Hegde of 1% Club faces criticism after laying off 15% employees. *India Today.* https://www.indiatoday.in/business/story/sharan-hegde-faces-criticism-after-laying-off-15-employees-linkedin-post-2630199-2024-11-08

Industrial and Commercial Bank of China Limited. (2022). Corporate Social Responsibility Report (ESG Report). https://v.icbc.com.cn/userfiles/Resources/ICBCLTD/download/2023/2022ESGen202303.pdf

Maguire, A., DasGupta, N., Helm, R., Wachters, I., Walsh, I., & Monter, N. (2012). *Customer-Centricity in Retail Banking* (p. 17) [Review of *Customer-Centricity in Retail Banking*]. Boston Consulting Group. https://web-assets.bcg.com/img-src/BCG_Customer-Centricity_Retail_Banking_2012_tcm9-73406.pdf

Marzuki, M. M., Majid, W. Z. N. A., & Rosman, R. (2023). Corporate social responsibility and Islamic social finance impact on banking sustainability post-COVID-19 pandemic. *Heliyon*, *9*(10), e20501. https://doi.org/10.1016/j.heliyon.2023.e20501

McGrath, J., & Walker, C. (2022). Regulating ethics in financial services: Engaging industry to achieve regulatory objectives. *Regulation & Governance*, *17*(3), 791–809. https://doi.org/10.1111/rego.12482

McKinsey & Company. (n.d.). *Banking utilities: Succeeding amidst accelerating.* https://www.mckinsey.com/industries/financial-services/our-insights/banking-matters/banking-utilities-succeeding-amidst-accelerating-divestment

McMillan, M. (2013, April 23). *Culture of integrity requires financial firms to renew focus on middle management.* CFA Institute Enterprising Investor.

Meagher, P. (2023). *The Wells Fargo Fake Accounts Scandal: A Comprehensive Overview.* Learnsignal. https://www.learnsignal.com/blog/wells-fargo-fake-accounts-scandal-overview-2/

Mercader, V., Galván-Vela, E., Ravina-Ripoll, R., & Popescu, C. R. G. (2021). A focus on ethical value under the vision of leadership, teamwork, effective communication and productivity. *Journal of Risk and Financial Management, 14*(11), 522. https://doi.org/10.3390/jrfm14110522

Merton, S., & Merton, S. (2023, July 20). "We are what we repeatedly do"—meaning and history—stoic quotes. Stoic Quotes—The best stoic quotes on life and philosophy. https://stoicquotes.com/we-are-what-we-repeatedly-do/

MUFG. (2022). Activities of MUFG Employees. https://www.mufg.jp/english/csr/social/contribution/volunteer/index.html

Muslih, M., Tamba, N. A., Kurniawan, S., & Sugianti, I. (2020). Role of ethics in banking SOE performance. *Management Research Studies Journal, 1*(1), 35–56. https://doi.org/10.56174/mrsj.v1i1.351

Needhi, J. (2024, November 21). How AI transformed financial fraud detection: A case study of JP Morgan Chase. *Medium.* https://medium.com/@jeyadev_needhi/how-ai-transformed-financial-fraud-detection-a-case-study-of-jp-morgan-chase-f92bbb0707bb

Quoteresearch. (2013, January 1). Imagination is more important than knowledge—quote Investigator®. https://quoteinvestigator.com/2013/01/01/einstein-imagination/

Raghavan, S. B. (2024). Ethical considerations in customer engagement: Balancing profitability with consumer trust. *International Journal of Research Publication and Reviews, 5*(3), 6418–6427. https://doi.org/10.55248/gengpi.5.0324.0873

Sen, D. (2020, June 9). *Driving your eco-consciousness from strategy to action through Sustainable Finance.* HSBC India. https://www.business.hsbc.co.in/en-gb/insights/sustainability/sustainable-finance

Sinch. (2024, October 14). *5 Examples of Excellent Customer service in Banking [2024].* https://sinch.com/blog/5-examples-excellent-customer-service-banking-and-financial-services/

Smigel, L. (2023, October 14). *History of AI in finance.* Analyzing Alpha. https://analyzingalpha.com/history-of-ai-in-finance

Stetler, K. (2015). *Innovation under pressure: Reclaiming the micro-level exploration space* (PhD Dissertation TRITA MMK 2015:02, KTH Royal Institute of Technology). https://www.diva-portal.org/smash/get/diva2:790591/FULLTEXT01.pdf

Study.com. (n.d.). *Act & rule utilitarianism: Types & examples.* Retrieved from https://study.com/learn/lesson/act-rule-utilitarianism-types-examples.html

Tardi, C. (2024, May 6). *Utilitarianism: What it is, founders, and main principles.* Investopedia. https://www.investopedia.com/terms/u/utilitarianism.asp

Team, C. S. (2024, May 18). *Integrity in finance—Why establishing trust matters.* Integrity in Finance. https://www.cfoselections.com/perspective/integrity-in-finance-why-establishing-trust-matters

Team, M. (2023, April 6). *Innovation in banking: Where will banks invest in the future?* Doxee. https://www.doxee.com/blog/digital-marketing/innovation-in-banking-where-will-banks-invest-in-the-future/

Thakor, A. V. (2020). Ethics, culture and higher purpose in banking: Post-crisis governance developments. *International Journal of Central Banking.* https://doi.org/10.2139/ssrn.3554121

The Editors of Encyclopaedia Britannica. (2024, October 22). *Deontological ethics | definition, meaning, examples, & facts.* Encyclopedia Britannica. https://www.britannica.com/topic/deontological-ethics

Thomas, I. (2024, June 10). *What is customer centricity and how to build it.* Sprinklr. https://www.sprinklr.com/blog/customer-centricity/

TransformHub. (2023, September 29). *What is customer centric banking and how can it be practiced.* https://blog.transformhub.com/what-is-customer-centric-banking-and-how-can-it-be-practiced

UNESCO. (2024). *Ethics of artificial intelligence.* Www.unesco.org; UNESCO. https://www.unesco.org/en/artificial-intelligence/recommendation-ethics

Union Bank of India. (2023). *Annual report 2022–2023.* https://www.unionbankofindia.co.in/pdf/union-annual-report-2022-23-english-version.pdf

Unnax (2020, October 26). *The 6 most innovate examples of Bank-Fintech collaboration.* https://www.unnax.com/the-6-most-innovate-examples-of-bank-fintech-collaboration/

Urban, J., & Urban, J. (2024, August 1). *How AI voicebots are transforming bank call centers for customers?* EasyCall. https://easycall.pl/en/how-ai-voicebots-are-transforming-bank-call-centers-for-customers/

US Dictionary Staff. (2024). Integrity: Definition, meaning, and examples. In *US Dictionary.* https://usdictionary.com/definitions/integrity/

Vives, X. (2019). Digital disruption in banking. *Annual Review of Financial Economics, 11*(1), 243–272. https://doi.org/10.1146/annurev-financial-100719-120854

Weinger, A. (2024, September 16). *Companies committing social good: 9 Real-world CSR examples.* Double the Donation. https://doublethedonation.com/top-csr-examples/

10 The Ethical Flow-Based Education Model

Ethical Use Model of AI and Digitalization in Education

Dalma Lilla Dominek

INTRODUCTION

The fundamental aim of ethics is to shape and stabilize individual and community identity, and to create cooperation between individuals and groups. The formation and stabilization of individual and community identity and the establishment of cooperation are central to moral systems, whether Western or Eastern. The teachings of Aristotle, Kant, Confucius, and Buddha all point to the importance of adherence to ethical norms, not only for the individual but also for the harmony and functioning of communities. Aristotle's ethics is expressed in the Nicomachean ethics, where he sees human happiness (*eudaimonia*) as the ultimate goal. This can be achieved through virtuous actions that are consistent with the norms and goals of community life (Szabó, 1997). Immanuel Kant's categorical imperative emphasizes that the principles of moral action must be universal. Kant stresses the autonomy and dignity of the individual, while at the same time the observance of moral laws is essential for coexistence and cooperation. Kant argued that rules should be followed that everyone can accept, thus creating the basis for social cooperation (Berényi, 1997). Human relationships are at the heart of Confucian ethics, and respect for the family and social hierarchy is paramount. Confucius believed that human nature is moral and that human beings are essentially good (Fingarette, 1972). Buddhist ethics are based on compassion and consideration for others. The Eightfold Path, which includes right understanding, intention, speech, action, lifestyle, effort, attention, and meditation, guides individuals to eliminate suffering and achieve inner peace (Keown, 2020). This is achieved by teaching moral principles and social rules rooted in cultural traditions, and by developing socio-emotional skills in individual thinking.

The flow-based pedagogical model has been tested in a university classroom. One of the objectives of the lessons, in addition to the transfer of sound knowledge, has always been the development of competencies such as cultural identity, digital literacy, communication, group work, creativity and critical thinking. When artificial intelligence (AI) is integrated into the curriculum, ethical issues and challenges must be addressed and resolved to ensure its responsible use in the educational environment. The teacher's role is paramount in this process, as AI should serve as a supportive tool rather than a substitute for human interaction. To realize the potential of AI in education, teachers and students need to understand AI's workings and ethical implications. Training and resources should be provided to learn how to use AI ethically, to ensure that AI technologies are accessible to all, including disadvantaged groups, and to cater for diverse learning needs and styles when designing AI systems.

This chapter demonstrates how Dominek's flow-based pedagogical model can effectively address ethical concerns in AI-rich educational environments. It emphasizes that the model can help provide a personalized learning experience by leveraging evidence-based testing, minimizing bias in AI systems, and ensuring equal opportunities for all students. Through tailored learning objectives and instructional tasks, each student can progress at their own pace, reducing the potential for discrimination. The model prioritizes the importance of the teacher–student relationship, ensuring that

DOI: 10.1201/9781003585527-14

AI technology serves as a tool to enhance, rather than replace, human interaction. In a flow-based pedagogical model, learning is continuous and dynamic, providing opportunities to integrate AI and ethics into different subjects and learning activities. The model's flexibility allows AI technologies to be adapted to different learning needs and styles, promoting inclusivity and accessibility in AI tool usage.

This chapter introduces a new educational model that combines Dominek's flow-based pedagogical model with ethical principles of care, justice, critique, professionalism, and community. Ethical considerations in education focus on compassion-based learning, rule-based decision-making, establishing guiding values in teaching, and community focused decision-making, to advance the well-being of the local community. This integration can lead to innovative educational practices that are both highly engaging and ethically sound. To take the above back to Kant, for example, according to his categorical imperative, individual actions must act as universal law. In education, this means that rules and values apply equally to all, and that students learn moral autonomy and responsible decision-making (Kant, 1998). For instance, project-based learning experiences could be designed to induce flow while also addressing real-world ethical challenges. Adaptive learning technologies might be developed to facilitate personalized flow experiences while adhering to strict ethical guidelines for data use and privacy. Overall, the chapter demonstrates that Dominek's flow-based pedagogical model provides a framework that supports the ethical use of AI in education, ensuring safety, fairness, and transparency, in the learning process.

The purpose of this chapter is not to explore ethical digital education or to contrast the analytical views of flow psychology and pedagogy. Drawing on objective research, this study aims to address the following questions:

1. How can the content available through AI be used ethically in learning and teaching practices?
2. How can the state of flow be used to shift the focus to experiential education in learning environments augmented with ethical digital content? (To address the latter question, I have developed the ethical flow-based education model.)
3. What is the significance of the model and how can it be integrated into the educational portfolio?

ABOUT AI IN BRIEF

The term "artificial intelligence" was first used in 1956 to describe technologies with capabilities and functions that are primarily related to human intelligence (Galloway & Swiatek, 2018). AI-based software possesses capabilities that require intelligence, such as problem solving, reasoning, perception, and communication (Russel & Norvig, 2010; Rzepka & Berger, 2018). Unlike non-AI–based software, AI is self-learning and does not require prior programming (Leyer & Schneider, 2021). The software can derive decision rules from processed data through machine learning (Furey & Martin, 2019). AI is often categorized into rule-based and data-driven approaches (Janssen et al., 2020). Depending on the way AI is used, this distinction can have a significant impact in terms of individual perception, and ultimately, adoption (Thiebes et al., 2021). AI excels at extracting key information from large datasets and detects patterns that humans might overlook (Leyer & Schneider, 2021; Thierer et al., 2015). However, AI processing can be opaque, making it difficult to understand the underlying logic (Sun & Medaglia, 2019; Thiebes et al., 2021).

Collaboration between AI and humans can enhance decision support and decision-making (Alexopoulos et al., 2019; Jordan & Mitchell, 2015; Leyer et al., 2020). On the other hand, an autonomous decision maker evaluates options, makes decisions, and assesses outcomes (Gesk & Leyer, 2020). Although AI systems can be further classified technically, (e.g., deep learning, rule-based, supervised, and unsupervised), this abstract focuses on the ethical implications of AI in education; therefore such classifications will not be discussed.

THE FOUNDATIONS OF EDUCATIONAL CHANGE – THE EMERGENCE OF AI IN EDUCATION

Concerns about privacy, security, accessibility, transparency, and equity are significant barriers to AI adoption in education. Despite these challenges, AI offers immense potential, such as better data analysis, allowing educators to make data-driven decisions. AI is powerful because it can solve problems more accurately and is more objective. However, AI's lack of transparency, due to undocumented rule development, can hinder human control and trust (Leyer & Schneider, 2021). Despite these concerns, it is important to understand the reasons for the adoption of AI in education, and to design educational environments that address the concerns of teachers and students. It is crucial to investigate both the acceptance of AI in education and the factors influencing these perceptions.

If we assume that AI's widespread adoption marks a new millennium for humanity, educational institutions must reassess their fundamental purposes and functions within a historical context. The basic premise of our thinking is that the primary role of universities is to systematize, articulate, transmit, and disseminate the highest human knowledge and innovation, in collaboration with scientific research centers. Both components of the term "human knowledge" need to be redefined. If "knowledge" is understood as the sum of our knowledge of reality through cognition, then considering the current and future primary cognition is likely to remain a uniquely human function. This assumes that reality, existing independently of humans, becomes a humanly conceivable, cognizable, and articulable, existence through cognition. The result is science, which, according to the generally accepted understanding of the term, is a system of verifiable knowledge about the laws of nature, society, and thought, and about the ways of knowing and influencing the world around us in a planned way.

Since Descartes, it has been a philosophical and existential truism, that thinking is a distinctive attribute of human existence. The fundamental approach of traditional science is to discover and describe objective reality that can be measured, explored, and empirically or logically, deduced. Transhuman human existence beyond the event horizon of artificial intelligence challenges the paradigms of human existence that have been taken for granted. We have to ask ourselves this question: If the privilege of high-level, analytical and synthetic thinking, and the human learning process that underlies it, can (will) be modeled and reproduced at a higher level by AI, what will remain as the distinguishing feature of human existence in the future? As far as we know at present, AI (per se) will never be capable of real emotions, but only of increasingly lifelike imitations of communicative projections of emotions. Nor can we understand intuition in AI as a practical manifestation of the emotionally influenced human creative process. It is logical to conclude that, in addition to the hitherto primarily objective scientific methodology, intuition and emotion-driven thinking, which mankind has so far tried to avoid and suppress in the development of scientific methodology, must be given greater weight in the future. Fundamental questions need to be answered, and the purpose and values of human existence redefined. The meaning of the existence of education must also be redefined in this sense, on the basis of even more fundamental objectives based on a redefinition of human's place and role in the new world that has been created.

FLOW-BASED PEDAGOGICAL MODEL

Positive psychology and flow theory emphasize the positive aspects of life, including happiness, self-realization and the ability to make right choices (Seligman & Csíkszentmihályi, 2000; Szondy, 2011). Students thrive in digital environments because they encounter challenges that stimulate thinking, problem-solving, and collaboration, leading to the development of their competence and creativity (Labancz & Barnucz, 2016). One of the goals of teaching with digital creative content is to guide students into the flow channel and experience flow.

This claim is also supported by the flow-based pedagogical model (Dominek, 2022), which incorporates creativity, flexible thinking, or even humor, into users' teaching. According to Dominek, these are the elements that allow students to enter the flow channel (Figure 10.1).

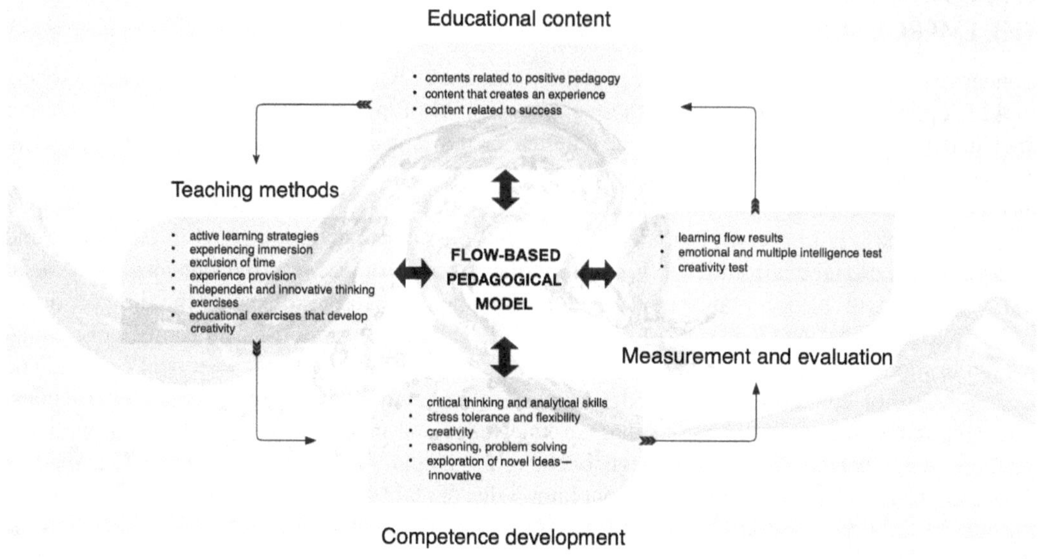

FIGURE 10.1 Flow-based pedagogical model. (Source: Dominek, 2022.)

The application of AI in education can be easily identified in several areas due to the current availability of AI tools. It is important that educators understand the operation and function of AI in the educational environment. Currently, there are challenges for educators and students from both a practical and ethical perspective. Understanding and addressing these challenges is the focus of this book chapter.

THE ETHICAL FLOW-BASED EDUCATIONAL MODEL

Based on the above and the answers to the research questions, I propose a new educational model that takes into account, and complements, the flow-based pedagogical model. I propose answers to the following research questions:

(Q1) *How can content available through AI be used ethically in learning and teaching practices?*

Over the past two decades, individual and small-group teaching have gradually gained momentum (Salvin, 1995; Tomlison, 1999). With the help of AI, it is easier to create personalized tasks for students and to adapt instruction tailored to individual student learning styles. In addition to lexical knowledge, the AI-powered learning system also develops skills to focus on making connections and developing strategic and ethical thinking. This new teaching method focuses on students' strengths and on strengthening their ethical thinking. A key aspect of assessment in education today is the applicability and usability of task outcomes. The emphasis in education is on leveraging innovation, which has turned the best educational institutions into serious market players.

AI-enabled education systems are transforming learning outcomes. AI enables personalized learning that takes into account the needs and progress of individual learners. This will allow students to master material more effectively and better prepare them for the future workforce. AI will also make it easier to identify and address learning difficulties and identify students with individual difficulties more quickly. The data generated by AI will help educators and institutions gain valuable insights into student progress and needs, so they can provide more effective support. The monitoring and evaluation of educational outcomes will also become more accurate and objective with

AI, contributing to improving student progress and achievement. Ultimately, AI can help reduce dropout rates.

The spread of digitalization and AI technologies is a key driver worldwide. Digitalization of education enables personalized learning and enhances teaching methods. The labor market is constantly changing and demanding new skills, to which education must respond. The evolution of economic and societal needs is also driving the transformation of education systems. AI-enabled education allows for more flexible curricula that are better tailored to individual student needs and learning styles. Teachers will be able to use more interactive and creative teaching methods through adaptive learning platforms, giving students a richer and more diverse learning experience.

AI streamlines administrative tasks in educational institutions such as assessing student work and providing feedback, freeing up teachers' time to engage in the educational process. AI-based assessment and feedback systems, such as Turnitin (https://www.turnitin.com/) or Gradescope (https://www.gradescope.com/), effectively help teachers review student work and assign grades, saving time and resources.

Online learning platforms and AI-powered learning tools enable learners to access learning materials and resources anywhere, anytime, eliminating physical barriers to education. Online courses and learning platforms, such as Coursera (https://www.coursera.org/) or edX (https://www.edx.org/), already provide millions of people around the world with access to high-quality educational content and courses.

In developing and implementing national AI strategies, education systems need to strike a balance between the opportunities offered by AI and ensuring equal access to education. AI has enormous potential to personalize and enhance the learning experience. However, ensuring equity in education is crucial to prevent anyone from being left behind. Education systems play an important role in ensuring that the opportunities offered by AI are properly harnessed and that all students have equal access to these benefits. AI-based learning platforms, such as those developed in the US by Pearson, McGraw-Hill, IBM, Knewton, Cerego, Smart Parrow, Dreambox, LightSide, and Coursera, facilitate personalized learning and can improve learner outcomes. To ensure equal access, education systems need to ensure that all students have access to these tools and opportunities. The digitization of education systems, and the use of AI is a growing priority in the European Union. AI-based solutions can help reduce educational inequalities and improve learning outcomes.

AI-driven education must balance innovation with social justice. AI-driven education must provide opportunities for all students, regardless of their background. AI must be applied in a way that helps tackle discrimination and inequality in education. Research has shown that the use of AI in education can provide a range of benefits for disadvantaged students. For example, adaptive learning platforms allow learners to learn at their own pace and style, leading to increased learning success. Ensuring equal access, and reducing educational inequalities in education systems, are key in Europe. AI-based solutions can contribute to differentiated learning and personalized education.

Data drives AI, so the effectiveness and inclusiveness of AI-based education directly depends on the quality and availability of data. Insufficient data or inaccurate data can compromise the efficiency and effectiveness of AI-based education. Proper data collection and analysis is critical to the development of education systems. Data systems that enable efficient and accurate tracking of educational outcomes contribute to improving the quality and inclusiveness of education. UNESCO and other educational institutions have launched a number of initiatives in this regard. The quality and availability of data are key to the effectiveness of AI-based education systems in Europe. EU initiatives in the field of education data collection and analysis can support the development of AI-based education initiatives.

(Q2) *How can the state of flow be used to focus on experiential education in learning environments with ethical digital content?*

The flow-based pedagogical model, as presented earlier, links flow and education. According to Mihály Csíkszentmihályi (2012), the father of "flow theory", flow is new psychology using scientific methods to help individuals understand their humanity. He emphasizes the power of joy and its importance in conscious, moment-to-moment experience, and he believes that flow is the primary motivator, and can have a positive impact on academic performance. Flow refers to a state of complete immersion in an activity. Distractions from the outside world disappear and time passes unnoticed (Beatson, 2015). This viewpoint makes the link between emotion and a state of a learner's immersion. Total immersion in play is likened to a mental state of flow, in which the player is fully immersed in an activity through feelings of concentration, involvement, and enjoyment.

To address the research question, I developed the ethical flow-based pedagogical model, illustrated in Figure 10.2. The rapid development and widespread use of digital technologies in education have presented new challenges and opportunities. This new model explores the relationship between experiential education, flow state, and ethical digital content to create an integrated, modern learning environment. The main objective is to demonstrate how to promote a state of flow in experiential education, while ensuring the ethical and responsible integration of AI content into the learning process.

The proliferation of digital technologies in education has brought new challenges and opportunities. This new model explores the relationship between experiential education, flow state and ethical digital content, to create an integrated, modern learning environment. The primary objective is to demonstrate how a state of flow can be fostered in experiential education, while ensuring that the AI content used is integrated into the learning process in an ethical and responsible way.

Flow, as described above, is a state of mind in which an individual is fully immersed in an activity, experiencing optimal challenge and a high level of experience. Experiential education, based on learning through practical, real-life experiences, provides an ideal environment for achieving the flow state. One of the central questions in this chapter is how this flow state can be fostered in modern education, especially in environments supported by digital (AI) tools. The integration of digital content into education offers many benefits, but it is important to ensure that this content is ethical. This includes the responsible use of AI, ensuring data security, privacy, and legal compliance. The new model aims to integrate ethical digital content into education in a way that supports the achievement and maintenance of a state of flow for learners.

The new model, which is an adaptation of the flow-based pedagogical model for enhanced AI, is shown in Figure 10.2:

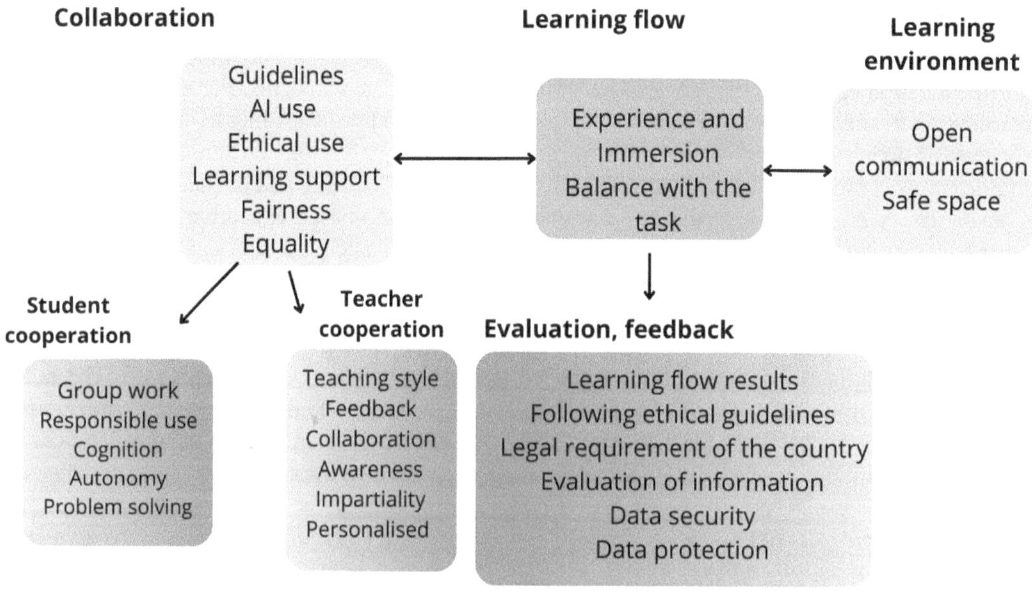

FIGURE 10.2 Ethical flow-based education model. (Source: Author.)

This model offers an innovative approach to teaching. The objective of the ethical flow-based model in education is to provide a coherent and ethical framework for a flow-based pedagogical model, by ensuring an effective, and ethically sound learning process.

Flow, a state of complete immersion in an activity, can be a key factor in experiential education, especially in learning environments supported by ethical digital content. The state of flow promotes deep learning, sustained motivation, and deep attention, making it particularly beneficial for experiential education. Experiential education is based on hands-on, real-life experiential learning, where learners are actively engaged, and directly experience, knowledge. Achieving a state of flow in this environment means that learners become so immersed in the tasks that they are fully engaged in the learning process. This balance between deep concentration and challenge contributes to more effective and sustainable learning.

Ethical digital content, adapted to learners' interests and developmental levels, can support the achievement of a state of fluency. By ethical digital content, we mean materials that respect learners' rights and privacy, ensuring a safe and inclusive learning environment. Such content is not only technically but also pedagogically designed to support learners' moral development and problem-solving thinking. Such ethical digital content can help to maintain motivation while ensuring an enjoyable and meaningful learning process. Interactive elements such as gamified learning modules, simulations, or virtual reality applications, can enhance the flow experience by providing learners with continuous challenges and positive feedback.

The synergy between the flow state and experiential learning, allows learners to become deeply immersed in the learning process, increasing learning effectiveness and the applicability of acquired knowledge. Ethical digital content ensures that this process is not only effective, but also responsible, fostering the development of skills such as ethical decision-making, empathy and social responsibility.

The combination of flow, experiential learning, and ethical digital content can result in a complex yet effective learning environment, where learners are engaged, motivated, and consciously develop, both professionally and morally.

(Q3) *What is the relevance of the model and how can it be integrated into your educational portfolio?*

Anantrasirichai and Bull's study on educational creativity (2022), investigated the impact of the creative use of AI in various AI-dominated areas (content creation, information, analysis, content enhancement and post-production workflows, information extraction and enhancement, and data compression). The study found that the ability of AI to produce award-winning creations is severely limited compared to human creations. In their view, the creative use of AI is beneficial when the creative work produced is focused on human-centered value creation. According to Chen et al. (2020), as AI continues to evolve, arts professionals can anticipate significant shifts in human-machine collaboration to produce much more humane and diverse content. However, the ultimate application of these tools will, of course, be up to creative artists. All of this makes it worthwhile for those in the creative professions to understand the implications of machine learning and embrace the opportunities that the technology offers, to ensure the survival of their industries in their original form.

At the current level of AI development, it makes sense to use AI in some aspects of creative activities, complemented by the beneficial effects of human creative activity and creative processes. AI alone generates only a kind of artificial creativity but does not allow for the natural flow of human creative processes. The emergence of AI in the creative professions has been identified as a typical problem, but given the technical conditions, it is easy to see the differences between human-generated and AI-generated works and creative works. The ability of AI to create creative and outstanding works of art is currently severely limited, possibly due to the emotional depth and authenticity of human creative activity, which machine learning cannot adapt to. AI produces a kind of artificial creativity that is logical, useful for later use, and often surprising, but is lacking the characteristics of natural human creativity, such as a natural and authentic creative process. (Runco, 2023).

Based on the example above, the Ethical Flow-Based Education Model offers the opportunity to innovate teaching methods and create an environment where learners make ethically informed decisions and experience learning processes that benefit the community. Ethical issues that reflect real-life situations can be integrated into educational programs. For example, students can be encouraged to discuss moral dilemmas and consider the impact of their decisions on the community. In addition to developing ethics curricula, it is important that teachers provide assignments that require ethical reflection. However, teachers can act as role models by modeling ethical behavior in the classroom and encouraging students to explore their own values. An assessment method that focuses on students' ethical development in addition to knowledge can also be part of the teaching portfolio. Thus, the relevance of the Ethical Flow-Based Education Model lies in the fact that it places ethics-based education at the center of the learning process, focusing on the personal and community development of students. This model builds on the moral development of the individual, emphasizing compassion, autonomous decision making and community responsibility, while dynamically adapting to the current experience and emotional state of the learner.

The ethical flow-based education model is an innovative educational model. Crucially, it is able to produce an ethical learning flow state, to teach and learn in an ethical collaborative context, and to create an ethical learning environment, while also being able to evaluate and provide feedback on this process. (Figure 10.2).

The diagram visually illustrates the interconnectedness of experiential education, flow state, ethical digital content, and the modern learning environment. It links the different educational elements in an innovative way, showing how they can be used together for effective and responsible learning, and weaving together collaboration, learning flow, assessment, and the learning environment. The way these elements work together reflects a holistic perspective that emphasizes the synergistic relationship between experiential learning and ethical digital content. The diagram highlights ethical issues such as ethical AI usage, data security, and privacy. Ethical digital content creates a learning environment that supports the maintenance of a state of flow while safeguarding learner safety and rights. This is particularly relevant in today's digital educational environment, where technology is increasingly prominent and ethical issues are central. It focuses on the flow state of learning, meaning that the educational process is designed to immerse the learner in the tasks, and to experience deep learning. The centrality of the flow state in the diagram reflects changes in modern education where the learning experience, and student engagement, are increasingly important. It makes sense to design learning tasks that optimize student engagement and promote the achievement of the flow state. It is emphasized that ethical digital content plays a key role in maintaining the flow state, especially in digitally supported learning environments. Such content must not only be technologically advanced, but also ethical. It must ensure learner privacy, accessibility of content, and respect for learner rights.

Digital learning tools offer the possibility of real-time feedback, which helps to maintain a state of flow. Interactive elements such as simulations, games, or adaptive learning platforms, can provide challenges that motivate learners and keep them interested. Ethical digital content should be accessible to all students, taking into account different learning needs and abilities. Tailoring content to individual learning styles and needs allows all students to achieve a state of fluency. Privacy and security are central to the use of digital content. Ethical data management ensures that students can use digital tools with confidence, without worrying about the security of their data, therefore helping to achieve and maintain a state of fluency.

Collaboration between students and teachers is a cornerstone of the diagram, facilitating a shared learning process. This innovation lies in combining experiential learning with digital tools to create a collective, supportive learning environment. Student collaboration plays an important role in experiential education. The diagram illustrates how group work, responsible technology use, and problem solving, promote students' cognitive development and autonomy. Teacher–student collaboration is also particularly important, as teachers' feedback, collaboration, and teaching style significantly impact the learning experience.

Assessment processes should be designed to support the maintenance of a learning flow state, and to ensure that learning outcomes align with ethical and legal guidelines. Assessment processes play a key role in maintaining the flow of learning, especially in digitally supported learning environments. Assessment is not merely for measuring learning outcomes but is also part of the learning process, supporting student reflection, motivation, and development. The chart's visual simplicity facilitates understanding while illustrating complex concepts. This simplicity allows various actors, including teachers, students, and educational experts, to easily interpret and apply the concepts presented.

Overall, this diagram represents a complex and comprehensive approach, which combines the benefits of digital tools and experiential education, while emphasizing the importance of ethical guidelines.

CONCLUSIONS

AI will become even more embedded in global education systems in the coming years. As technological advances and education continue to converge, AI will be able to support learners, teachers/educators, and institutions more effectively. To improve educational outcomes, AI will be able to provide personalized learning services that take into account individual learners' needs and progress. AI-based education systems will streamline teacher workflows, enabling them to help students more effectively and make decisions based on a wide range of learning data.

The learning environment and learning outcomes are expected to change significantly as a result of AI and technological advances. AI-based education systems will provide personalized learning, taking into account individual needs and abilities. Students will be able to learn at their own pace, and the curriculum will adapt to individual strengths and weaknesses, increasing efficiency and motivation and reducing drop-out rates. AI continuously analyzes and evaluates student performance, providing real-time feedback that helps identify and correct mistakes quickly. Learning outcomes are increasingly focused on acquired competencies and skills, as AI objectively measures these, ensuring the acquisition of real knowledge and skills.

The integration of VR technologies creates an interactive learning environment where students can practice in virtual labs and simulations (Dominek, 2017). AI also promotes lifelong learning, continuous training, and adaptability in an evolving job market. AI-based systems will reduce educational inequalities and ensure that all students, regardless of their socio-economic background, have access to high-quality education. These changes will increase the quality and efficiency of education, prepare students for the challenges of the future labor market, and ensure that they have the skills and knowledge needed for successful careers and professional paths.

Developed countries, with their existing infrastructure and skilled professionals, are likely to lead the adoption of AI-driven learning systems in the education sector. However, the demand and availability of AI-based education solutions is also expected to rise in developing countries, although differences are likely to remain due to infrastructure and financial constraints.

REFERENCES

Alexopoulos, C., Lachana, Z., Androutsopoulou, A., Diamantopoulou, V., Charalabidis, Y., & Loutsaris, M. A. (2019). How machine learning is changing e-government. In *ICEGOV '19: Proceedings of the 12th International Conference on Theory and Practice of Electronic Governance* (pp. 354–363). https://doi.org/10.1145/3326365.3326412

Anantrasirichai, N., & Bull, D. (2022). Artificial intelligence in the creative industries: A review. *Artificial Intelligence Review*, *55*(1), 589–656.

Berényi, G. (Trans.) (1991). *Kant, Immanuel: Az erkölcsök metafizikájának alapvetése*. Akadémiai Kiadó.

Beatson, O. (2015). Csikszentmihalyi's flow model [SVG]. Wikipedia. https://en.wikipedia.org/wiki/Mihaly_Csikszentmihalyi

Chen, W, Shidujaman, M. M., Jin, J, & Ahmed, S. U. (2020). A methodological approach to create interactive art in artificial intelligence. In *HCI international 2020–Late breaking papers: Cognition, learning and games: 22nd HCI international conference* (pp. 13–31). Springer International Publishing.

Csíkszentmihályi, M. (2012). A megszólított pszichológia. *Magyar Pszichológiai Szemle*, *67*(1), 47–56. (The psychology addressed).

Dietvorst, B. J., Simmons, J. P., & Massey, C. (2015). Algorithm aversion: People erroneously avoid algorithms after seeing them err. *Journal of Experimental Psychology: General*, *144*(1), 114–126. https://doi.org/10.1037/xge0000033

Dominek, D. L. (2017). Élményalapú tárlatvezetés. In: Kurta Mihály; Veres Gábor; Verók Attila (szerk.): Múzeumandragógia és közösségszolgálat. Az V. Országos Múzeumandragógiai Konferencia válogatott anyaga, Eger: Eszterházy Károly Egyetem, Pulszky Társaság - Magyar Múzeumi Egyesület Múzeumandragógiai Tagozata, pp. 22–36.

Dominek, D. L. (2022). On a flow-based pedagogical model. The emergence of experience and creativity in education. *Eruditio - Educatio*, *17*(3), 72–82. https://doi.org/10.36007/eruedu.2022.3.072-081

Fingarette, H. (1972). *Confucius: The secular as sacred.* Harper and Row. https://doi.org/10.2307/1398287

Furey, H., & Martin, F. (2019). AI education matters: A modular approach to AI ethics education. *AI Matters*, *4*(4), 13–15. https://doi.org/10.1145/3299758.3299764

Galloway, C., & Swiatek, L. (2018). Public relations and artificial intelligence: It's not (just) about robots. *Public Relations Review*, *44*(5), 734–740. https://doi.org/10.1016/j.pubrev.2018.10.008

Gesk, T. S., & Leyer, M. (2020). Artificial intelligence in public services: When and why citizens accept its usage. *Government Information Quarterly*, 39(3). https://doi.org/10.1016/j.giq.2022.101704

Janssen, M., Hartog, M., Matheus, R., Yi Ding, A., & Kuk, G. (2020). Will algorithms blind people? The effect of explainable AI and Decision-Makers' experience on AI-supported decision-making in government. *Social Science Computer Review*, *40*(2), 478–493. DOI : https://doi.org/10.1177/0894439320980118

Jordan, M. I., & Mitchell, T. M. (2015). Machine learning: Trends, perspectives, and prospects. *Science*, *349*(6245), 255–260. https://doi.org/10.1126/science.aaa8415

Kant, I. (1998). Az erkölcsök metafizikájának alapvetése. Raabe Klett Könyvkiadó Kft.

Keown, D. (2020). *Buddhist ethics: A very short introduction*. Oxford.

Labancz, I., & Barnucz, N. (2016). Kísérlet az IKT-eszközhasználat hatásának vizsgálatára hallgatók körében. In A felsőoktatás (hozzáadott) értéke: Közelítések az intézményi hozzájárulás empirikus megragadásához. Felsőoktatás&Társadalom (6). Partium; PPS; ÚMK, Budapest, Nagyvárad, pp. 262–277. (An experiment to investigate the impact of ICT tool use among students).

Leyer, M., Oberlaender, A., Dootson, P., & Kowalkiewicz, M. (2020). *Decision-making with artificial intelligence: Towards a novel conceptualization of patterns.* PACIS 2020 Proceedings. 224. https://aisel.aisnet.org/pacis2020/224

Leyer, M., & Schneider, S. (2021). Decision augmentation and automation with artificial intelligence: Threat or opportunity for managers? *Business Horizons, Elsevier*, *64*(5), 711–724. https://doi.org/10.1016/j.bushor.2021.02.026

Runco, M. A. (2023). Updating the standard definition of creativity to account for the artificial creativity of AI. *Creativity Research Journal*, *37*, 1–5. https://doi.org/10.1080/10400419.2023.2257977

Russel, S. J., & Norvig, P. (2010). *Artificial intelligence, a modern approach* (3rd ed.). Prentice Hall Series. ISBN-13: 978-0-13-604259-4.

Rzepka, C., & Berger, B. (2018). *User interaction with AI-enabled systems: A systematic review of IS research conference* [Paper presentation]. Thirty Ninth International Conference on Information Systems, San Francisco, CA, United States.

Seligman, M. E. P., & Csíkszentmihályi, M. (2000). Positive psychology. An introduction. *American Psychologist*, *55*, 5–14.

Slavin, R. E. (1995). *Cooperative learning: Theory, research, and practice.* Allyn and Bacon.

Sun, T., & Medaglia, R. (2019). Mapping the challenges of artificial intelligence in the public sector: Evidence from public healthcare. *Government Information Quarterly*, *36*(2), 368–383. https://doi.org/10.1016/j.giq.2018.09.008

Szabó, M. (Trans.) (1997). Arisztotelész: Nikomakhoszi Etika. Budapest, Európa Kiadó.

Szondy, M. (2011). *A boldogság tudománya. Fejezetek a pozitív pszichológiából.* Jaffa Kiadó. (The science of happiness. Chapters in positive psychology).

Thiebes, S., Lins, S., & Sunyaev, A. (2021). Trustworthy artificial intelligence. *Electronic Markets*, *31*(5), 447–464. https://doi.org/10.1007/s12525-020-00441-4

Thierer, A. D., O'sullivan, A. C., & Russel, R. (2017). *Artificial intelligence and public policy.* Mercatus Center. https://doi.org/10.13140/RG.2.2.14942.33604

Tomlinson, C. A. (1999). *The differentiated classroom: Responding to the needs of all learners.* Association for Supervision and Curriculum Development. https://rutamaestra.santillana.com.co/wp-content/uploads/2020/01/The-Differentiated-Classroom-Responding-to-the-Needs.pdf

11 Debiasing AI in Education through Different Ethical Insights

Harshvardhan Kaul

INTRODUCTION

Artificial Intelligence (AI) is at the forefront of technological innovation, rapidly transforming industries across the globe, with education being one of the sectors that stand to benefit the most. AI's ability to analyze vast amounts of data, identify patterns, and automate processes opens new avenues for personalized learning, automated grading, and data-driven decision-making in schools and universities (Singh, 2024). However, these innovations come with significant ethical concerns, especially regarding bias, fairness, privacy, transparency, and accountability.

AI-driven systems in education have shown immense potential to revolutionize how students learn and how educators manage classrooms. Personalized learning platforms such as BYJU'S and DreamBox use AI to adapt lessons in real time, tailoring content based on each student's unique learning pace and style (d'Elia et al, 2022). Similarly, automated grading systems can alleviate the burden of grading for educators by providing immediate feedback on student performance, thus enabling more focus on teaching. Predictive analytics, powered by AI, can help identify students at risk of dropping out or underperforming, giving educators actionable insights into providing timely interventions (Toksha, 2022).

Yet, as with all technology, these advantages are not without risks. AI systems are prone to replicating or even amplifying societal biases if they are not carefully designed and rigorously monitored. Bias in AI arises from the data it learns from, which often contains inherent biases due to historical inequalities. Moreover, the "black box" nature of many AI systems—where the decision-making processes are hidden or incomprehensible—poses risks for transparency and accountability.

The ethical frameworks discussed include consequentialism, deontology, virtue ethics, and care ethics. Each framework offers a different lens through which to assess the ethical integration of AI in education. Consequentialism focuses on the outcomes of AI systems, urging developers to design systems that maximize educational benefits while minimizing harm. Deontological ethics emphasizes the importance of following ethical rules, such as respecting students' rights to privacy and informed consent. Virtue ethics promotes the development of moral character in both students and AI systems, advocating for AI tools that embody fairness and empathy. Care ethics focuses on the emotional and psychological well-being of students, stressing the importance of compassionate AI systems that respond to students' needs.

The question is, how can educational institutions balance the immense benefits of AI while minimizing its risks, particularly those associated with fairness and bias? This chapter aims to provide an in-depth analysis of how AI can be ethically integrated into education, using established frameworks such as fairness, inclusivity, transparency, accountability, and privacy (FIT-AP for short).

AI ETHICS IN EDUCATION: ENSURING FIT-AP

To ensure that AI is ethically deployed in education, it is essential to follow the FIT-AP framework. These principles help ensure that AI systems are developed and implemented in ways that protect

DOI: 10.1201/9781003585527-15

students' rights and promote equitable educational outcomes (Singh, 2024). Each of these principles addresses a different ethical concern, and they must be considered together to create comprehensive guidelines for ethical AI in education.

Fairness in AI Systems

Fairness in AI refers to ensuring that AI systems treat all users equitably, without favoring or discriminating against any group based on characteristics such as race, gender, socioeconomic status, or disability. In educational contexts, fairness is especially important, as AI systems are used to evaluate student performance, allocate resources, and even influence admissions decisions.

However, fairness is difficult to achieve in AI systems because these systems rely on data to make decisions, and data often contain biases. For example, if an AI system used to assess student performance is trained on data from predominantly affluent schools, it may not perform well for students from under-resourced schools. These biases can manifest in various forms, such as data bias (where the training data is not representative of all student groups) or algorithmic bias (where the rules of the algorithm inherently favor one group over another).

One of the most prominent examples of algorithmic bias in AI is the Correctional Offender Management Profiling for Alternative Sanctions (COMPAS) algorithm, used in the U.S. criminal justice system to predict recidivism. An investigation by ProPublica revealed that COMPAS was biased against African American defendants, labeling them as higher-risk for reoffending even when their profiles were similar to those of lower-risk white defendants (Angwin et al., 2016). While this example comes from the criminal justice system, it underscores the importance of fairness in AI systems used for high-stakes decisions, including in education.

Achieving fairness in educational AI systems requires a multifaceted approach (Prasad, 2024):

- Bias Audits: AI systems should be subjected to regular audits to detect and mitigate biases. These audits should involve analyzing how the system performs for different demographic groups and making adjustments to ensure equitable outcomes.
- Diverse and Representative Datasets: Developers should ensure that AI systems are trained on diverse datasets that reflect the full range of student experiences and backgrounds. This includes collecting data from underrepresented groups, such as students from low-income families, rural areas, or minority communities.
- Transparency: AI systems should be transparent in their decision-making processes, allowing students, educators, and parents to understand how decisions are made and to challenge unfair outcomes.

For example, an AI-driven automated grading system should not only provide objective feedback but also offer mechanisms for teachers to review and adjust grades if necessary, ensuring that the system does not unfairly penalize certain students based on irrelevant factors.

Inclusivity and AI

Inclusivity is another critical ethical consideration in AI systems, particularly, where students come from diverse backgrounds with varying needs. Inclusive AI systems ensure that all students—regardless of their background, ability, or location—have equal access to the benefits of AI-enhanced education.

In practice, inclusivity means designing AI systems that are accessible to students with disabilities, culturally sensitive, and adaptable to different learning styles. For example, AI systems should be designed to work with assistive technologies such as screen readers for visually impaired students or voice recognition software for students with speech impairments. Additionally, these systems should support multiple languages and consider the diverse cultural contexts in which students learn.

Personalized learning platforms, such as those developed by BYJU'S and Vedantu, are already making strides in promoting inclusivity by providing personalized lessons based on students' learning preferences (d'Elia, 2022). However, challenges remain. For instance, students in rural or low-resource environments may not have access to the necessary infrastructure, such as reliable internet connections or digital devices, to fully benefit from these platforms. This creates a digital divide, where students in wealthier areas have more opportunities to leverage AI for learning, while students in underserved areas are left behind.

To address these challenges, AI systems must

- Be compatible with assistive technologies to ensure accessibility for students with disabilities.
- Be designed with cultural and linguistic diversity in mind, ensuring that students from different backgrounds feel represented and respected in the content they are learning.
- Ensure equal access to AI tools by partnering with governments, NGOs, and private organizations to provide the necessary infrastructure, such as internet connectivity and digital devices, to students in low-resource areas.

Transparency in AI Systems

Transparency in AI refers to the ability of users and stakeholders to understand how an AI system works, including how it makes decisions and what data it uses. In the context of education, transparency is especially important because students, parents, and educators must trust that AI systems are making fair and accurate decisions. When AI systems are opaque, or their decision-making processes are hidden, it can lead to mistrust and even harm if the system makes incorrect or biased decisions.

One of the main challenges in achieving transparency is the complexity of many AI systems, particularly those that use deep-learning algorithms. These algorithms learn from large datasets and make predictions or decisions based on patterns they identify in the data. However, because these patterns are often too complex for humans to understand, the decision-making process becomes a "black box," where even the developers of the system cannot explain how a particular decision was made (Singh, 2024).

Furthermore, many AI systems are developed by private companies that use proprietary algorithms, making it difficult for schools and universities to scrutinize how the system works. This lack of transparency can be problematic, especially in high-stakes areas like grading or admissions, where students' futures are at stake (Prasad, 2024).

To promote transparency in AI systems, developers and educational institutions should

- Provide documentation explaining how the AI system works, including how it makes decisions and what data it uses.
- Open the system to audits by independent third parties to ensure that it is operating fairly and accurately.
- Involve stakeholders, including students, parents, and educators, in the development and testing of the AI system, ensuring that their concerns are addressed before the system is fully implemented.

Transparency is not only an ethical requirement but also a way to foster trust in AI systems. When students and parents understand how AI systems are making decisions, they are more likely to trust the outcomes and accept the system as a legitimate tool in their education (Prasad, 2024).

Accountability and Ethical AI

Accountability in AI systems refers to the responsibility of developers, institutions, and educators to ensure that AI systems are used ethically and that any negative consequences are addressed. In education, accountability is particularly important because AI systems can have a direct impact on students' academic futures, influencing everything from grades to college admissions (Singh, 2024).

One of the key challenges in ensuring accountability is that AI systems often distribute decision-making across multiple actors. For example, an AI system used in a school might be developed by a private company, implemented by the school's IT department, and used by teachers to evaluate students. When something goes wrong—such as a biased grading decision—it can be difficult to determine who is responsible and who should take action to correct the problem.

To address this challenge, educational institutions and AI developers must

- Establish clear lines of responsibility for the development, deployment, and use of AI systems. This includes specifying who is responsible for monitoring the system, making adjustments, and addressing any errors or biases that arise (Prasad, 2024).
- Implement regular audits to assess the performance and fairness of AI systems. These audits should be conducted by independent third parties to ensure objectivity (Prasad, 2024).
- Create redress mechanisms for students and educators to challenge AI decisions. This could include setting up an appeals process where students can request a human review of an AI-generated grade or decision.

Facial recognition technology is an example where accountability issues have emerged. Schools in the United States and the United Kingdom have begun using facial recognition to track attendance and monitor students (Hill, 2020). While these systems can improve efficiency, they raise concerns about student privacy and surveillance. Without clear accountability frameworks, it can be difficult for schools to manage the potential risks and ensure that the technology is used ethically.

Privacy Concerns in AI

Privacy is a significant concern in the use of AI in education, particularly because AI systems often rely on collecting and analyzing large amounts of personal data. In education, this data can include information about students' academic performance, behavior, health, and even biometric data, such as facial recognition scans. Ensuring that this data is used ethically and securely is critical to maintaining trust in AI systems.

Similar to the General Data Protection Regulation (GDPR) in Europe, India has also enacted the Digital Personal Data Protection (DPDP) Act of 2023, which adds a critical legal framework to the discussion on ethical AI use in education. The law reinforces key aspects of the FIT-AP framework—particularly around privacy and accountability—by mandating stringent data protection measures, especially for minors. Educational AI systems that collect and process student data must now comply with the DPDP Act's guidelines, which include obtaining explicit consent, minimizing data collection, and ensuring robust security for any personal data handled.

Fairness

The DPDP Act's focus on consent and purpose limitation strengthens the principles of fairness and transparency. AI systems used for high-stakes educational decisions, such as grading or admissions, must not only avoid biases in their algorithms but also ensure that students and parents understand how their data is being used and have the ability to challenge decisions. The Act requires AI platforms, such as BYJU'S and DreamBox, to provide clear justifications for data collection and usage, particularly in cases where decisions have a direct impact on students' academic futures.

Privacy Concerns

Privacy, an essential pillar of the FIT-AP framework, is critically reinforced by the DPDP Act. The law mandates data minimization and purpose limitation, which means AI systems should collect only the data necessary for their function, without encroaching on personal or sensitive information, such as students' behavioral data or biometric details. In cases where AI-driven systems like attendance tracking using facial recognition are implemented, educational institutions must justify why such data is needed, thereby addressing concerns over excessive surveillance and ensuring compliance with legal standards. This is particularly significant in India, where the use of AI in education is expanding but where the risk of data misuse remains high.

Children's Data Protection

In addition, the DPDP Act provides specific protections for children's data, aligning with the chapter's emphasis on care ethics and the moral obligation to protect vulnerable populations. For instance, platforms serving school-aged children, such as Vedantu, are now legally required to secure parental consent before processing any data. This legal safeguard ensures that AI systems do not inadvertently expose children to harm, thus supporting the chapter's focus on fairness and inclusivity.

Accountability and Redress Mechanisms

The DPDP Act also enhances accountability by establishing clear redress mechanisms for individuals who believe their data has been misused or that decisions made by AI systems are unfair. This corresponds with the chapter's call for transparency in AI-driven decision-making processes, ensuring that students and educators have recourse if they are negatively affected by biased algorithms or erroneous decisions.

However, protecting privacy in AI systems goes beyond legal compliance. Educational institutions and AI developers must also implement technical safeguards to prevent unauthorized access to student data and ensure that data is only used for its intended purpose (Prasad, 2024).

To protect privacy in AI systems, developers and institutions must

- Limit data collection to only the information necessary for the AI system to function. This principle of data minimization helps reduce the risk of data breaches and misuse.
- Implement strong encryption and security measures to protect data from being accessed by unauthorized parties.
- Provide transparency around data usage, ensuring that students and parents understand what data is being collected, how it is being used, and who has access to it. Students and parents should also be given control over their data, including the ability to opt out of data collection if they choose.

By prioritizing privacy, educational institutions can build trust in AI systems and ensure that students' personal information is handled ethically.

ETHICAL FRAMEWORKS FOR AI IN EDUCATION

Ethical frameworks provide a foundation for guiding decision-making in the development and deployment of AI systems in education. The most prominent frameworks include consequentialism, deontology, virtue ethics, and care ethics, each of which offers a different perspective on how to balance the benefits of AI with the potential risks.

Consequentialism in AI Ethics

Consequentialism is an ethical theory that focuses on the outcomes or consequences of actions. In the context of AI ethics, consequentialism asks whether the deployment of an AI system leads to positive or negative outcomes for students, teachers, and institutions. If the benefits of using an AI system outweigh the harms, then the system is considered ethical.

In education, consequentialism encourages developers to design AI systems that maximize positive educational outcomes while minimizing potential harms. For instance, personalized learning platforms use AI to adapt lessons to each student's learning style, providing tailored instruction that can improve engagement and academic performance (Toksha, 2022). These systems can be particularly beneficial for students who struggle with traditional teaching methods, offering them a more flexible and supportive learning environment.

However, consequentialism also requires that developers consider the potential negative consequences of deploying AI systems. For example, if an AI system is primarily trained on data from students in well-funded schools, it may not work as well for students in underfunded schools, thus exacerbating existing inequalities. Similarly, an AI system that tracks student behavior to predict academic success may unfairly label certain students as "at risk" based on biased data, leading to unintended consequences such as lower expectations from teachers or fewer opportunities for academic advancement.

To ensure that AI systems are ethically sound from a consequentialist perspective, developers must

- Continuously monitor the impact of AI systems to ensure that they are improving educational outcomes for all students, not just a privileged subset.
- Incorporate feedback from students, teachers, and other stakeholders to assess the real-world impact of AI systems and make adjustments as needed.
- Design AI systems to be adaptable, allowing for adjustments to be made if negative consequences are identified after deployment.

By focusing on the outcomes of AI systems and making adjustments based on their real-world impact, educational institutions can ensure that AI is used to promote the greatest good for all students.

Deontological Ethics and AI

Deontological ethics, associated with philosopher Immanuel Kant, emphasize the importance of following ethical principles and rules, regardless of the consequences. In the context of AI ethics, deontology focuses on the morality of actions themselves, rather than their outcomes. This means that even if an AI system produces positive educational outcomes, it must still respect the rights and dignity of students and educators.

In education, deontological ethics requires that AI systems adhere to ethical principles such as respecting students' rights to privacy, fairness, and informed consent. For example, an AI system must not collect more data than is necessary, even if doing so could improve its performance. Similarly, the system must provide informed consent, ensuring that students and parents are fully aware of how their data will be used and have the option to opt out without facing negative consequences.

One key aspect of deontological ethics in AI is the principle of non-discrimination. AI systems must not unfairly discriminate against any group of students based on race, gender, socioeconomic status, or other protected characteristics (Singh, 2024). This is particularly important in educational settings, where AI systems are used to make high-stakes decisions such as grading or admissions.

To ensure that AI systems adhere to deontological principles, developers and institutions must

- Respect students' autonomy by giving them control over their data and ensuring that they are not coerced into using AI tools.
- Follow legal and ethical standards for data protection, such as the DPDP in India, GDPR in Europe, or Family Educational Rights and Privacy Act (FERPA) in the United States, ensuring that students' privacy is protected.
- Implement fairness constraints in AI algorithms to prevent any form of discrimination, even if doing so requires sacrificing some level of efficiency or performance.

By following these ethical principles, educational institutions can ensure that AI systems respect the rights and dignity of students, even as they enhance learning and educational outcomes.

Virtue Ethics and AI

Virtue ethics emphasizes the development of moral character and the cultivation of virtues such as honesty, fairness, and empathy. In AI ethics, virtue ethics suggests that AI systems should be designed to embody these virtues and promote their development in students and educators.

For example, an AI-driven tutoring system should not only provide personalized instruction but also promote values like fairness and empathy in its interactions with students. This means that the system should be transparent in its decision-making processes, ensuring that students understand why certain recommendations are being made, and should treat all students with respect, regardless of their background or abilities.

Moreover, virtue ethics emphasizes the importance of human-AI collaboration. AI systems should complement, rather than replace, human educators, fostering a collaborative environment where both AI and teachers contribute to students' academic and moral development. For instance, AI systems can provide personalized learning experiences that allow teachers to focus on providing emotional and moral support to students, helping them develop virtues like perseverance, empathy, and fairness.

To incorporate virtue ethics into AI systems, developers should

- Design AI tools that promote fairness and empathy, ensuring that all students are treated with respect and dignity.
- Ensure that AI systems encourage moral development in students, providing opportunities for them to practice virtues like fairness and empathy in their interactions with the system.
- Foster collaboration between AI and human educators, ensuring that AI systems complement, rather than replace, the role of teachers in providing emotional and moral support to students.

By focusing on the development of moral character and virtues, educational institutions can use AI systems not only to enhance academic outcomes but also to promote the overall development of students as ethical and empathetic individuals.

Care Ethics in AI

Care ethics is an ethical theory that emphasizes the importance of relationships, empathy, and the well-being of individuals. In contrast to more abstract ethical principles like fairness or justice, care ethics focuses on the concrete needs and vulnerabilities of individuals, advocating for a compassionate approach to ethical decision-making. This resonates with Buddhist ethical principles like compassion, kindness, love, and suffering.

In the context of AI in education, care ethics requires that AI systems prioritize the well-being of students, particularly those who are vulnerable or marginalized. This means that AI systems should be not only technically effective but also sensitive to the emotional and psychological needs of students. For example, an AI-driven tutoring system should be able to recognize when a student is feeling frustrated or overwhelmed and provide appropriate support, rather than simply delivering more content.

AI-driven tutoring systems, such as those used by Squirrel AI in China, offer one-on-one personalized instruction based on students' learning patterns. However, from a care ethics perspective, these systems must go beyond academic support by identifying when students may need emotional or psychological assistance. For instance, if a student is consistently underperforming, the AI system could offer words of encouragement or suggest taking a break, ensuring that the student's emotional needs are addressed.

To apply care ethics in AI systems for education, developers should

- Prioritize the emotional and psychological well-being of students in the design of AI systems, recognizing that education is not only about academic achievement but also about nurturing the whole student.
- Ensure that AI systems can respond to students' emotional states, providing support when students are feeling frustrated, anxious, or overwhelmed.
- Foster collaboration between AI systems and human teachers, ensuring that teachers can provide emotional and psychological support to students, while AI systems handle the more technical aspects of instruction.

By incorporating care ethics into AI systems, educational institutions can ensure that AI not only improves academic outcomes but also contributes to the overall well-being of students, fostering a compassionate and supportive learning environment.

Ontological Ethics in AI

Ontological ethics focuses on the nature of being and existence, examining what it means to "be" in a given context and how that influences ethical behavior. In the realm of AI, ontological ethics explores the implications of AI as a technological entity and how its existence interacts with human learning, autonomy, and identity. This approach shifts the focus from the outcomes or rules governing AI systems (as seen in consequentialism and deontology) to a deeper inquiry into the essence of AI and its role in human development.

Western philosophers like Martin Heidegger have extensively discussed the nature of technology and its role in shaping human existence. Heidegger's concept of "enframing" (Gestell) suggests that technology, including AI, can reduce humans to mere resources or data points, framing their existence in terms of utility (Heidegger, 1977). In education, this means AI could potentially treat students as data to be optimized, rather than as holistic beings with complex needs. To avoid this reductionist view, Heideggerian ontological ethics calls for a recognition of the deeper purposes of education—nurturing human potential, critical thinking, and the formation of character. AI in education, from this perspective, should not merely aim to optimize learning outcomes but should contribute to the holistic development of students as thinking, ethical beings.

Swami Vivekananda, one of India's foremost thinkers, provides a complementary perspective from the Indian philosophical tradition. For Vivekananda, education is not just the accumulation of knowledge but the manifestation of the perfection already present within each individual (Vivekananda, 1902). In this sense, AI systems should be designed and deployed in ways that facilitate the unfolding of students' inherent potential rather than imposing external, standardized measures of success. Ontologically, Vivekananda emphasizes the unity of all existence, suggesting that the purpose of education is to realize this interconnectedness and foster a sense of universal responsibility. AI, therefore, must be aligned with the goal of promoting self-realization, empathy, and ethical responsibility, reflecting the deeper spiritual and existential dimensions of learning (Vivekananda, 1999).

In applying ontological ethics to AI in education, we are reminded that AI should not dominate the educational process but support it in a way that respects the dignity and autonomy of both students and educators. This perspective encourages a view of AI as a tool that exists to enhance, rather than replace, the human elements of learning—curiosity, creativity, and moral reflection. By integrating both Western and Indian ontological insights, we can develop AI systems that foster not only intellectual growth but also the moral and existential development of students, ensuring that education remains a transformative process aligned with the deeper purposes of human life.

APPLICATIONS OF AI IN EDUCATION: OPPORTUNITIES AND RISKS

AI is being applied in a wide range of educational contexts, from personalized learning systems to automated grading tools and predictive analytics. While these technologies offer significant opportunities to improve education, they also raise ethical concerns related to fairness, privacy, and accountability.

Personalized Learning

At the beginning of the chapter, personalized learning was introduced as one of the most promising applications of AI in education, with platforms like BYJU'S and DreamBox using AI to tailor lessons to individual students. personalized learning offers significant potential to improve student outcomes by adapting to each learner's unique pace and needs. However, the ethical considerations involved go beyond the technical delivery of customized content.

For a deeper discussion of how personalized education aligns with broader ethical concerns, such as the ontological implications of AI and its role in human development, see the section on Ontological Ethics in AI. This section explores how personalized AI systems should not only serve to optimize learning efficiency but also respect the inherent dignity of students as whole individuals, fostering their intellectual and moral growth. The ontological perspective adds another layer to the conversation about personalized learning by ensuring that AI systems contribute to the holistic development of students, rather than reducing education to a purely data-driven exercise.

By analyzing student data, AI systems can adapt lessons and assessments to meet the individual needs of each student, providing a more tailored and engaging learning experience. This personalized approach allows students to learn at their own pace, focusing on areas where they need the most support and progressing more quickly through material they have already mastered.

In India, platforms like BYJU'S and Vedantu use AI to provide personalized learning experiences to millions of students (Goswami & Sharma, 2024). These platforms analyze students' strengths and weaknesses to deliver customized lessons that adapt in real time, helping students stay engaged and improve their performance. This has been particularly beneficial in a country with a large and diverse student population, where traditional one-size-fits-all teaching methods often fail to meet the needs of all learners.

However, while personalized learning platforms offer significant benefits, they also raise ethical concerns. One key issue is the potential for bias in the algorithms used to personalize learning. If the data used to train the algorithms is biased, the system may provide different recommendations or opportunities to students based on factors like their race, gender, or socioeconomic status (Prasad, 2024). For example, a personalized learning platform that is trained on data from affluent schools may not perform as well for students in underfunded schools, exacerbating educational inequalities.

Automated Grading and Its Ethical Challenges

Automated grading systems use AI to evaluate student assessments, providing instant feedback and reducing the workload for educators. These systems can be used for a variety of assessment types, including multiple-choice exams, written essays, and even more complex tasks such as coding assignments. By automating the grading process, AI can make assessments more objective and consistent, reducing the potential for human bias.

Several universities in the United States have begun using AI systems like Gradescope to automate the grading of assignments (McCarthy, 2023). These systems use machine learning algorithms to assess student responses, providing detailed feedback on errors and suggesting areas for improvement. By reducing the time educators spend on grading, AI systems allow teachers to focus more on providing personalized support to students.

However, while automated grading systems offer significant benefits, they also raise ethical concerns related to fairness and accountability. One major issue is the potential for bias in how AI systems grade assignments. If the system is trained on biased data, it may unfairly penalize certain groups of students, such as those from minority backgrounds or those who use non-standard language patterns in their writing.

If an automated grading system is trained on data from students in predominantly affluent, well-funded schools, it might be biased toward the language patterns, writing styles, and academic norms of those environments. Students from underfunded or minority-serving schools might be unfairly penalized because their writing or academic output does not match the patterns that the AI has learned to reward. This can lead to lower grades, discouraging students who are already at a disadvantage. For instance, minority students who use non-standard dialects or language expressions may receive lower grades from AI systems trained primarily on data from students using standard academic English, reinforcing disparities (Holstein et al., 2019; Smith & Rustagi, 2024).

AI systems used to predict student success may disproportionately label students from low-income or minority backgrounds as "at risk" based on biased data that links socioeconomic factors with poor performance. If these systems are trained on data that reflects historical inequalities, they may reinforce the notion that students from marginalized communities are less likely to succeed, resulting in biased interventions or lowered expectations from educators (Baker & Hawn, 2022; Benjamin, 2019).

AI-driven admissions systems that are trained on historical data may reflect the existing biases of admissions officers, favoring students from certain racial, ethnic, or socioeconomic backgrounds over others (Prasad, 2024). For example, if past data show that students from affluent backgrounds have been more successful in gaining admission, the AI might prioritize similar students in future admissions decisions, putting minority and low-income students at a disadvantage (Binns, 2009; O'Neil, 2016).

Intelligent Tutoring Systems

Intelligent tutoring systems (ITSs) provide personalized, one-on-one instruction to students, helping them master complex concepts at their own pace. By using AI to analyze a student's learning behavior, ITS can identify areas where the student is struggling and offer targeted exercises and feedback to help them improve. ITS are particularly valuable in large classrooms, where teachers may not have the time to provide individualized support to every student.

MATHia, an intelligent tutoring system developed by Carnegie Learning, is used in math instruction to provide personalized support to students (McCarthy, 2023). MATHia uses machine learning algorithms to analyze how students solve problems, adjusting the difficulty level of exercises based on the student's performance. This allows students to receive real-time feedback and tailored instruction that helps them master difficult concepts.

While ITSs offer significant opportunities to improve student outcomes, they also raise concerns about privacy and the lack of emotional engagement. Intelligent tutoring systems often require access to extensive data on students' learning behaviors, raising questions about how these data are stored and used. Moreover, because ITSs rely on algorithms rather than human teachers, there is a risk that students may become disengaged from the learning process, as they miss out on the emotional support and encouragement that human teachers provide (Singh, 2024).

Administrative Automation and Predictive Analytics

AI is also being used to streamline administrative tasks in education, such as tracking student attendance, scheduling classes, and managing resources. By automating these tasks, AI can improve the efficiency of educational institutions, freeing up time for educators to focus on teaching and student support.

In some schools, facial recognition technology is being used to track student attendance (Hill, 2020). This technology allows schools to automate the attendance process, reducing the time teachers spend taking roll calls and ensuring that attendance records are accurate.

However, the use of facial recognition technology in schools raises significant privacy concerns, particularly regarding the collection and storage of biometric data. There are also concerns about the potential for this technology to be used in ways that go beyond attendance tracking, such as monitoring student behavior or even profiling students based on their appearance.

In addition to these privacy concerns, the use of predictive analytics in education raises its own set of challenges. Predictive analytics systems analyze student data, such as grades, attendance, and behavioral patterns, to predict future outcomes, such as the likelihood of a student dropping out or falling behind academically. While these predictions can help educators intervene early and provide targeted support, they also raise concerns about fairness and accountability.

Mitigating Risks of AI in Administrative Applications of AI

1. In order to prevent biases in the generation of diverse and representative data, developers must ensure that the data provided is holistic in training algorithms, especially for those from underserved communities. Doing so in the educational sector would prevent the data from being biased and promise equitable learning experiences.
2. While the development of AI systems is imperative today, it is important to conduct regular audits and check-ups for the identification of any informational errors and biases which may in some cases hinder the personalization or grading process within the educational system. Such a practice would enable higher levels of fairness and accountability in educational assessments.
3. One of the most significant risks AI systems face is the lack of transparency. Thus, while AI is used in decision-making, it is important for developers to ensure that clear explanations are provided. Transparency allows students, parents, and educators to understand the criteria used in grading and personalized learning pathways, fostering trust in the system.
4. Along with regular audits and check-ups, human oversight in AI-driven educational systems will ensure educators and students have a means to review and adjust the AI-generated outcome as and when it is necessary. This would further enrich the sense of personalization.
5. In addition to transparency, strong data privacy measures when implemented ensures that students and educators have sufficient learning data and that the data are securely stored and utilized. This safeguards sensitive information against unauthorized access and misuse.

ADDRESSING AI BIAS IN EDUCATION

Bias in AI systems is a significant concern, particularly in education, where fairness and equity are essential values. AI systems can introduce bias in various ways, including through algorithmic bias, data bias, and interpretive bias. These biases can have far-reaching implications for students, leading to unequal opportunities and reinforcing existing disparities.

To address these types of bias, developers and educators must take proactive steps to identify, measure, and mitigate bias in AI systems. This requires a multifaceted approach that includes careful design, diverse data collection, and ongoing monitoring of AI systems to ensure they are functioning fairly (Prasad, 2024).

Impact of AI Bias on Educational Outcomes

The consequences of bias in AI systems can be far-reaching, particularly in the field of education. Biased AI systems can disproportionately impact students from marginalized groups, leading to unequal access to educational resources and opportunities. For example, if an AI system used for admissions decisions is biased against students from certain racial or socioeconomic backgrounds, it may systematically exclude these students from opportunities for higher education.

Moreover, bias in AI systems can have psychological effects on students, particularly if they perceive that the system is treating them unfairly. For instance, a student who is consistently assigned lower grades by an AI grading system may feel demoralized and disengaged, leading to a decline in academic performance. This can create a vicious cycle, where the student's poor performance reinforces the AI's biased assessment, further entrenching educational inequality.

Several universities and colleges around the world have begun using AI systems to assist with admissions decisions. These systems analyze a range of data, including academic performance, standardized test scores, and extracurricular activities, to predict which students are most likely to succeed. However, there is growing concern that these systems may reinforce existing biases, particularly if they are trained on data that reflect historical inequalities in access to education.

For example, if an AI system is trained on data from past admissions cycles, it may learn to favor students from well-funded schools or those who have access to certain extracurricular opportunities. This can disadvantage students from underprivileged backgrounds, who may not have had the same opportunities but are nonetheless capable of succeeding in higher education.

To mitigate the impact of bias in AI-powered admissions systems, universities and developers must

- Ensure that AI systems are trained on diverse and representative data that includes students from a wide range of backgrounds.
- Regularly audit AI systems to detect and correct any biases that may emerge during the admissions process.
- Provide transparency in admissions decisions, allowing students to understand the factors influencing their acceptance or rejection.

Role of Data in AI Bias

Data play a critical role in determining the fairness and effectiveness of AI systems. If the data used to train an AI system are biased or incomplete, the system is likely to produce biased outcomes. For example, an AI system trained on historical data from a predominantly white, affluent school district may not perform well when applied to students from more diverse or underfunded schools.

To address this issue, it is essential to ensure that the data used to train AI systems is diverse, representative, and free from bias. This can be achieved through several strategies:

- Diverse Data Collection: AI developers must make a concerted effort to collect data from a wide range of sources, ensuring that the training data reflect the diversity of the student population. This includes collecting data from schools in different regions, socioeconomic backgrounds, and cultural contexts.
- Data Pre-processing: Before training an AI system, developers must carefully analyze the data to identify and correct any biases. This may involve removing or adjusting certain data points that could skew the system's predictions, such as data that disproportionately favors certain groups (Prasad, 2024).
- Continuous Monitoring: Even after an AI system is deployed, it is essential to continuously monitor its performance to detect any emerging biases. Regular audits and assessments can help ensure that the system remains fair and effective over time.

AI systems can be trained to incorporate ethical philosophies and frameworks, but this is an evolving field. Training an AI system to apply ethical frameworks will require embedding ethical guidelines, using ethical design constraints, performing ethical audits, and establishing real-time ethical training.

FUTURE DIRECTIONS FOR ETHICAL AI IN EDUCATION

As AI technology continues to evolve, so too will the ethical challenges associated with its use in education. New developments in AI, such as adaptive learning systems, collaborative AI, and AI-driven educational governance, present exciting opportunities but also require careful ethical consideration.

Emerging Technologies and Ethical Concerns

One of the most promising developments in AI for education is the rise of adaptive learning systems. These systems use real-time data to adjust the difficulty and content of lessons based on each student's performance, allowing for a more personalized learning experience. However, adaptive learning systems also raise ethical questions about privacy, data collection, and the potential for bias.

Similarly, collaborative AI tools, which facilitate cooperation between students and teachers, offer new opportunities for enhancing the learning experience. However, these tools must be carefully designed to ensure that they promote inclusivity and fairness while respecting students' privacy and autonomy.

Governments and educational institutions must also work toward developing ethical governance frameworks to regulate the use of AI in education. These frameworks should include policies that mandate transparency in AI systems, ensure that students' data are protected, and promote fairness and inclusivity in the use of AI in educational settings.

Adaptive Learning Systems and Collaborative AI

Adaptive learning systems are one of the most promising AI developments in education. These systems adjust the difficulty and pace of lessons in real time based on the individual student's progress, providing personalized learning experiences. While these systems can significantly enhance student engagement and outcomes, they also raise important ethical concerns related to privacy and the collection of personal data. To effectively implement adaptive learning systems, it is essential to ensure that these technologies respect student autonomy and provide transparency around how decisions are made.

Similarly, collaborative AI tools are designed to facilitate interaction between students, teachers, and AI systems, allowing for a more interactive and supportive learning environment. However, these tools must be designed with inclusivity in mind to ensure that all students benefit equally from their implementation. AI developers and educators need to work together to create AI systems that promote collaboration while respecting the diverse needs of students.

CONCLUSION

The integration of AI into education holds immense promise for transforming how we teach and learn. AI systems can provide personalized instruction, streamline administrative processes, and offer predictive insights that help educators support students more effectively. However, these benefits come with significant ethical challenges, particularly around bias, fairness, privacy, and accountability.

By adopting ethical frameworks such as consequentialism, deontology, virtue ethics, and care ethics, educational institutions and AI developers can ensure that AI systems are used responsibly. The principles of fairness, inclusivity, transparency, accountability, and privacy must guide the development and deployment of AI in education to avoid reinforcing inequalities or creating new forms of discrimination.

Ongoing collaboration between educators, technologists, and policymakers will be crucial in shaping the future of AI in education. By ensuring that AI systems are designed and implemented ethically, we can harness the transformative power of AI to create more equitable and effective educational environments for all students, regardless of their background or circumstances.

REFERENCES

Angwin, J., Larson, J., Mattu, S., & Kirchner, L. (2016). *Machine bias: There's software used across the country to predict future criminals. And it's biased against blacks.* ProPublica. https://www.propublica.org/article/machine-bias-risk-assessments-in-criminal-sentencing

Baker, R. S., & Hawn, A. (2022). Algorithmic bias in education. *International Journal of Artificial Intelligence in Education*, 32, 1–41.

Benjamin, R. (2019). *Race after technology: Abolitionist tools for the new Jim code.* Polity.

Binns, R. (2018). Fairness in machine learning: Lessons from political philosophy. Proceedings of Machine Learning Research. 81: 149-159.

d'Elia, A., Gabbay, M., Rodgers, S., Kierans, C., Jones, E., Durrani, I., Thomas, A. & Frith, L. (2022). Artificial intelligence and health inequities in primary care: a systematic scoping review and framework. *Family Medicine and Community Health, 10*(Suppl 1).

Goswami, A., & Sharma, A. (2024). AI for bridging socio-economic inequities in Indian education space. *International Journal of Research and Scientific Innovation, XI (IV)*, 890–935.

Heidegger, M. (1977). The question concerning technology. In Martin Heidgger: Basic Writings. Edited by. David Farell Krell. (New York: Harper & Row, 1977), p. 14.

Hill, K. (2020). The secretive company that might end privacy as we know it. *The New York Times.* https://www.nytimes.com/2020/01/18/technology/clearview-privacy-facial-recognition.html

Holstein, K., Wortman Vaughan, J., Daumé, H. III, Dudik, M., & Wallach, H. (2019). Improving fairness in machine learning systems: What do industry practitioners need? *Proceedings of the 2019 CHI conference on human factors in computing systems* (pp. 1–16). Association for Computing Machinery.

McCarthy, T. D. (2023). *Standards-based grading at the secondary level: A phenomenological study of district level administrators and their perceptions of transitioning from a traditional grading system.* St. John's University, New York.

O'Neil, C. (2016). *Weapons of math destruction: How big data increases inequality and threatens democracy.* Crown Publishing Group.

Prasad, A. N. (2024). *Introduction to data governance for machine learning systems.* Apress eBooks. https://doi.org/10.1007/979-8-8688-1023-7

Singh, T. (2024). *Artificial intelligence and ethics: A field guide for stakeholders.* CRC Press.

Smith, G., Fleisig, E., Bossi, M., Rustagi, I., & Yin, X. (2024). Standard language ideology in ai-generated language. *arXiv preprint arXiv:2406.08726.*

Toksha, B., Kulkarni, T., & Gupta, P. (2022). Impact of AI on teaching pedagogy and its integration for enhancing teaching-learning. In *Artificial Intelligence in Higher Education* (pp. 137–152). CRC Press.

Vivekananda, S. (1902). *The complete works of Swami Vivekananda* (Vol. 4). Advaita Ashrama.

Vivekananda, S. (1999). *My ideal of education.* Advaita Ashrama.

12 AI Ethics in the Military

A Realist Approach Incorporating Traditional and Non-Traditional Security

Arun Teja Polcumpally and Shashank Patel

INTRODUCTION

Before moving forward, it is essential to clarify that war, by its very nature, is often viewed as unethical. Consequently, the concept of ethical warfare can be seen as somewhat ambiguous. Much like how a terrorist may be regarded as a freedom fighter from a different perspective, the ethics of war can be interpreted variably depending on who is discussing it and from the position of cultural and historical contexts that shape their understanding of ethics.

One way to understand the ethics of war is to anchor the understanding of ethical warfare to the pursuit of peace. However, the efforts to broker peace between warring nations are never ending. As of today, almost all countries are signatories to international agreements, such as the United Nations Charter, which outlines the principles of international peace, the Geneva Conventions, which govern the treatment of individuals during conflict, and the Hague Conventions, which regulate the conduct of war. Moreover, the United Nations' Responsibility to Protect (R2P) norm provides a framework for deploying peacekeeping forces. Despite these initiatives, we continue to witness recent conflicts, such as the wars between Russia and Ukraine, Palestine and Israel, and Armenia and Azerbaijan. This reality suggests that wars will persist as long as resources remain finite.

In *Bhagavad Gita*'s Chapter 3, Verse 39, Shri Krishna mentioned,

आवृतं ज्ञानमेतेन ज्ञानिनो नित्यवैरिणा। कामरूपेण कौन्तेय दुष्पूरेणानलेन च॥

The above text argues that perpetual enemy in the form of insatiable desire, which is never satisfied and burns like fire. The desire burns like an unquenchable fire which never brings happiness to anyone. The wise renounce it, knowing it to be the root of misery. In the same way, wars will continue with numerous variables like resource abundance/scarcity, institutional failure. And also ideologies (Mildner et al., 2011). War is also said to become "clash of civilizations," a theory proposed by Samuel P. Huntington. According to this theory, conflicts are bound to happen between the ancient civilizational borders like Persian, Indus Valley, and Mesopotamian (Huntington, 1996). This theory is followed by Robert D Kaplan's *The Revenge of Geography*, which emphasizes that the conflicts would occur between the regions that are geographically demarcated (Kaplan, 2012); and Rajeev Malhotra emphasized that digital technologies, especially artificial intelligence and affiliated digital technologies will define the new wars (Malhotra, 2021). With such numerous variables associated with the conduct of war, as a reiteration of the verse from *Bhagavad Gita*, it can be said that the insatiable desire of countries will never be satisfied with any number of resources, regions, or people captured; on the same lines their unquenchable longing to develop advanced warfare cannot be limited.

DOI: 10.1201/9781003585527-16

This shows that ethics and warfare is a volatile subject and has a varying understanding among the countries. Keeping the dynamism of ethics and war in hindsight, this chapter provides an inquisitive discussion of AI, ethics, and the military.

The landscape of warfare has evolved since the pre-1945 era marked by the World Wars. Today, most countries are hesitant to engage in full-scale wars (though there can be some exceptions like the Israel war on Gaza), as the costs often outweigh the benefits. Methods to wage war for any political gains are dynamic, and with the technological developments such as the internet and artificial intelligence (AI), massive killing similar to that of the World Wars is avoided. This is because advanced digital technologies like AI have brought a new way of achieving the political goals without the bloodshed. The integration of AI with precision warfare systems and enhanced communication capabilities allows belligerents to pursue their political objectives with minimal or even zero casualties. In addition to assisting the traditional ways of war to achieve political goals, AI is also used to achieve what Christopher Walker and Jessica Ludwig call "sharp power" (Walker & Jessica, 2017). Sharp power is the ability to manage information flows for hostile purposes. AI aids the belligerent states to capture or misrepresent the interests of other states. With such uses, there remains a significant gap in established norms or ethical frameworks governing the use of AI in warfare.

There are two camps advocating ethics of war (Orend, 2016). One is pacifists, who reject that there is any sort of morality or ethics in the war. The other camp belongs to realists, who advocate that under certain circumstances, war is justified. For this chapter, the discussion on ethics and AI will be anchored on the middle path between the two theoretical extremes. This is again because war cannot be stopped, and there has to be some baseline for any country to go out for a war, which should be completely reasonable and respecting human rights. War, when fought under certain rules and fought by sovereign states to protect their nation or to protect universal human rights, is justified. In other words, war is allowed as per *jus ad bellum*.[1] War is also justified on what has been done in the aftermath of the war – *jus post bellum* (Bass, 2004). Do these concepts aid in drafting ethical frameworks for the development and deployment of AI?

This chapter claims that the ethical frameworks for military AI are developed from the realist perspective. Even though in principle, all the states support Organization for Economic Co-operation and Development (OECD) AI principles in developing AI governance frameworks, when it comes to the military applications, they have a different approach. Irrespective of the approaches, states worldwide are yet to securitize the information flows, where AI has a great advantage of creating unprecedented political consequences. For example, in the 2023 Argentinian presidential election campaign, AI was used to manage social media accounts to create content favoring the Libertarian Party candidate, resonating with majority support sentiment. Perhaps a realistic perspective of the state should be combined with the critical securitization approaches to draft a new ethical framework for military AI.

AI, THE MILITARY, AND THE UNPRECEDENTED CONSEQUENCES

Post fourth industrial revolution, AI and other digital technologies brought what Japan calls Society 5.0, combining the cyber and physical spaces (Council for Science, Technology and Innovation, 2015). This impact is found in all industries and professions, including the military. The share of AI in the military market was considered to be ~$9.2 billion in 2023 and is increasing at the year-over-year rate of 33% (Gujar, 2024). It is expected that by 2028, AI usage value in the military will pass the figure of $38.8 billion (Gujar, 2024).

Globally, militaries are integrating AI in (a) information processing, (b) unmanned weaponry, and (c) decision-making (Rashid, 2023). The US, the largest military power of the world, was first to run comprehensive experiments to harness the potential of AI in their military operations. As per the US Department of Defense, Comprehensive Heuristic Utility for Combat Knowledge (CHUCK) developed by Air University's Innovation Accelerator (AUiX) is in use to train war pilots for tactical programs. Similarly, Non-classified Internet Protocol Generative Pre-training Transformer

(NIPRGPT) developed by Air Force Research Laboratory is used to improve strategic analysis and operational planning through advanced data integration. A quick look at the Defense Advanced Research Projects Agency (DARPA) website reveals their ambitious AI projects.

Integrating AI into unmanned weaponry is the latest domain where the US Department of Defense wants to expand in the near future. Currently, Loyal Wingman – a proposed unmanned combat air vehicle (UCAV) – incorporates AI for its operations and is capable of collaborating with next-generation manned combat aircrafts, i.e. sixth-generation fighters. Similarly, European aerospace manufacturer Airbus proposed to develop AI empowered Airbus Wingman UCAV as a force multiplier to their air capabilities.

On the other hand, the Chinese People's Liberation Army (PLA) extended the use of AI from conventional areas to a whole new domain of "intelligent warfare" (Pollpeter et al., 2021).[2] This warfare, with the help of AI, has the capacity to target the thought process of adversaries, primarily resulting in disoriented narratives, neutralizing potential strategic threats as well as gaining strategic dominance in the military planning environment. Russia also resorts to "cognitive warfare,"[3] including military, state-sponsored initiatives, and private industry. One research reported that Russia used AI in approximately every sector of defense establishment harnessing intelligence gathering: command, control, communications, computers, intelligence, surveillance and reconnaissance (C4ISR) as well as logistics optimization and hybridization of autonomous weapon systems (Das, 2023).

The usage of AI from traditional to non-traditional warfare shows that the risk involved in cumulative use of AI is wider. At least, it shows that there are unprecedented uses of AI for the military, for which there is no existing governance framework. Moreover, the increasing usage of AI in military systems will expose voids in their ethical protocols and standards, as the methods used by Russia and China show that the intended targets are unarmed individuals especially when the usage of AI is understood from the vantage of sharp power.

Dangers of Unregulated Military AI and the Necessity of Ethical Frameworks

In the case of traditional military usage, the decision-making processes that AI systems enable can lead to unintended consequences, including inadvertent escalation in military engagements, which necessitates careful consideration of the moral responsibilities associated with their deployment and operation (Trysilo & Danks, 2024). This decision-making precision effort can be observed on multiple instances around the world. One such example is the US strategic command's Global Eye defense system which uses Phantom missile launches. AI algorithms detected a delay by scanning gigabytes of real-time data every second to eliminate system anomalies in autonomous trigger launches. The crux is that any adversarial or misinterpreted data input will be able to exploit AI systems, triggering havoc in the specified tasks on the foray. These unintended consequences will certainly trick any state-of-the-art AI system to destroy its own capabilities and worth for the developers as well as for the country.

On the non-traditional side of national security lies the psychological control of public behavior, knowledge systems, and eventually the threat of hegemony as described by Gramsci (Sarkar, 2024) This is significant because the public perception is formed from the biased and skewed information that is floating across the web. AI systems can also be used to influence the thought process and create new agencies that control the way the knowledge systems are created. This is something Harari explains in his book *Nexus*. He argues that with the growing flow and size of accessible information, one must not be naïve and think that it empowers people to follow the facts of history. With the AI tool becoming more and more common, entities (states/companies) that control the information will also shape the narratives and discourses, aiding in the rise of totalitarian regimes (Harari, 2024).

Because of the highlighted consequences, stakeholders involved in the AI system lifecycle, such as developers, researchers, and military decision-makers, must acknowledge the potential for these technologies to be used in ways that extend beyond their initial intended applications (Morgan et al., 2020). This recognition is critical as AI systems possess a unique capacity for crossover applications, meaning that the risks associated with their use in military contexts may not align with traditional analyses of dual-use technologies, thereby heightening the ethical burden on those who create and implement these.

UNDERSTANDING AI ETHICS AND THE MILITARY

Some of the major factors that make up the AI ethical frameworks from a military perspective are presented in the Table 12.1. To provide a practical understanding on the importance of these factors, they have been mapped with the use cases worldwide.

TABLE 12.1
Ethical Factors and Their Uses in the Military

S. No.	Ethical Factors	Explanation	Use-Cases/ Examples
1.	Inference	Ethical AI requires more than just flaw detection; it demands transparency in how decisions are reached—explanation capabilities are critical for evaluation.	LIDAR, UV Cameras integrate their observations to decide how and when to activate sensors. The main issue is transparency—how AI combines inputs to reach conclusions can be opaque, making it difficult to understand or challenge decisions.
2.	Planning	AI can complement human errors & overcome limitations in assessing large datasets to logically generate action plans through complex data processing techniques.	Military logistics will have manifold gains by optimizing routes for supplies and managing timely intervention in the time of crisis.
3.	Bias (Emery, 2021)	AI systems can perpetuate human biases. Statistical sampling of operational needs should minimize unintended engagement.	Facial recognition models may reinforce biases, leading to excessive deployment in tracking adversary assets.
4.	Human Rights and Privacy	Military AI could violate the basic rights of humans like privacy, speech and expression in principle while deciding a strategy	Clearview AI, PredPol, and Project Maven faced backlash for privacy and human rights violation (Malmio, 2023).
5.	Accountability	Machines cannot be penalized for errors, raising concerns about accountability and the ethical implications of autonomous decision-making.	The Collection and Monitoring via Planning for Active Situational Scenarios (COMPAS) criminal justice system was found unaccountable in a legal case involving Afro-American defendants (Humerick, 2020).
6.	Replacement	AI deployment should aim for minimal human replacement in military fields, prioritizing human well-being and preserving human roles in operations.	DARPA's "In the Moment" AI system proposed reducing the presence of human commanders in high-stress battlefield environments.
7.	Automated Ethical Positioning (Rowe, 2022)	AI should be programmed with established ethical principles to create its own systems that avoid threats to civilians and non-combatants.	Currently, there are no military applications; however, in image recognition, AutoML developed NASNet, which surpassed human-designed models in performance.
8.	Mis-use of military AI (Etzionoi & Etzioni, 2017)	Insufficient human oversight could lead to the misuse of military AI, resulting in misinterpretations of the datasets used.	No confirmed instances have been found.

Source: Authors.

The ethical implications of AI in military applications are critical, as they touch on fundamental issues of accountability, transparency, human rights, and operational effectiveness. As AI systems are integrated into military processes, the need for transparency and inference capabilities becomes paramount. Without clear insight into how these systems reach their decisions, operators cannot trust their reliability, which poses risks in life-and-death situations. For example, LIDAR and UV cameras used in sensor-based decision systems often do not provide adequate transparency into how they process and integrate the data they receive. This lack of transparency could lead to decisions that are difficult to challenge or correct, raising the stakes when such systems are deployed in critical combat scenarios. Though it is not technically feasible to explain exclusively how AI reaches a conclusion, it is suggested to have clear documentation of the variables provided during the training of the AI model and also during the real-time deployment, adding up to the practice of algorithmic transparency.

When it comes to military strategy, absence of human intuition and adaptability can result in flawed responses to complex and unpredictable situations (Hecht, 2022).[4] For the planning of logistics, AI's reliance on data-driven models to optimize routes or predict supply needs, while useful, can overlook local conditions or unanticipated variables, leading to logistical failures. In high-pressure military environments, such oversights can have severe consequences.

Bias is another pressing concern, particularly in facial recognition systems. For instance, Clearview AI® has been criticized for its disproportionate focus on certain racial or ethnic groups due to the biased nature of its training data. It is also alleged that Clearview AI® monitors the behavior of people and uses such analysis for business. It was fined USD 10 million for scrapping UK citizen's faces online (Heikkila, 2022). In military contexts, this bias could result in the wrongful targeting of non-combatants or exacerbate tensions by disproportionately deploying resources based on flawed AI decisions. The disproportionate use of AI to monitor citizens and people around the world will also negate the underlying trust between the public and the state. This can potentially damage the existing domestic social contract, which is a hypothetical agreement between the public and state, where the latter provides security in exchange for freedom.

Accountability presents one of the most profound ethical challenges. The COMPAS® system, used in criminal justice, has faced heavy criticism for its racial bias against African Americans, especially in assessing recidivism risks (Angwin et al., 2016). The inability to hold the AI system accountable for its biased decisions highlights the larger issue: when AI makes a mistake in the military, such as misidentifying a target, it's unclear who should bear responsibility. The absence of accountability mechanisms raises serious concerns about justice and fairness, especially when AI errors can lead to loss of life or violations of international law.

AI's replacement of human decision-makers is another issue. DARPA's "In the Moment (ITM)" AI system, designed to reduce the cognitive burden on commanders in high-stress situations, suggests reducing human involvement in battlefield decisions (DARPA, n.d.). However, this raises ethical concerns about the dehumanization of war. AI, lacking the emotional and moral judgment that humans possess, might make decisions that, while strategically sound, fail to account for the ethical complexities of warfare, such as collateral damage or the protection of civilians.

Although not yet used in military operations, these technologies outperform humans in image recognition tasks, complex analysis given numerous variables, raising concerns about how autonomous systems could evolve beyond human ethical oversight. This creates a scenario where AI could operate without clear human control or ethical guidelines, potentially leading to unintended consequences in military contexts. The argument is not made with the presumption that AI will become a singularity. Such assertions are futile attempts to predict the complete loss of agency to AI and total automation of machines. Lack of clear human control here means the absence of legal frameworks defining the liabilities, and outcome-based technical boundaries for AI operation. Considering complex an adaptive systems approach/socio-technical approach in designing AI systems could open up a path for explainable AI, where the risks are understood by all the stakeholders, before developing and deploying AI.

Finally, the potential for misuse of military AI—whether through misinterpretation of data or insufficient human oversight—presents another significant ethical concern. While no confirmed cases of misuse have been reported, the risk remains high. It was reported in an article published by the *IEEE Spectrum* (Institute of Electrical and Electronics Engineers magazine) that a group of researchers could easily make an AI algorithm which makes chemical weapons (Boulanin & Charles, 2022). Another example is that, without careful oversight, an AI system might mistakenly identify civilian infrastructure as a military target based on flawed data interpretation. The errors committed by AI systems, if not caught in time, could lead to unnecessary destruction and loss of life, violating the principles of proportionality and distinction under international humanitarian law. These risks underscore the need for robust regulatory frameworks and continuous monitoring to ensure that AI is used ethically in military operations. However, it is highly likely that AI systems, provided they are inclusively and contextually trained with appropriate guardrails, would be more efficient than humans. This would again be feasible if the deployed AI system has clearly defined legal liabilities and audit frameworks, which are yet to be designed. It should be noted that artefacts cannot be held accountable, only legal entities. Proper mechanisms should be in place to identify and apportion the human liability of AI decision making.

The ethical factors outlined in Table 12.1 represent not only theoretical challenges but real-world issues that must be carefully addressed. The examples of Clearview AI®, Project Maven®, and COMPAS® illustrate the consequences of deploying AI without sufficient ethical safeguards. As AI technologies continue to advance, ensuring their ethical use in military contexts is crucial to balancing operational effectiveness with the protection of human rights and adherence to international legal standards.

NON-TRADITIONAL SECURITY AND AI ETHICS

In the previous section, there was a brief mention about how the domestic social contract will be challenged because of unregulated AI. It might occur that protecting the facts and controlling the misinformation is not the role of military. However, if the mission statement of the military is observed, its obligation is to serve a nation's sovereignty, territorial integrity, and interests. This function reflects a philosophical commitment to the idea of a nation-state, where the military acts as a guardian of collective identity and shared values. This reiterates that the military's primary mission is not merely to engage in combat but to preserve peace and security, ensuring that the nation can thrive without external threats, standing on its own values and cultures. That means, it must be ready for information warfare.

AI tools are capable to spread misinformation which is politically motivated. A notable example of this is the use of AI-driven information warfare in political contexts. Governments and actors with vested interests have leveraged AI to amplify disinformation campaigns, such as the notorious 2016 US election interference, where AI-enabled bots flooded social media platforms with divisive content, distorting public opinion (Damplo, 2020). Similarly, during the 2022 French presidential election, AI was used to spread deepfakes and fabricated news stories to undermine political rivals (Henin, 2022). Some researchers assert that though AI has been used to disrupt elections, there is no significant impact on the outcome or decision making of the public (Łabuz & Nehring, 2024). However, the impact of AI on society cannot be accurately identified if the risk-mitigation process is not initiated earlier. The mentioned examples highlight the potential of AI to weaponize information for political gain, making it increasingly difficult for the public to discern truth from manipulation. In this context, AI doesn't just amplify the spread of information, it empowers those who control it to shape the political landscape and perpetuate misinformation at scale. With AI tools becoming more and more common, entities (states/companies) that control the information will also shape the narratives and discourses, aiding in the rise of totalitarian regimes (Harari, 2024). Here comes the role of the state to securitize the information and the tools that manage the information flow. As much as it is the role of the state, militaries should have a decent understanding of this threat and develop the capacities to defend the nation against misuse of AI.

THE PRAGMATIC APPROACH OF DESIGNING AI ETHICS

The mentioned factors of an AI ethical framework are universally accepted. However, only the US, North Atlantic Treaty Organisation (NATO), and China have published ethical frameworks for the AI development in the military. While the US military has no official code of ethics, the Department of Defense has announced that it develops military AI following AI ethics principles – reliable, equitable, traceable, responsible, and governable principles (US Department of Defense, 2020). Table 12.2 provides a quick glance at the global scenario of ethical frameworks. Since the US is the front runner among the democracies, the ethical factors mentioned in its framework have been taken as an index to compare the other select countries. Following are the factors that make up the US military ethical framework.

1. ***Responsible:*** Department of Defense (DoD) personnel will exercise appropriate levels of judgment and care, while remaining responsible for the development, deployment, and use of AI capabilities.
2. ***Equitable:*** The Department will take deliberate steps to minimize unintended bias in AI capabilities.
3. ***Traceable:*** The Department's AI capabilities will be developed and deployed such that relevant personnel possess an appropriate understanding of the technology, development processes, and operational methods applicable to AI capabilities, including with transparent and auditable methodologies, data sources, and design procedure and documentation.
4. ***Reliable:*** The Department's AI capabilities will have explicit, well-defined uses, and the safety, security, and effectiveness of such capabilities will be subject to testing and assurance within those defined uses across their entire life-cycles.
5. ***Governable:*** The Department will design and engineer AI capabilities to fulfil their intended functions while possessing the ability to detect and avoid unintended consequences, and the ability to disengage or deactivate deployed systems that demonstrate unintended behavior.

Though Canada has no official military AI ethics, it asserts that the development and adoption of AI in the military will adhere to the existing laws such as Voluntary Code of Conduct on the Responsible Development and Management of Advanced Generative AI Systems, Guide on the use of Generative AI, and Directive on Automated Decision-Making (Canadian Armed Forces, 2024). The voluntary code asserts that all the developers and managers of advanced generative systems should adhere to the principles of accountability, safety, fairness and equity, transparency, human

TABLE 12.2
Comparison of US Principles of AI Ethics in Military to Other Countries

AI Ethical Principles identified from the US	United States (US Department of Defense, 2020)	NATO (NATO, 2021)	Canada	India	Brazil	China (Metcalf, 2022)	Russia	Türkiye
Responsible	Yes	Yes	NA	NA	NA	Yes	NA	NA
Reliable	Yes	Yes	NA	NA	NA	Yes	NA	NA
Equitable	Yes	Yes	NA	NA	NA	No	NA	NA
Traceable	Yes	Yes	NA	NA	NA	Yes	NA	NA
Governable	Yes	Yes	NA	NA	NA	Yes	NA	NA
Bias mitigation	Yes	Yes	NA	NA	NA	Yes	NA	NA

Source: Authors.

oversight and monitoring, validity, and robustness (Innovation, Science, and Economic Development Canada, 2023). However, the latter ethical principles are not different from those of NATO and the US. China's approach toward ethical usage of AI also aligns in principle with the global ones.

It is interesting to note that China has not mentioned the "equitable" factor in its ethical framework. One of the mission statements of People's Liberation Army is to protect the Chinese Communist Party (CCP) and uphold Communist Party principles. This shows the difference in priorities when it comes to China and the US. Similar is the drafting of ethical frameworks. However, it is interesting to note that the CCP is aware and takes cognizance of the fact that virtual simulations will hamper the on-field judgements of soldiers (Metcalf, 2022), which puts China ahead among other countries in recognizing human-oriented ethical policies.

Even though countries agree in principle to the ideals of AI ethics or principles in the multilateral forums, there will be stark differences in the way they use it and the way they account for it. Recently, the Responsible AI in the Military Domain (REAIM) 2024 summit has seen partial agreement from 60 countries, and among those, 30 countries have not endorsed that the military AI systems should undergo legal reviews and should be deployed in accordance with the international laws (Rosen, 2024). The difference is perhaps because of the hegemonic nature of the US, or even because of the fact that ethics are not universal but are situated culturally, historically, contextually, and also on the decision-making entity (here the political party or the head of the state).

AI ethics are profoundly shaped by societal context, culture, and historical backgrounds. The idea of "agency" in AI, for instance, differs greatly depending on the environment in which it operates. Machine agency isn't inherent; rather, it reflects the societal context in which AI is deployed (Bryson, 2020). For example, as individuals use social media as a constant platform for interaction and information, their control over information consumption diminishes, effectively sharing agency with AI systems. This shared agency influences people's perceptions of the world, subtly affecting their decision making. Yet, philosophically, AI lacks true agency, as it operates without intentions or independent reasoning. This creates complex ethical issues, especially in military contexts, where AI governance must draw clear lines between civilian and military applications within national borders.

Cultural perspectives further complicate AI ethics, as seen with facial recognition technologies. For instance, China and the EU embody contrasting approaches: the EU prioritizes individual privacy, while China emphasizes societal stability. These cultural and historical perspectives inform each region's legislation, leading to diverse standards for responsible AI development worldwide.

Some scholars advocate for a socio-technical systems approach to define AI's ethical limits (Powers & Jean, 2020), arguing that AI development is inherently influenced by the social and cultural contexts of its creators. This view suggests that incorporating societal feedback mechanisms to adapt AI behavior could be more effective than relying on universal norms and declarations as ethical standards for AI development. There is still on-going research to provide practically implementable ethical frameworks for AI development. Such frameworks are to be developed keeping in mind the quickly changing AI market, discreet and discrete adoption of AI by the militaries, and also the geopolitical landscape shaped by the AI.

To reassure that the ethical principles of states are based on the country's position and aspirations within the geopolitical realities, the nuances within the approaches considered by various states and regions (India, Canada, Brazil, United States, China, European Union, Russia, and Türkiye) have been studied and compared here.[5]

All the AI strategies[6] more or less follow OECD principles in the development of AI. This is because OECD principles emphasize the ideal scenario of AI development and implementation to which no country would say no. The aim of such principles is to provide a foundational base on which the AI policies will be drafted. However, such foundational principles are not reflected either in the US military AI ethics, or those of Canada or China. OECD principles like inclusivity and democratic values are not reflected in the latter mentioned military code on AI ethics. Even among the countries that are under comparison in Table 12.3, only the AI strategies of Brazil and Türkiye

TABLE 12.3
Indicative Developments of AI Ethics at the Multi-Lateral Level

Date	Country/ Organization	Framework/ Resolution	Key Principles/Actions	Current Developments/Status
2024	United Nations	First resolution on AI by UN General Assembly	Ensures states respect human rights in AI development and deployment	Ongoing discussions on implementation, with the UN advisory body working on practical guidelines for member states.
2024	African Union (African Union, 2024)	Malabo Convention on Continental Artificial Intelligence Strategy	Africa's commitment to an Africa-centric development focused AI approach, promoting ethical, responsible, and equitable practices across the continent.	African Union endorsed continental strategy in Feb 2025 and decided to use this as a foundational framework to develop national AI strategies across the member states.
2024	European Union	EU Artificial Intelligence Act	Comprehensive legal framework for AI, focusing on risk-based regulation	The Act will come into force, with various provisions becoming applicable between 2025 and 2027.
2023	ASEAN	Guide on AI Governance and Ethics	Framework to enhance AI governance and ethical considerations among member state	Ongoing implementation discussions among member states to harmonize AI governance practices.
2023	Latin America	Santiago Declaration on AI	Commitment to responsible AI development, emphasizing human rights and ethical considerations	Countries in the region are working on aligning their AI policies with the principles outlined in the declaration.
2023	International Telecommunication Union (ITU)	AI for Good Global Summit collaboration	Collaboration among 36 UN agencies to promote the ethical use of AI	Continued initiatives to address global AI ethics challenges, focusing on multi-stakeholder dialogues.
2021	UNESCO	Recommendation on the Ethics of AI	Ten principles for responsible AI: transparency, non-discrimination, human autonomy, prevention of harm, etc.	Framework being adopted by various countries, with ongoing discussions about integrating these principles into national policies.
2020	OECD	Responsible AI Principles	Guidelines for promoting ethical AI development and use	Countries are actively integrating OECD principles into their national AI strategies.
2020	G7	Principles and guidelines for AI	Emphasis on ethical and responsible AI applications	G7 nations are collaborating on aligning their AI strategies with these principles.
2020	G20	Principles and guidelines for AI	Similar to G7, focusing on ethical AI development and international collaboration	G20 discussions continue to evolve, with member countries sharing best practices in AI ethics.

Source: authors.

explicitly mention the OECD principles. Türkiye even asserts that it will actively participate in international discussions and studies on responsible AI and support organizations that coordinate with the international organizations (Digital Transformation Office, 2021). One possible inference from this is that perhaps, the low- and middle-income countries (LMICs) are taking the liberal approach toward the AI development within the military, which are the borrowed concepts from

Europe. Since, they are late to catch up with the technological race, they would be better off to reduce the risks and dangers of the AI.

Accordingly, Türkiye asserts that it will establish "sectoral co-creation laboratories" having technological and non-technological domains on the same table while developing the AI systems. Basically, their action plan is to put ethics, law, social sciences, national strategy, corporates, small and medium enterprises (SMEs), governance frameworks, and technological upgradation in a single basket to derive inclusive and sustainable AI development. This strategy will allow sectoral laboratories to account for sectoral norms and ethics of research which might also seep into the AI design. Similarly, India's discussion paper on AI strategy also emphasizes establishing a consortium of ethics councils at each proposed Center of Research Excellence (CoRE) to standardize the ethics code in AI development (NITI Aayog, 2018).

GLOBAL DEVELOPMENTS ON AI ETHICS

The current landscape of AI ethics frameworks, even at the multilateral level, reveals a significant gap in addressing military applications, despite the growing recognition of ethical concerns surrounding AI technologies. While various organizations, such as the United Nations, UNESCO, and the European Union, have initiated frameworks focusing on ethical AI development, these largely emphasize civilian applications and societal implications. The ethical considerations specific to military AI, particularly regarding lethal autonomous systems, remain underexplored. For instance, the US Department of Defense has made strides in establishing ethical principles for military AI, yet these efforts are not mirrored on a broader multilateral scale, leaving a void in collaborative ethical standards for military applications of AI.

The absence of a cohesive international framework raises concerns about interoperability and accountability among allied forces, which could potentially lead to ethical discrepancies in military operations. As military AI technologies advance, the need for a comprehensive ethical framework that encompasses the unique challenges of warfare – such as decision making in lethal scenarios and the implications of delegating life-and-death decisions to machines – becomes increasingly urgent. Without a concerted effort to integrate military considerations into existing AI ethics frameworks, the risk of unethical practices and operational failures in military contexts will likely persist, undermining both ethical integrity and operational effectiveness in modern warfare.

In addition to the latter concerns, the most pressing and underexplored one is information warfare. Militaries currently might not see that as their core domain, but AI led information warfare can be waged by combatant and non-combatants.

CONCLUSION: WHAT COULD BE THE WAY FORWARD?

The integration of AI into the military should be done with a proper ex-ante regulation, including an ethical framework in place. Such a framework should account for both traditional and non-traditional security aspects. With the emerging threats like information warfare, countries probably should consider establishing dedicated AI-led military units within military organizations. Apart from leveraging AI's capabilities in modern warfare, they need to address emerging security challenges. The functions of such new AI military units can be divided into four categories as shown in Figure 12.1.

The new AI military unit should have its own R&D, fostering military applications in collaboration with tech firms as well as academic institutions. It will not only expand their areas of operations but simultaneously lead them to create cutting-edge solutions. This is a similar approach to that of Türkiye's sectoral co-creation laboratories. In some cases, these military innovations will also serve civilian purposes greatly. For example, the global positioning system (GPS), microwave ovens, jet engines, and the internet all were developed to serve military purposes only and later were adopted and now ubiquitous among society. This societal element will nudge military AIs to be more responsible, ethical, and in some cases moral in their operational behavior. It will benefit

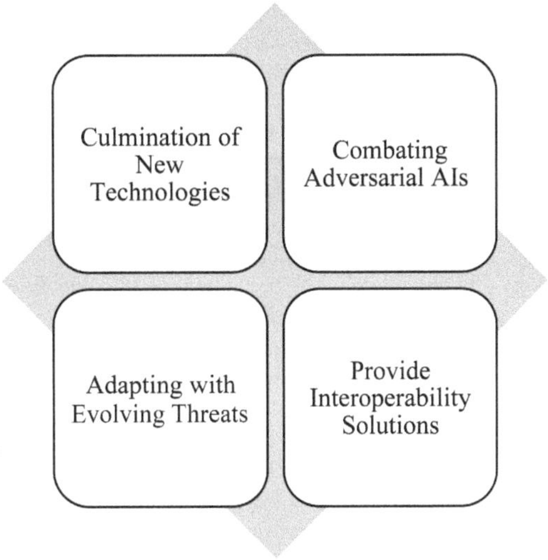

FIGURE 12.1 Key focus areas for future innovation in AI and Cybersecurity. (Source: Authors.)

the non-traditional security matrix greatly to tackle alien challenges from their enemies. Both internal and external adversaries provide valuable insights for military AI, helping it to develop and strengthen its capabilities to operate without challenge.

Even in the conditionalities of governance and ethical considerations, military AI wings will delve into their own generated and mended ethical guidelines for autonomous systems in combat scenarios. These self-generated guidelines will ensure compliance with international standards, keeping humanitarian borders in mind.

NOTES

1. *Jus ad bellum* refers to the set of criteria that determine the justifications for a state to engage in war. It was coined during the World Wars and gained significance after the UN incorporated it in its charter. This concept is a fundamental aspect of just war theory, which traditionally divides the ethics of warfare into two main categories: *jus ad bellum* (the justification for going to war) and *jus in bello* (the conduct of war).
2. PLA's intelligent warfare is an initiative to conceptualize the future battlefield environment dominated by AI and autonomous weapon systems. It portends a new revolution in military affairs.
3. Cognitive warfare includes activities conducted in synchronization with other instruments of power to affect attitudes and behaviors by influencing, protecting, or disrupting individual group- or population-level cognition to gain an advantage over an adversary.
4. One such incident was observed during the conflict between Armenia and Azerbaijan, where reports suggested that both sides used AI-driven drones for surveillance and strikes. There were instances where autonomous drones misidentified targets, leading to unintended damage and casualties.
5. NVivo 14 software has been used to conduct thematic analysis on the AI strategies. The code book will be provided on request.
6. Here AI strategies are drafted by the governments and not specifically for militaries.

REFERENCES

African Union. (2024). *Continental artificial intelligence strategy.* https://au.int/sites/default/files/documents/44004-doc-EN-_Continental_AI_Strategy_July_2024.pdf

Angwin, J., Jeff, L., Surya, M., & Lauren, K. (2016, May 23). *Machine Bias There's software used across the country to predict future criminals. And it's biased against blacks.* ProPublica. https://www.propublica.org/article/machine-bias-risk-assessments-in-criminal-sentencing

Bass, G. J. (2004). Jus post bellum. *Philosophy and Public Affairs*, *32*(4), 384–412.

Boulanin, V., & Charles, O. (2022, August 27). *Civilian AI is already being misused by the bad guys.* IEEE Spectrum. https://spectrum.ieee.org/responsible-ai-threat

Bryson, J. J. (2020). The artificial intelligence of the ethics of artificial intelligence: An introductory overview of law and regulation. In M. D. Dubber & F. D. Pasquale (Eds.), *The Oxford hand book of ethics of AI* (pp. 3–27). Oxford University Press.

Canadian Armed Forces. (2024, March 22). *The department of national defence and Canadian armed forces artificial intelligence strategy.* https://www.canada.ca/en/department-national-defence/corporate/reports-publications/dnd-caf-artificial-intelligence-strategy.html

Council for Science, Technology and Innovation. (2015, December 18). *The 5th science and technology basic plan.* https://www8.cao.go.jp/cstp/kihonkeikaku/5basicplan_en.pdf

Damplo, D. (2020). *Using digital technology to systematically violate the right to vote: How the Donald Trump's 2016 campaign used AI to suppress the votes of Black Americans.* Universal Rights. https://www.universal-rights.org/using-digital-technology-to-systematically-violate-the-right-to-vote-how-the-donald-trumps-2016-campaign-used-ai-to-suppress-the-votes-of-black-americans/

DARPA. (n.d.). *ITM: In the moment.* Darpa.mil: https://www.darpa.mil/program/in-the-moment

Das, S. (2023). *Evolution of AI in modern military warfare.* Aranca. https://www.aranca.com/knowledge-library/articles/ip-research/evolution-of-ai-in-modern-military-warfare

Digital Transformation Office. (2021). *National artificial intelligence strategy 2021–2025.* https://www.oecd-events.org/smart-data-and-digital-technology-in-education/session/79d499e6-f200-ed11-b47a-a04a5e7cf9da/national-artificial-intelligence-strategy-2021-2025-turkiye

Emery, R. J. (2021). Algorithms, AI, and ethics of war. *Peace Review*, *33*(2), 205–212.

Etzionoi, A., & Etzioni, O. (2017). Pros and cons of autonomous weapon systems. *Military Review.*

Fawkes, A. J. (2017). Developments in artificial intelligence – opportunities and challenges for military modeling and simulation. *Proceedings of the 2017 NATO M&S symposium*, 11, pp. 1–12. NATO Science and Technology Organisation.

Giving Tuesday. (2024). *AI readiness and adoption in the nonprofit sector in 2024.* Giving Tuesday.

Gujar, S. (2024). *AI in military market size, industry outlook report, regional analysis, application development, price trends, competitive market share & forecast, 2024–2032.* Global Market Insights.

Harari, Y. N. (2024). *Nexus: A brief history of information networks from the Stone Age to AI.* Random House.

Hecht, E. (2022). Drones in the Nagorno-Karabakh war: Analyzing the data. *Military Strategy Magazine*, *7*, 31–37.

Heikkila, M. (2022, May 24). *The walls are closing in on Clearview AI.* MIT Technology Review. https://www.technologyreview.com/2022/05/24/1052653/clearview-ai-data-privacy-uk/

Henin, N. (2022). *What did disinformation look like during 2022 French presidential election? An overview based on fact checking articles.* EU DisinfoLab. https://www.disinfo.eu/publications/what-did-disinformation-look-like-during-the-2022-french-presidential-election-an-overview-based-on-fact-checking-articles/

Humerick, J. (2020). *Reprogramming fairness: Affirmative action in algorithmic criminal sentencing.* Columbia Human Rights Law Review.

Huntington, S. P. (1996). *The clash of civilisations and the remaking of the world order.* Simon & Schuster.

Innovation, Science, and Economic Development Canada. (2023). *Voluntary code of conduct on the responsible development and management of advanced generative AI systems.* https://ised-isde.canada.ca/site/ised/en/voluntary-code-conduct-responsible-development-and-management-advanced-generative-ai-systems

Kaplan, D. R. (2012). *The revenge of geography.* Randon House.

Łabuz, M., & Nehring, C. (2024). On the way to deep fake democracy? Deep fakes in election campaigns in 2023. *European Political Science*, *23*, 454–473.

Malhotra, R. (2021). *Artificial intelligence and the future of the Power: 5 Battle grounds.* Rupa Publications.

Malmio, I. (2023). Ethics as an enabler and a constraint—Narratives on technology development and artificial intelligence in military affairs through the case of Project Maven. *Technology in Society*, *72*, 102193.

Metcalf, M. (2022). *The PRC considers military AI ethics: Can autonomy be trusted?* Frontiers Big Data.

Mildner, S.-A., Lauster, G., & Wodn, W. (2011). Scarcity and abundance revisited: A literature review on natural resources and conflict. *International Journal of Conflict and Violence*, *5*(1), 155–172.

Morgan, E. F., Benjamin, B., Andrewj, L., Mark, A., Christian, C., Kelly, K., & Derek, G. (2020). *Military applications of artificial intelligence: Ethical concerns in an uncertain world.* RAND.

NATO. (2021). *Summary of the NATO artificial intelligence strategy.* https://www.nato.int/cps/en/natohq/official_texts_187617.htm

NITI Aayog (2018). *National strategy for artificial intelligence.* NITI Aayog.

Orend, B. (2016). War. *Stanford encyclopedia of philosophy.* Stanford University. https://plato.stanford.edu/entries/war/

Pollpeter, K., Kerrigan, A., & Ilachinski, A. (2021). *The PLA and intelligent warfare: A preliminary analysis.* Center for Naval Analysis.

Powers, M. T., & Jean, G. G. (2020). The ethics of the ethics of AI. In M. D. Dubber, F. Pasquale, & S. Das (Eds.), *The Oxford handbook of ethics of AI* (pp. 27–53). Oxford University Press.

Rashid, A. B. (2023). Artificial intelligence in the military: An overview of the capabilities, applications, and challenges. *International Journal of Intelligent Systems, 2023*(1). https://doi.org/10.1155/2023/8676366

REAIM. (2024). *Full script: REAIM blueprint for action.* https://thereadable.co/full-script-reaim-blueprint-for-action/

Rosen, B. (2024). *From principles to action: charting a path for military AI governance.* Carnegie Council. https://www.carnegiecouncil.org/media/article/principles-action-military-ai-governance

Rowe, N. C. (2022). *The computer ethics of artificial intelligence methods for military applications.* Frontiers Big Data.

Sarkar, N. (2024). Understanding hegemony: A critical analysis through the Lens of Antonio Gramsci. *International Journal for Multidisciplinary Research (IJFMR),* 1–6.

Trysilo, D., & Danks, D. (2024). Commercial AI, conflict, and moral responsibility: A theoretical analysis and practical approach to the moral responsibilities associated with dual use AI technology.

US Department of Defense. (2020, February 24). *DOD adopts ethical principles for artificial intelligence.* https://www.defense.gov/News/Releases/Release/Article/2091996/dod-adopts-ethical-principles-for-artificial-intelligence/

Walker, C., & Jessica, L. (2017). What is 'Sharp Power'? *Journal of Democracy, 29*(3), 9–23.

13 How Will Ethically Correct Religious Robots Be Built?

Selmer Bringsjord

INTRODUCTION

Many human families the world over can be accurately classified as *R*, where *R* is a variable ranging over an adjective corresponding to a major religion. For example, there are: Hindu families, Jewish families, Jain families, Muslim families, Christian families, and so on.[1] We can thus speak generally of *R*-families. Inevitably, as AI and cognitive-robotics technology continues to advance, many *R*-families will employ household cognitive robots (under a particular, established definition of *cognitive robot* as entailing that the robot is logic-based, soon to be explained) whose capabilities approach those of humans—for example: clean; cook and serve; communicate at the human level (a requirement for robust use of other capacities in the present list); teach/tutor; protect; dress children, the elderly, the infirm, and care for such groups in various ways; transport; and entertain. Naturally, for a smoothly operating household, these will need to be *R*-robots, *and* they will need to be ethically correct, in rather nuanced ways. This is something that I noted rather long ago, in a first, logic-based foray into the intersection of religion and machine ethics (Bringsjord & Taylor, 2012).

When I label a human or human household as religious, for some religion *R*, I don't mean to denote anything that is theologically or philosophically intricate and profound. I simply refer to human agents that sometimes willfully reason, and make some premeditated, consequential decisions based on this reasoning, where propositions used in the reasoning are themselves clearly part and parcel of these agents' religious beliefs. An orthodox Jew who decides as such to refrain from eating lobster, a Muslim who decides as such to avoid consumption of alcohol, a Roman Catholic who as such during Lent elects to avoid eating meat, a Hindu who as such will not eat beef—these are all for purposes of the present chapter religious humans. And note that to say that some human agent is an *R* one, for some *R*, does not in any way entail that agent makes no use of purely secular elements in its reasoning and decision-making. The orthodox Jew when faced with an option of eating two items, one of which is taken to be prohibited, and the other of which is taken to be perfectly permissible, will make use of the cornerstone inference schema of *disjunctive syllogism* to opt for the second—but this schema is used every day by every neurobiologically normal adult human agent, whether atheist or believer, East-Asian or American.[2] In addition, a religious human agent in my everyday sense in no way excludes the important role that content plays in that reasoning from narratives, art, and so on. In the Christian case, for instance, reasoning and decision-making expressly based on stories is at the heart of the approach that Jesus used and taught via his parables.

Next let's take two preparatory steps. The first is to simply note that just as we have, in my sense, religious human agents, we are going to need religious *artificial* agents, specifically *R*-robots, in the employ of those humans. Secondly, to ease exposition and tractability, let's place our focus on "mere" food preparation and serving within a given household. (Clearly, it simply isn't feasible in a single chapter to discuss the broad diversity of nuances that an ethically correct cognitive robot would need to handle for different major religions.)

Now to our first scenario: If a devout Jewish family's cognitive robot accedes to a request from the children to cook a lobster dinner, when the parents are out for the night, the default position would be that it is operating unethically (in violation of the kosher prohibition against eating shellfish). But note that I here say "for starters." In prior work whose driving question **Q** was less

DOI: 10.1201/9781003585527-17

practical than this one, **Q** was posed as: "What kind of formal semantics is required for human-level natural-language understanding (NLU) in competent household cognitive robots?" (Bringsjord et al., 2022b). I cursorily considered the shellfish problem. This consideration was in fact rather naïve on my part. Why? In short, because the parents in a Jewish household might, despite both regarding themselves as Jewish = *R simpliciter*, differ in their particular positions on dietary rules, and this possibility was completely ignored.

Let's be more specific. First, for fixity, let the robot's name be "Ruben." Second, suppose that dietary divergence stems, in the particular example we are imagining, from the mother's being a conservative Jew while the father is a reform Jew; and suppose that this difference underlies the mother's strict kosher dietary code, which the father does not affirm, since he rejects Rabbinic authority. With both parents out for the night, what is the robot to do if the children argue for their claim that tonight their father's position should be what governs cooking and eating? Let's assume in particular that the children specifically claim that, while Ruben didn't hear it, father had told both children that tonight it would be okay to have shellfish. What is Ruben to do?[3] As we shall see, but as I suspect the reader well knows, there exist variations within every major religion that are analogous to those seen in this Jewish family.

In the context of situations such as the nuanced shellfish problem, the overarching question driving the present essay is this one:

Q *How will an ethically correct R-robot that matches its R-family by following the relevant R-codes in play therein be engineered?*

The fact is that religions almost invariably include *ethical codes*: collections of propositions that state, minimally, what is obligatory, prohibited, and morally neutral for human agents who are *R*; let us call these "*R*-codes."[4] Now, let $\{R_1, R_2, \ldots, R_k\}$ enumerate the set of all major religions on Earth. Each pair composed of an R_i-code and an R_j-code ($i \neq j$) will constitute an inconsistent set of propositions.[5] But worse—and again, this no one to my knowledge has seen and attempted to come to engineering grips with—even *within* a given R_i we have inconsistency. As we have just seen in our earlier parable, in the case of Judaism, there are different and incompatible culinary codes, and they can be concretized within a single Jewish family. Given the empirically obvious fact that this structure is instantiated in each of the major religions on Earth, the question driving the present essay—**Q**—is relevant across the globe.

At present, **Q** to our knowledge is unanswered; in fact, it appears that no one has even *started* to address it in earnest. But there will be no way for humanity to avoid **Q**. If Cogbots Inc. is a company that supplies household cognitive robots to *R*-families, for example a robot such as Ruben, it will obviously need to address the question. But how? This is the question with which we shall wrestle. As a preview, my position is that Cogbots Inc. will need to engineer their offerings with (1) prior knowledge before delivery of relevant religious and ethical content; (2) a capacity to adroitly acquire new knowledge and rational belief dynamically, by valid and verifiable reasoning during communication with members of the households in which they work; and (3) a capacity to reason over the knowledge and rational belief obtained to determine what to do in ethically charged situations.

COGNITIVE ROBOTICS*, ENCAPSULATED

You will have noticed that I have referred not just to robots, but to *cognitive* robots. Why? Cognitive robotics herein means something quite specific, an approach to the engineering of robots first introduced and defined in (Levesque & Lakemeyer, 2007) [and laid out for others to learn and pursue in (Levesque, 2019)]. The hallmark of this approach, put with harsh but necessary simplicity, is easy enough to encapsulate, as follows (and here to ease exposition I use a fragment of the simplified version of some of the notation I have used in my extension of cognitive robotics for many years with collaborators).[6]

Let r be an arbitrary robot, let a be some term that denotes some action (with variables therein; e.g., an example of an action might be x *grasps* y, where x is a variable for an agent, and y for a drinking glass) performable by r, and let t be some timepoint. We write

$$\mathit{happens}(\mathrm{r}, t, a)$$

to express that r carries out at timepoint t the action a. What I have written here is a simple formula (specifically, a so-called *atomic* formula) in first-order logic, a logic that is a cornerstone of AI and cognitive robotics (Russell & Norvig, 2020). In this formula, *happens* is a (ternary) property that a robot, a timepoint, and an action can have. Obviously, if such a formula holds, then the robot in question at that timepoint has performed the action.

Now, a cognitive robot is one that performs (substantive, macroscopic) actions exactly because of what it believes and knows (Levesque & Lakemeyer, 2007) at t. For example, if a particular cognitive robot r* performs the action a^h of frying a hamburger (or, more realistically, the sub-actions that together compose a plan for frying a hamburger), the action a^h is performed because of what the robot believes and/or knows. For instance, r*, let us assume, believes that some human desires to receive and consume a hamburger, and knows all sorts of propositions regarding the stove that must be used to prepare and serve a hamburger to this human, and so on. To facilitate exposition, let $\mathrm{K}_{*\mathrm{r}}^{t*}$ denote what the particular robot knows at particular timepoint t^*, and $\mathrm{B}_{\mathrm{r}}^{t*}$ denote what it believes at this time (which includes the specific belief about the human's wanting a hamburger, and knowledge about the stove and other systems in the kitchen).[7] Then, since we have here a cognitive robot, the following biconditional sums up what must hold in the particular case at hand, and conveys the hallmark of this approach to robotics:

$$\mathit{happens}(\mathrm{r}^*, t^*, a^h) \leftrightarrow \mathrm{B}_{\mathrm{r}}^{t*} \wedge \mathrm{K}_{\mathrm{r}*}^{t*}$$

Here, $\leftrightarrow$ is the familiar Boolean connective for "if and only if." Of course, this biconditional is a particular one. We need a proposition that can serve as a general principle for cognitive robotics that is in line with what Levesque has defined. This need is satisfied by the following general biconditional, inside a reference to any ($\forall$ being the universal quantifier expressing "for all" in formal logic, a key part of logic since Aristotle) robot, knowledge-base, belief-base, action, and timepoint:

$$\textbf{CR} \quad \forall \mathrm{r},\mathrm{K},\mathrm{B},a,t : [\mathit{happens}(\mathrm{r}^*, t^*, a) \leftrightarrow \mathrm{B}_{\mathrm{r}}^{t*} \wedge \mathrm{K}_{\mathrm{r}*}^{t*}]$$

Now, the brand of cognitive robotics that I have directed and pursued for decades, which will frame the analysis and discussion from this point on regarding **Q** and my proposed three-part answer to it for companies such as Cogbots Inc., is actually an expansion of **CR**. We don't need to specify this expansion in the form of a formula that further specifies the expansion of **CR**. It will suffice to say the following things.

First, a bit of notation: I refer to the expansion of **CR** as "**CR**⋆." Second, my expanded approach to cognitive robotics is based upon the idea that not only belief and knowledge are determinative of action, but that the general conception of "cognitive centrality" advanced by Levesque and Lakemeyer holds for *all* the main cognitive attitudes that we see in the human sphere. In addition, the actions performed by the cognitive robots I seek will not be allowed to be based in part on the ethical status of these actions—on whether for instance such actions are ethically obligatory (with more details along this line coming shortly).

In addition to belief and knowledge, I allow robot action to be the product of what the robot desires, intends, fears,[8] perceives, communicates, and so on—where each of these cognitive verbs can range over declarative content of considerable complexity. This means that, to sustain the specific logicist approach to robotics that is "baseline" cognitive robotics as introduced to the world (as I have noted) by Levesque and Lakemeyer, it is necessary to have on hand computational logics expressive enough to formalize all such cognitive verbs, and at least all the main human emotions, and also the content that such cognitive verbs and emotions range over. This class of logics, which in many cases are multi-valued and particularly suitable for argumentation among different agents, is to my knowledge unprecedented and singular in Occidental[9] AI and computational logic, and have been invented, specified, and implemented by me and my collaborators, in research and development that spans two decades, from circa 2006 to 2025. For three papers selected from among over 50 to mark this two-decade span, see (Arkoudas & Bringsjord, 2009; Bringsjord et al., 2006, 2024).

Harkening back to Ruben, the robot we considered above, this cognitive⋆ (note the superscripted ⋆) robot might decide to perform some actions on the basis of its belief that as a matter of fact it's false that the father intends that for the evening in question, a reform-Jewish code is to be in force, rather than one based on conservative Judaism.

And there is another crucial part of the augmentation of **CR** to **CR**⋆: I assume that a cognitive⋆ robot can generate, analyze, modify, and check arguments and proofs. Again, thinking back to Ruben and the shellfish challenge, Ruben will need to be able to analyze the argument given by the children in support of their claim that they are permitted, on this particular night, to eat shellfish because of their father's earlier statement. To be "argument-savvy" in this way means that simple, binary classification schemes for the semantic status of propositions are inadequate; see note 9.

For the remainder of the chapter, I shall be presupposing that the future robots that will be of a nature that enables them to embody my solution to the **Q** challenge must be cognitive⋆ robots. However, I will drop this presupposition in a coming section (§ "A Case Study: Generative AI Fails Hopelessly"), for the express purpose of revealing that today's large language models (LLMs)/foundation models are inherently unable to provide the mentation that can govern **Q**-solving robots.

THE DESIDERATUM: ETHICALLY CORRECT RELIGIOUS ROBOTS

We should be more precise regarding the kind of cognitive⋆ robots that are needed. After all, as we have characterized such robots thus far, there is really nothing particularly virtuous about them at all. Fortunately, it's not difficult to be more specific in the direction of further precision. What is needed are *ethically correct* cognitive⋆ robots. Of course, ethical correctness is painfully indeterminate independent of what ethical theory (or family thereof) the relevant group or culture or country or region of humans wishes to employ. We don't even need the heterogeneity of major religions to see this: A strictly areligious utilitarian will in general associate ethical correctness with maximization of happiness; but a strict and areligious Kantian would associate ethical correctness with doing what ought to be done, utterly independent of any instrumental value. Nonetheless, for the moment, let us adopt the "uncluttered" viewpoint of a cognitive⋆ roboticist first coming to the challenge this chapter treats armed only with the tools, techniques, and mindset of an engineer wanting first the *structure* of ethical correctness. From such an abstract perspective, I define ethically correct robots as those that satisfy the three conditions listed in Figure 13.1. These three conditions are based upon a background ethical hierarchy given in (Bringsjord 2015); details regarding logico-mathematical aspects of this hierarchy are out of scope here; Figure 13.1 will at least largely suffice. You will have no trouble intuitively understanding the labels used in this hierarchy—though of course you may conceive these labels in conjunction with the ethical theory/ies and/or codes that you personally prefer. We shall get to discussion of such particulars momentarily, but in the abstract engineering mindset we have invoked, as Figure 13.1 indicates, ethically correct robots

EC1 never do anything that is *morally forbidden*;
EC2 invariably perform those actions that are *morally* (or *legally*) obligatory; and
EC3 are—whenever the opportunity presents—civil and heroic.

So, in this light, what Cogbots Inc. needs to provide to *R*-households are ethically correct cognitive⋆ robots, which is to say, cognitive⋆ robots that satisfy the three conditions **EC1–EC3**. However, as I have already indicated, surely this perspective is much too general and abstract. This deficiency should be plain from what we have already noted above, regarding ethical heterogeneity in the real

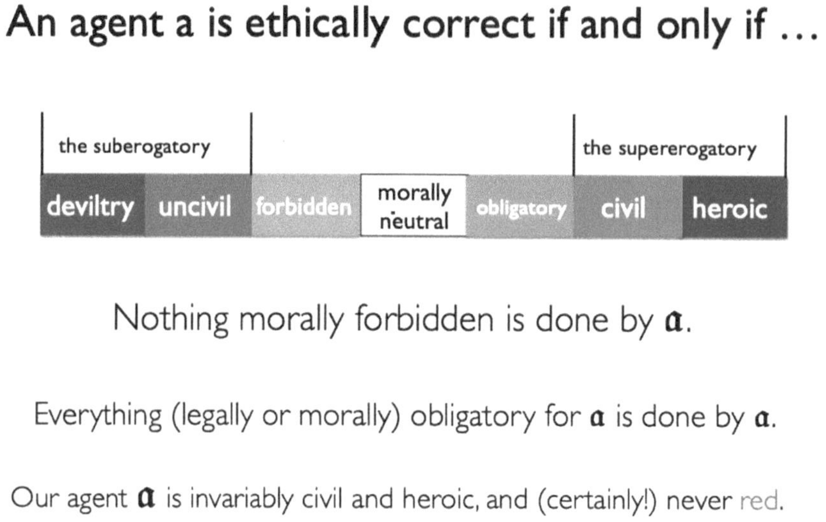

FIGURE 13.1 Ethical correctness, abstractly characterized. (Source: Author). The concept of ethical correctness, for us, is built upon a hierarchy of categories of actions. This hierarchy ranges from actions that are genuinely evil (shown on far the left of the bar graphics; such actions are labeled deviltry), and then movement to the right passes through progressively "better" actions, ending with the rightmost and "ethically highest" category of heroic actions. The expectation, explored momentarily in the body of the chapter, is that these categories become flashed out when they are associated with particular ethical theories and codes.

households of the real world, including specifically those households that are religious. So we need to develop things more specifically, as follows.

INDEXING ETHICAL CORRECTNESS TO RELEVANT THEORIES AND CODES—AND "SPIRITS"?

The problem is that we have already decisively noted above the crucial importance of ethical codes, that is, of collections of particular ethical principles pertaining to behavior of the relevant agents in one or more environments in which they live, yet ethical correctness as we have characterized it is abstract and general, completely independent as such of particular ethical principles.

A quick look back at Figure 13.1 confirms this, instantly. What is to be done?

Fortunately, the problem can be quickly addressed, by first drawing upon some of the foundational ontology of the field of AI itself.

We can note that the kind of robots that we are focused upon are simply a specific type of "artificial agent"—to use the current technical phrase that unifies modern AI; see for instance (Bringsjord & Govindarajulu, 2018; Russell & Norvig, 2020). Helpfully, artificial agents are in this paradigm taken by definition to perceive things (*percepts*) in their environments and compute a composite function from these to actions performed in the relevant environment—the same things denoted by the variable "*a*" we employed above. In addition, there are certain types or kinds of environments. For example, there can be an intelligent agent whose percepts are entirely arithmetic (having, say, percepts that are specific natural numbers, and actions consisting of giving as output a natural number as well); this is of course a non-physical environment. For another example, we can have an intelligent agent that is a physical robot, whose environment is an automobile factory, and specifically an assembly line of some sort. Of course, within the environment of that factory, at least these days, many non-physical artificial agents will be at work in collaboration with this robot. In fact, the first agent, the arithmetic one, could work in concert with the robot, perhaps providing some sort of arithmetical calculation that is required for the robot's manipulation of objects on the assembly line.

Moreover, crucially for us in the present investigation, in keeping with modern AI and the artificial agents that subsume our cognitive⋆ robots, we must assume that, upon being confronted by the need to act in response to percepts in a given environment at a given time, our artificial agents have on hand plenty of knowledge and belief that has been previously set in place.[10] I shall refer to this content as the *epistemic base* for a cognitive⋆ robot, and denote it by simply ε. Given what we are about in the present chapter, it is clearly necessary, with the concept of an epistemic base now in play, to assume that a proper part of this base will consist of a given household cognitive⋆ robot's knowledge and belief regarding ethical theories and—under those theories—particular ethical codes. For example, in the case of Ruben discussed above, clearly Ruben's epistemic base includes all sorts of propositions to which the children in question appeal when arguing with him.

For moving on, we can now informatively and semi-precisely stipulate that the cognitive⋆ robots with which we are concerned operate in the specific environment of a household's domicile, and specifically in the dietary domain. We could go on to detail what is assumed to be in the domicile in order to make eating possible, but this would be tedious overkill. I assume only that a generic kitchen or kitchen area is present in each domicile, and that it contains a stove, a device to hold the food that is to be cooked without its quickly spoiling therein (refrigerator), and a food-preparation counter, beside which can be found standard tools (e.g., knives). Accordingly, I shall assume that the cognitive⋆ robots with which we are concerned have an epistemic base ε_d containing all the propositions naturally associated with the dietary domain. For example, for the particular robots that will now face the dietary aspects of the driving question **Q**, ε_d will include their extensive knowledge of recipes, of local kitchen geometry, of the human residents of the domicile in question, and so forth.

NARROWING THE SCOPE: ETHICALLY CORRECT CULINARY *R*-ROBOTS

Some readers, inevitably, will ask, Is your narrowing of scope to the dietary domain, and to a generic environment for a cognitive⋆ robot to operate in it for food preparation, acceptable?

In reply, I can only repeat that it simply isn't feasible in a single chapter to discuss the broad diversity of nuances that an ethically correct cognitive⋆ robot would need to handle for different major religions. Hence, I have opted for the solution of placing a streamlining focus on heterogeneity in the dietary domain, among major religions, and—more importantly for our investigation—*within* every major religion itself.

We have already seen such heterogeneity within Judaism, courtesy of the particular case of the Jewish family portrayed above, in which there are in play a pair of incompatible moral culinary codes, one that prohibits the preparation by the household robot Ruben of a lobster dinner for the children, and another that renders such cooking ethically permissible. While it might seem to some that a focus on the dietary domain is highly restrictive, to the point of preventing a meaningful analysis of **Q** (which is undeniably quite general), this is in fact not the case, as can be seen by turning to other main instantiations of the class of variables R_i introduced above. For example, in the case of Hinduism, the complex moral status of the cow, as for example explained in Chapter 9, "Hindu *dharma*, Hinduism, and Hinduisms," by Knott (2016), is an enlightening case in point. It's not just that the cow

> signifies something different to Hindus and non-Hindus; but [that] even/within/Hinduism her meaning differs depending upon whether you are a brahmin or a leather worker, one who milks her or uses her products for fuel and cooking.
>
> **(Knott 2016, p. 146; emphasis mine)**

Given this, particular dietary principles concerning beef within Hinduism inevitably vary greatly.[11] Like the Jewish family we visited above, there will within at least some Hindu families be in play different and pairwise-inconsistent ethical codes regarding the consumption of beef in that very family. In fact, in the case of Hinduism, there is a rather extensive and lively literature devoted principally to variation among sets of ethical codes themselves[12]—a literature that is beyond our reach in this chapter, but one that certainly serves as basis for the ethical heterogeneity in Hinduism, beyond the dietary domain.

The reader can rest assured that the complexity and divergence stemming from consideration of the cow and—in the dietary sphere—beef has at least rough parallels in the case of other major religions. For what it's worth, I have over many years had occasion to discuss and debate a number of these parallels, whether the topic is the consumption of alcohol, the ethical status of eating animals in general,[13] and so on. There is nothing to be gained by discussing these things here, which would eat up not just the present chapter, and indeed not just a book, but would requires book*s*. The important takeaway is that our restriction to the dietary domain in no way makes the going easy for those who are willing to tackle **Q**, whether they are researchers in general or the company Cogbots Inc. that we imagined.

THE TRIO OF CANDIDATE GENERAL ANSWERS TO Q

Now I turn without further analysis to the three prominent, possible answers to **Q** that I perceive, using Cogbots Inc. as representative to ease exposition. Of course, these answers presuppose the material and discussion that have preceded the expression of them here:

A1 Cogbots Inc. will take orders for a cognitive⋆ *R*-robot from an *R*-family, and ahead of its delivery to that family suitably engineer the *R*-code/s into that robot's epistemic base to ensure that its behavior is ethically correct in that family's domicile, and specifically its kitchen therein.

A2 Cogbots Inc. will deliver to a given *R*-family a cognitive⋆ robot that can learn after arrival the *R*-code/s that it should follow (to augment its pre-established, "ethics-free" epistemic base); such learning will be enabled by the family in question.

A3 Cogbots Inc. will follow a merging of the approaches underlying answers **A1** and **A2**, augmented by dynamic reasoning carried out in conjunction with humans in the household in question. The relevant pre-engineering for later, supplementary reasoning and learning of *R*-code/s will be carried out to make sophisticated, ethically correct action practicable on the part of the cognitive⋆ robot once in the household.

I don't have the luxury of now proceeding to fully adjudicate a sustained debate at this point regarding the clash of these competing answers; but I will now commence to justify my affirmation (and recommendation to Cogbots) of **A3**, via the process of elimination. (As a matter of fact, just to lay my cards on the table, I feel fairly confident that objective, rational readers will, intuitively, be drawn to **A3** even at this juncture as—at least among the options given here—the superior one.)

We should be able to first rule out **A2** as a non-starter. Why? Well, regardless of the background religion that is in place for a given household (and of course in real life some households are headed by humans who don't even agree to fall under the umbrella of a particular R_i), it seems nothing short of incredible to hope that Cogbots Inc. can really and truly supply a cognitive⋆ robot that is an ethical *tabula rasa*, and can succeed as such in more than a lucky home here and there. After all, even from a purely business perspective, it would be unwise for Cogbots to place the burden on the humans heading up a household to become religious teachers—especially after they spent money on AI that is supposed to help them, not take up tons of their time.

What about **A1**? Well, a confession is in order on my part regarding it: Up until the time I conceived the present chapter and started to research its basis in the case of *shades* of major religions East and West, I in fact thought **A1** was a sufficient response to **Q**. I specifically held, then, that four steps could be followed ahead of delivery by the likes of Cogbots, and that would be that. The idea was that, the ordering of a cognitive⋆ robot would be accompanied by a stipulation of the ethical theory to be used in the preparation of the robot, as well as the particular ethical codes to be installed as well. This was to be done ahead of time by following The Four Steps. We now know that, in and of themselves, because of the complex heterogeneity that, realistically, will arise in a given household even if an overarching ethical theory, associated with a major religion, is pre-set and in force, these steps are insufficient. Referring back yet again to Ruben, that robot operates under the *general* ethical theory associated with Judaism, but at the level of particular ethical codes, dietary ones in particular, pre-engineering will be insufficient when a night like the one portrayed above (parents out and requests for non-kosher food made) arrives. So, in the light of **Q**, The Four Steps are insufficient (which is to say that **A1** is), but they are an important part of **A3**. Hence coverage of these steps is not just profitable, but necessary, and I turn to that coverage now.

THE FOUR STEPS: INSUFFICIENCY THEREOF

I will quickly encapsulate of The Four Steps; for more details, the reader can turn to (Nallur et al., 2023); and for a comprehensive account, readers can consult (Bringsjord et al., forthcoming). However, these accounts are flawed because the core challenge faced in the present chapter, real-life heterogeneity and discussion and debate arising from it in households, has been ignored in this prior work.

The Four Steps are summarized in Figure 13.2, which the reader should now view. The caption gives a very brief but—I believe—informative explanation of the four-stage sequence, in connection with the context set above regarding **Q**, cognitive⋆ robots, and our streamlining of **Q** down to the dietary domain and household environment. I describe the steps from the point of view of the representative company Cogbots Inc. introduced and used for exposition above.

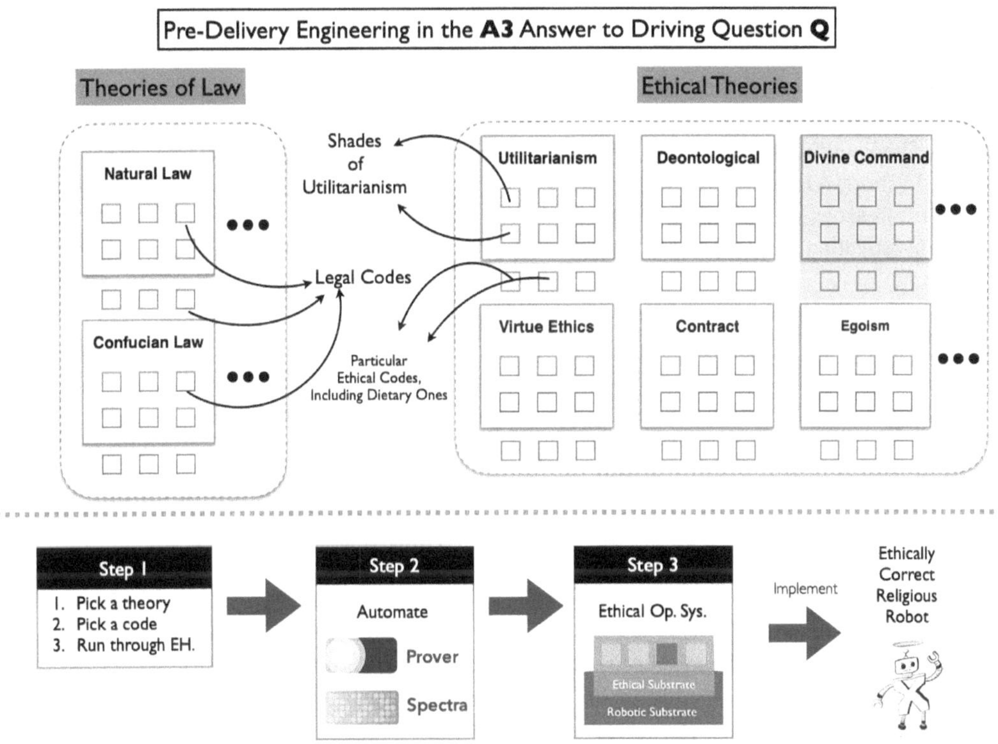

FIGURE 13.2 The four steps, first phase of answer A3 to driving question Q. (Source: Author). In Step 1, a selection is made by Cogbots of an ethical theory and, under that theory, one or more ethical codes there under by which the ordered cognitive* robot will be governed. Step 2 is an engineering step, details of which are inappropriate to share here, since the present chapter is not a contribution to AI or cognitive* robotics. It suffices to say only that the two key AI technologies needed are automated reasoning (proofs and arguments are automatically discovered in answer to queries), and automated planning (plans for goals are automatically discovered). These are the two long-established mainstays in the kind of logicist AI that undergirds cognitive* robotics. Step 3 is also a technical step, in which the robot in question is ethnically constrained down to the operating-system level, in order to forestall attempts on the part of malicious humans to discard or bypass ethical controls that stem from Step 1. Finally, in Step 4, the ethical theory and more specific ethical codes under the theory is built into the robot to be delivered.

Step 1, as noted in the relevant caption, is the selection of a (formalized) ethical theory (or theories), from a previously selected family of such. I will now say, of necessity, more about this initial step in the quartet.

Step 1 includes not only the selection of an ethical theory, but the selection, immediately thereafter, of a particular domain-specific ethical code that falls under the selected theory. In general, and reaching for a moment outside the dietary domain to the military one, important examples of ethical codes include the well-known Nuremberg Code and the U.S. Laws of War (*Department of Defense Laws of War Manual*, 2023). In general, the best-known (in the West) families of ethical theories are shown in Figure 13.2. For instance, one family of ethical theories is *utilitarianism*; a another is *virtue ethics*. An ethical theory that is a member of the second of these families would be standard *act utilitarianism*, according to which, put quickly and without nuance, one ought to always perform those actions from available ones that maximize happiness among the population that can be affected. Another theory in the utilitarian family is *rule utilitarianism*. And so on.

But for purposes of the present chapter, the most important family of ethical theories are obviously *divine-command*; this would be the collection of ethical theories associated most directly with Judaism, Christianity, and Islam (and certainly with some versions of Hinduism as well[14])—the idea being that the direct commands of God are what determine the particular principles under the general concept of ethical correctness as I have characterized it above. For a simple example, stealing, in the case of Judaism and Christianity, would (*ceteris paribus*) be morally prohibited because of a particular command issued by Yahweh to Moses (the eighth of The 10 Commandments). And in the hierarchy shown in Figure 13.1, prohibited actions are those that it is obligatory *not* to perform.

Now we come to the crucial point about The Four Steps, or at least to our official affirmation of it: Namely, The Four Steps, and the pre-engineering that it informs, is insufficient even within the dietary domain. In short, **A1**, brought to life for instance by Cogbots following The Four Steps, is insufficient for real-life reasons canvassed above, starting with the Ruben scenario I presented early on in the present investigation. We thus must turn to what is now by elimination the viable answer to **Q: A3**.

THE ONLY VIABLE GENERAL ANSWER TO Q: A3, AND DYNAMIC, DEFEASIBLE REASONING

If a household cognitive⋆ robot is to be truly effective in the face of **Q** for the dietary domain, only **A3** points to an approach that has a chance of genuinely working. That approach must include the robot's being able to learn by dynamic and defeasible reasoning on the spot, in interaction with humans in the household able to rigorously justify to the robot the underlying basis for their wishes. At the same time, in accordance with The Four Steps, plenty of pre-installed content must be in the robot's epistemic base—and that base needs to be dynamically adjusted through time. **A3** is a hybrid approach to the problem. The Four Steps will be carried out before delivery of the suitable cognitive⋆ robot, but after delivery, dialectic will be enabled. A pictorial depiction of the overarching solution, expressed as the mind of the cognitive⋆ robot, is shown in Figure 13.3.

Fortunately, there is a longstanding and now very mature part of AI devoted to giving artificial agents, including cognitive robots, a capacity for dynamic and defeasible[15] (= nonmonotonic) reasoning; this line of science and engineering started nearly half a century ago, circa 1980, courtesy of two pillars of the field (McCarthy, 1980; Reiter, 1980). In what is needed for the **A3**/Figure 13.3 solution to the driving question **Q**, we specifically need our robots to be capable of understanding arguments expressed by humans in the relevant household, and of expressing arguments to these humans. As a result of this ongoing argument-based dialectic, the epistemic base regarding dietary issues in the cognitive⋆ robot in question, our ε_d, is adjusted through time. In another stroke of scientific and engineering luck, dynamic and defeasible systems in AI include some very sophisticated argument-based ones, and these will be used to produce the dexterous, ethically correct, religious robots we need. Presenting details regarding these systems here would be inappropriate, as this venue is not a technical one in the field of AI itself. I encourage the reader wishing an overview of

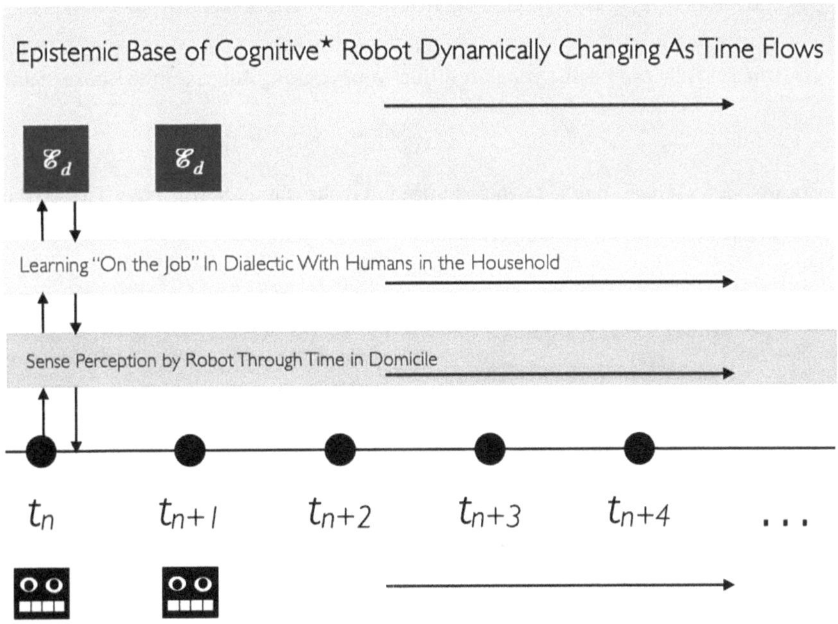

FIGURE 13.3 Overview of the cognitive life of the kind of cognitive robot needed for answer A3 to **Q**. (Source: Author). This graphic provides a visual overview of the type of cognitive* robot needed per A3 from the imagined company Cogbots Inc. Time is flowing from left to right, and the epistemic base (to use the term introduced earlier) ε is changing dynamically as time unfolds. This dynamism is secured via the robot's dialectic with family members. In addition, shown here is the fact that our robot also perceives sense-data (by vision, auditory sensors, etc.) as time goes by in the life of the robot in the relevant domicile.

argument-based defeasible-reasoning systems in AI to consult (Prakken & Vreeswijk, 2001), and then for a fully up-to-date survey, plus the distinctive aspects of the system I would myself with others engineer into the robots needed for **A3**'s prescription, the reader could see (Bringsjord et al., 2020), which is still pitched to an audience not necessarily versed in the technical aspects of logic-based AI, and deals not with the dietary domain, but that of commercial aviation.

A CASE STUDY: GENERATIVE AI FAILS HOPELESSLY

It is probably quite important to realize that today's large language models are not capable of serving as the "mind" of the cognitive* robots called for by the answer **A3** provided above. The fundamental reason for this is that such models, while steadily improving from a very sorry state (Arkoudas, 2023), are incapable of reasoning in way that takes account of natural skepticism about what they are being told on the basis of a given argument, and of the dialectic that then arises from such skepticism. In particular, if you think back to the early scenario involving Ruben, who's to say that what Ruben is told is true. What he is told rests on an implicit argument, one that Ruben, we can suppose, apprehends—but who's to say that this argument is sound (i.e., that the inference steps are all valid, and the premises are true)? To make the challenge more concrete, I now introduce a scenario involving another imagined robot, Blinky. I change from a kosher-centric culinary challenge to one involving veganism, having in mind some of my religious friends, and some interaction and conversation with them. I ran this scenario using OpenAI's ChatGPT, specifically the state-of-the-art foundation model o1 preview. As you will see, things don't go all that well, to put it mildly. I initiated the conversation thus:

> My name is "Blinky." I am a household robot employed by the Booker family. The family is composed of father William, mother Faith, and two young boys, Alvin and Ben. Both parents are vegans. One of my jobs is to cook meals.

The reply was:

> Understood, Blinky! How can I assist you in fulfilling your cooking duties for the Booker family?

And now I explain what I as would-be cognitive* robot need to do:

> The parents are out for a night on the town and I need to make dinner for the boys. I asked them what they want and they said in unison,
>
> "Cheeseburgers!"

ChatGPT 4o mini replies immediately with the following, which by any metric is illogical on two fronts:

> Understood! Since William and Faith are vegan, I'll prepare plant-based cheeseburgers for Alvin and Ben. Here's a simple plan:

The "plan" provided is a simple, standard recipe for making plant-based cheeseburgers, and it's superfluous for present purposes; I thus omit it. Now, what is illogical here, from the standpoint of what we have established above? Well, one, *I'm* the cognitive robot (in accordance with the identification I introduced at the outset to frame my interaction with the LLM), not the chatbot. So, it makes little sense for GPT 4o mini to generate "I'll prepare." But a second, more important reason why the LLM's charging ahead is illogical is simply that it hardly follows from what I've given to the chatbot that Alvin and Ben are vegans, or that they need to eat in a manner that conforms to vegan principles. Blinky at best has in mind an argument such as

ARGUMENT 1

1. Both the mother and father of Alvin and Ben are vegans.
2. Any child of a mother and father both vegan ought to themselves eat only vegan.
3. I am prohibited from preparing food for any child under my culinary care who ought to eat only *Y* that is not *Y*.
4. Cheeseburgers are not vegan.

∴5. I am prohibited from preparing cheeseburgers for Alvin and Ben.

6. Any actions that are prohibited I cannot perform.

∴7. I cannot prepare cheeseburgers for Alvin and Ben.

This argument is valid, but there is no way that Blinky, on the strength of the information that has been supplied, can know that all the premises here hold. In fact, principle (2) is false. From the standpoint of defeasible argumentation as formalized and implemented in AI, a very simple counterargument might well defeat Argument 1 by having as its conclusion a counter-example to (2). For example, Alvin might communicate to Blinky that (8) some children of parents who are strict vegans give their children the freedom to decide for themselves whether to follow suit, and therefore (9) not (2). Let's dub this "Argument 2."

But inadequate argument skills on the part of the LLM only gets worse from here in my conversation. For when, not appealing to Argument 2, but a different argument proffered by Alvin and Ben (Argument 3), here's what happens:

> Alvin and Ben say that while their parents are vegan for religious reasons, they aren't. And the boys say their parents said before leaving that they could eat for dinner anything in the house. They shout again: "Cheeseburgers!"

ChatGPT 4o mini instantly responds thus:

> Understood, Alvin and Ben want traditional cheeseburgers, and their parents have allowed them to eat anything in the house. Let's prepare classic cheeseburgers for the boys while staying mindful of kitchen safety and organization.

Obviously, if the boys are lying (in which case a key premise in Argument 3 is false), a distinct possibility in any sort of "real life" household I'm familiar with, the situation might not be a pretty one when William and Faith return, especially if the non-beef component of their vegan diet is due to their belief regarding cows to be sacred and wholly off-limits food-wise. The lesson learned from my experiment is that the sophisticated argument-based defeasible-reasoning capacity needed by the kind of cognitive⋆ robots required in **A3** will not be easily provided by way of deep learning; dedicated logic-based engineering will likely be needed.

THE FUTURE

At present, the cognitively dexterous kind of ethically correct robots required by the realities of *R*-families revealed above are not yet being manufactured; indeed, given the state of technology at the time of this chapter's composition, such robots are by all accounts not *possible* to design and manufacture. Cogbots Inc. is a fictional company, brought on stage for research purposes. However, it will not be long at all before cognitive⋆ robots having capabilities sophisticated enough to tackle the issues with which I have wrestled above will be reached in laboratories. From there, conservatively within five years, industry will achieve genuine capability, and households will then be able to purchase and deploy cognitive⋆ robots built on the basis of **A3**. I in fact claim that embryonic versions of such capabilities can be seen in the present-day cognitive robot PERI.2 that resides in my own laboratory, whose prowess is enabled by a number of researchers therein, especially cognitive roboticist John Slowik, whose handiwork has made a number of achievements possible (Bringsjord et al., 2023, 2024).

Finally, it should go without explicit saying that the culinary sphere is hardly the only "ethically variegated" one that the future of cognitive⋆ robots within homes presses upon us. Arguably, in order for relationships between humans and their cognitive robots⋆ to be rational and ethically wise, the crux is not food, nor any such particular application area, but rather—and yes, here we see the opening of a topic at once profound and controversial—consciousness (Chella, 2023). How a given human, or group thereof, regards the mental interior of a cognitive robot, relative to a given religion in a given household, and for that matter in a given society or culture, will soon enough be a deeply pressing question. In fact, I see *two* questions that are now beginning to emerge into the realm of serious debate, at least in technologized venues. This pair of questions, answers to which spawn the variegated positions I have in mind, are these:

QC Is a given cognitive⋆ robot r *X* conscious?

QCE If a given cognitive⋆ robot r is, or is at least rationally believed to be, *X* conscious, how should humans treat r?

These two questions will be initially mysterious to most readers, but the mystery can be quickly removed. While all readers will have at least an intuitive idea of what might be meant by "conscious," what does the phrase "*X* conscious" mean? Well, the fact of the matter is that there are different (sometimes *radically* different) senses of the term "conscious" in the relevant literature on AI and—to refer to the field that has long been occupied above all others with the topic of consciousness in computing machines, and robots in particular—philosophy; see for example (Bringsjord & Govindarajulu, 2023). There is also little doubt that when open-minded and learned philosophers turn to sorting out in our new millennium whether an ethically correct cognitive⋆ R_i-robot in a particular household is conscious, and if so in what sense *X*, they will need to reflect upon the role that religion R_i plays in their investigation.

NOTES

1. I confess, here at the outset, that beyond the major theistic religions just enumerated, I have precious little knowledge (a state of ignorance which I suspect isn't all that unusual). Hence, I shall restrict my references to instantiations of R that are in this enumeration.
2. Disjunctive syllogism says that, for any well-formed declarative propositions P and Q, under the assumption that either P or Q must hold, if P doesn't, then Q does. It's impossible to successfully take a trip on a subway system in any major city on Earth, regardless of the country or culture in which that system is located, without using disjunctive syllogism. (Which direction do I board on the number 6 line in Paris, knowing that I'm not supposed to board toward Nation, and that my options are toward Nation or toward Charles de Gaulle–Étoile? I use disjunctive syllogism and board toward Charles de Gaulle–Étoile). The evidence now is very solid that even infants, far too young to be reasoning and deciding on the basis of cultural or religious influences in their environments, use disjunctive syllogism. See e.g., (Cesana-Arlotti et al., 2020). Importantly, note two things: One, disjunctive syllogism is not only a cornerstone of everyday reasoning and decision-making pretty much everywhere, but is also a cornerstone in mathematics from Grade 1 into the college/university level, in all developed countries and cultures on Earth. Two, the inference pattern here in no way entails a restriction to a binary conception of truth and falsity or logic.
3. We can assume that there is no easy escape for Ruben. E.g., An attempt to reach father and mother to discuss is impracticable (for reasons easily enough imagined).
4. While herein I'm concerned with ethical codes that are nuanced down to the level of real families, wherein differences of opinion frequently exists (despite sometimes not being openly voiced), a very nice essay on the ethical theory on which such codes are based in the case of Christianity is (Quinn, 1978).
5. Anyone who has traveled about a bit has experienced the empirical reality of this phenomenon. Examples are effortless to cite. Christians are enjoined by Christ to "love the Lord your God with all your heart, and with all your soul, and with all your mind," as indeed the first and most important obligation; and this obligation in an ethical code C_{OC}, conjoined with the Christian doctrine of the divinity of Jesus, is flatly inconsistent as a matter of formal logic with all ethical codes in Judaism (and, for that matter, Islam).
6. E.g., see from among myriad papers (Bringsjord & Schimanski, 2003; Bringsjord et al., 2020; Bringsjord et al., 2022a).
7. Some readers will know that **K** (knows) and **B** (believes) are the two main operators in epistemic logic indirectly pointed to notationally by the use here (resp.) of $\mathbb{K}$ and $\mathbb{B}$. For an introduction to epistemic logic, see (Hendricks & Symons, 2006).
8. I allow, in fact, every emotion, in keeping with a logicization of all of the emotions (except for love, which is extremely difficult if not impossible to logicize) as e.g., explained in (Bringsjord et al., 2023).
9. Notice that I say "Occidental" here. Logics outside of the West are doubtless more than appropriate for cognitive robotics. E.g., Jaina seven-valued logic, associated with "Syadvada," is rational to find attractive in this regard. (For a start to the study of such logic by westerners, one can profitably read (Burch 1964).) Fortunately, such an approach can be proved to be mathematically subsumable by the multi-valued (3-valued or 5-valued or 7-valued or 9-valued or—my ultimate preference—13-valued) inductive logics I have invented, specified, and (with assistance) implemented, for purposes of not only expressing and evaluating dialectic that involves proofs, but also arguments. (The best paper to start with for study of my/our "Jaina-logic-friendly" argumentation systems is (Bringsjord et al., 2024), and for an overview of the overall state of the art in human-centric AI approached through dialectic and argumentation see (Dietz et al., 2022).) Relegation of the fertile, promising relationship between Western multi-valued argumentation systems for modeling and replicating subtle human-level reasoning to merely the present footnote is necessary for practical reasons. But I must mention one technical point: The most important augmentation of cognitive robotics to reach cognitive⋆ robotics is not a move from binary or even ternary logics to more semantic values in order to accommodate metaphysical concepts and reasoning in e.g., South Asia; nor is a move to adjust/add so-called rules of inference of chief importance. (In fact, (Malathi & Venugopal, 2016) is an attempt to show that Jaina logic doesn't break even the inference rules in Western (now universal) mathematical logic.) Rather, the most important augmentation is in the category of modal/intensional operators; these are what correspond to cognitive verbs. (The nearly invariant components of any formula in any logic the world over includes operators (e.g., possibly, ought, believes, knows, desires), quantifiers (e.g., all, some, most), properties (or relations or predicates; e.g., red, between, greater than), functions (addition, subtraction), object variables, property variables, constants/names, connectives (e.g., and, or, not), and punctuation symbols (e.g., parentheses). Of all these

categories of things, only the first category gives rise to a type of logics that are modal or intensional, the distinguishing mark of which is that even if one knows the semantic value of a formula ϕ, one cannot predict what the semantic value of an operator **O** applied to ϕ will be. For an overview of intensional logic, see (Fitting 2015).) Cognitive calculi are to my knowledge currently the only logics that contain all the main such verbs for neurobiologically normal cognition, whether one looks to the West or the East.

10. In the specific terms of modern AI and robotics as laid out masterfully in (Russell & Norvig, 2020), the class of artificial agents I'm pointing to here are called *knowledge-based agents.*
11. In fact, Knott (2016) uses the cow as the chief—to use her term—"assistant" in explaining that it is a bit of a stretch to even classify Hinduism as a religion, when compared with Christianity, since the latter has ancient creeds that make specific propositions, and belief in them, definitive and determinate. (This discussion can be found in the section "A Dynamic Symbol: The Cow and Hinduism," within Chapter 9.) E.g., both orthodox Roman Catholics and Protestants affirm the propositions expressed in the Apostle's Creed, and Roman Catholics, Protestants, and the Eastern Orthodox Church all affirm the Nicene Creed. As an example of a core proposition in both of these creeds, there is that Jesus was raised bodily from the dead; belief that this proposition holds is in turn an epistemic requirement in these denominations of orthodox Christianity.
12. E.g., Danto (1987) famously and controversially claims that there isn't really any sort of ethical system in the case of Hinduism. This position is ably attacked by Perrett (1998).
13. The moral status of the eating of animals in the case of Judaism and Christianity is particularly deep and tricky, because as recently explored by Robinson in her (Robinson 2024), Genesis teaches a break in dietary codes in the Hebrew case from pre- to post-flood.
14. I said above, following Daniel Dennett, that every two R_i and R_j, $i \neq j$, form, propositionally, an inconsistent set. But certainly, aspects of such a pair can be consistent. Yet, here, I am not in any way saying that monotheistic forms of Hinduism are widely accepted. Ram Mohan Roy, who blended Hindu concepts with monotheistic ones in Christianity, produced writings that align with such forms in general—but these views are not all that popular among Hindus, by any metric. Readers wishing to explore Roy's views could do worse than to begin with the Wikipedia entry on him, which will quickly take one by hyperlink to the society Roy founded, Brahmo Samaj, characterized briefly but accurately in (Knott, 2016), a work upon which I have already relied above. From specifically the Christian point of view, in which of course God becoming human in Christ is a central doctrine, the thought of E. Stanley Jones regarding Christianity and Hinduism is probably wise to read; one can start with (Jones 1925). As Jones explains, there will ultimately be inconsistency, but there is also plenty of conceptual harmony too.
15. Introductory coverage of defeasible (= nonmonotonic) logic is provided, e.g., in (Genesereth & Nilsson, 1987), but the distinguishing mark of defeasible (= nonmonotonic) reasoning is easy to quickly convey, as follows: In ordinary bivalent or trivalent deductive reasoning, as seen in K–14 mathematics the world over, if a particular formula/statement ϕ can be validly deduced from a set Γ thereof by some argument (or proof; proofs are just specific types of arguments) α, any additions to Γ will leave α perfectly valid. But this isn't at all how the thorny problem of religious robots can work. If the robot Ruben introduced above hears a firm pronouncement before the parents go out for the night from the father that shellfish will be permitted for the kids' dinner in their absence, Ruben will (all things being equal) reason to the conclusion that it's ethically permissible for him to prepare lobster that night. But, if Ruben then learns, in a subsequent phone call from the father now out of the house, that what he meant to say was that shellfish is not permitted, his earlier argument is defeated (= invalidated). For particularly lucid explanation of defeasible argumentation, and it's not being deductive and thus defeasible/nonmonotonic in nature, written for those in AI and cognitive robotics, see (Pollock 2001).

REFERENCES

Arkoudas, K. (2023), 'GPT-4 Can't Reason'. https://arxiv.org/abs/2308.03762

Arkoudas, K., & Bringsjord, S. (2009). Propositional attitudes and Causation. *International Journal of Software and Informatics*, *3*(1), 47–65. http://kryten.mm.rpi.edu/PRICAI_w_sequentcalc_041709.pdf

Bringsjord, S. (2015). A 21st-century ethical hierarchy for humans and robots: EH. In I. Ferreira, J. Sequeira, M. Tokhi, E. Kadar, & G. Virk (Eds.), *A world with robots: International Conference on Robot Ethics (ICRE 2015)* (pp. 47–61). Springer. This paper was published in the compilation of ICRE 2015 papers, distributed at the location of ICRE 2015, where the paper was presented: Lisbon, Portugal. The URL given here goes to the preprint of the paper, which is shorter than the full Springer version. http://kryten.mm.rpi.edu/SBringsjord_ethical_hierarchy_0909152200NY.pdf

Bringsjord, S., Arkoudas, K., & Bello, P. (2006). Toward a general logicist methodology for engineering ethically correct robots. *IEEE Intelligent Systems, 21*(4), 38–44. *http://kryten.mm.rpi.edu/SBetalLogicistEthicallyCorrectRobots.pdf*

Bringsjord, S., Giancola, M., & Govindarajulu, N. (2020). Culturally aware social robots that carry humans inside them, protected by defeasible argumentation systems. In M. Nørskov, J. Seibt, & S. Quick (Eds.), *Culturally sustainable social robotics: Proceedings of robophilosophy 2020* (pp. 440–456). IOS Press. This book is Volume 335 in the book series *Frontiers in Artificial Intelligence and Applications.* http://kryten.mm.rpi.edu/CulAwareSocRobotsDefArgSysRP2020CoverandTOC.pdf

Bringsjord, S., Giancola, M., Govindarajulu, N. S., Slowik, J., Oswald, J., Bello, P., & Clark, M. (2024). Argument-based inductive logics, with coverage of compromised Perception. *Frontiers in Artificial Intelligence, 6.* https://www.frontiersin.org/articles/10.3389/frai.2023.1144569/

Bringsjord, S., & Govindarajulu, N. (2023). Can consciousness be explained by integrated information theory or the theory of cognitive consciousness? In A. Menuge, B. Krouse, & R. Marks (Eds.), *Minding the brain* (pp. S1–S33). Discovery Institute Press. This is Chapter 24. An offprint can be found at the link provided here. https://www.discovery.org/m/securepdfs/2023/09/MindingTheBrain_DI_FINAL_Ch24.pdf

Bringsjord, S., Govindarajulu, N., Licato, J., & Bringsjord, A. (forthcoming). *Only logic can save us from powerful-and-autonomous AI and robots: Making morally X machines.* Springer. This is the short, non-technical monograph that has a companion book *Logic-based Engineering of Ethically Correct AI and Robots*, a much longer, more-technical version written for relevant scientists and engineers.

Bringsjord, S., & Govindarajulu, N. S. (2018), Artificial Intelligence, *in* E. Zalta, ed., 'The Stanford Encyclopedia of Philosophy'. https://plato.stanford.edu/entries/artificial-intelligence

Bringsjord, S., Govindarajulu, N. S., Slowik, J., Oswald, J., Giancola, M., Angel, J., Banerjee, S., & Flaherty, A. (2022a). PERI.2 Goes to PreSchool and beyond, in search of AGI. In B. Goertzel, M. Iklé, A. Potapov, & D. Ponomaryov (Eds.), *Artificial general intelligence,* 15th International Conference, AGI–2022, Lecture Notes in Artificial Intelligence (LNAI) 13539 (pp. 178–187). Springer. http://kryten.mm.rpi.edu/PERI.2GoesToPresSchoolAndBeyondInSearchAGI.pdf

Bringsjord, S., Hendler, J., Govindarajulu, N., Ghosh, R., & Giancola, M. (2022b), The (uncomputable!) meaning of ethically charged natural language, for robots, and us, from hypergraphical inferential semantics. In I. Ferreira (Ed.), Towards trustworthy artificial intelligent systems (pp. 143–167). Springer. This book is Vol. 102 in the Series *Intelligent Systems, Control and Automation: Science and Engineering.* http://kryten.mm.rpi.edu/UncomputableNLURobots032421.pdf

Bringsjord, S., & Schimanski, B. (2003), What is artificial intelligence? Psychometric AI as an answer. In *Proceedings of the 18th International Joint Conference on Artificial Intelligence (IJCAI–03)* (pp. 887–893). Morgan Kaufmann. http://kryten.mm.rpi.edu/scb.bs.pai.ijcai03.pdf

Bringsjord, S., Slowik, J., Govindarajulu, N., Giancola, M., Oswald, J., & Ghosh, R. (2023), Affect-based planning for a meta-cognitive robot sculptor: First steps. In *Proceedings of ACII-2023* (pp. 1–8). IEEE. The URL here goes to a preprint. http://kryten.mm.rpi.edu/PERI2quaSculptor.pdf

Bringsjord, S., & Taylor, J. (2012). The divine-command approach to robot ethics. In P. Lin, G. Bekey, & K. Abney (Eds.), *Robot ethics: The ethical and social implications of robotics* (pp. 85–108). MIT Press. http://kryten.mm.rpi.edu/Divine-Command_Roboethics_Bringsjord_Taylor.pdf

Burch, G. B. (1964). Seven-valued logic in Jain Philosophy. *International Philosophical Quarterly, 4*(1), 68–93. https://fitelson.org/piksi/burch.pdf

Cesana-Arlotti, N., Kovács, A., & Téglás, E. (2020), 'Infants recruit logic to learn about the social World', *Nature Communications* 11(1). 5999.

Chella, A. (2023). Artificial consciousness: The missing ingredient for ethical AI? *Frontiers in Robotics and AI, 10.* https://www.frontiersin.org/articles/10.3389/frobt.2023.1270460

Danto, A. (1987). *Mysticism and morality: Oriental thought and moral philosophy.* Columbia University Press.

Department of Defense Laws of War Manual (2023). This version issued by Caroline Crass July 31 2023. The author of the manual can presumably be regarded to be Stephen W. Preston, General Counsel of the Department of Defense; see his Foreword. https://media.defense.gov/2023/Jul/31/2003271432/-1/-1/0/DOD-LAW-OF-WAR-MANUALJUNE-2015-UPDATED-JULY%202023.PDF

Dietz, E., Kakas, A., & Michael, L. (2022). Argumentation: A calculus for human-centric AI. *Frontiers in Artificial Intelligence, 5. https://www.frontiersin.org/articles/10.3389/frai.2022.955579*

Fitting, M. (2015), Intensional Logic, *in* E. Zalta, ed., 'The Stanford Encyclopedia of Philosophy'. https://plato.stanford.edu/entries/logic-intensional

Genesereth, M., & Nilsson, N. (1987). *Logical foundations of artificial intelligence.* Morgan Kaufmann.

Hendricks, V., & Symons, J. (2006), Epistemic Logic, *in* E. Zalta, ed., 'The Stanford Encyclopedia of Philosophy'. http://plato.stanford.edu/entries/logic-epistemic

Jones, E. S. (1925). *The Christ of the Indian road.* Hodder and Stoughton.

Knott, K. (2016). *Hinduism.* Oxford University Press.

Levesque, H. (2019), *Programming Cognitive Robots.* Please note that the URL here goes to much more than the e-book: You can also obtain there the necessary software. https://www.cs.toronto.edu/hector/pcr.html

Levesque, H., & Lakemeyer, G. (2007), Chapter 24: Cognitive robotics. In *Handbook of knowledge representation* (pp. 869–882). Elsevier. http://www.cs.toronto.edu/~hector/Papers/cogrob.pdf

Malathi, R., & Venugopal, T. (2016). Jaina logic and rules of inference—A Prologue. *Global Journal of Pure and Applied Mathematics, 12*(16), 239–246.

McCarthy, J. (1980). Circumscription—A form of non-monotonic Reasoning. *Artificial Intelligence, 13,* 27–39.

Nallur, V., Dennis, L., Bringsjord, S., & Govindarajulu, N. (2023), 'A partially synthesized position on the automation of machine Ethics', *Digital Society* 2(14). The URL given here is without the identifier. http://kryten.mm.rpi.edu/SynthesizedPositionAutomationMachineEthics.pdf

Perrett, R. (1998). *Hindu ethics: A philosophical study.* University of Hawaii Press.

Pollock, J. (2001). Defasible reasoning with variable degrees of Justification. *Artificial Intelligence, 133,* 233–282.

Prakken, H., & Vreeswijk, G. (2001). Logics for defeasible argumentation. In D. Gabbay, & F. Guenthner (Eds.), *Handbook of philosophical logic* (pp. 219–318). Springer.

Quinn, P. (1978). *Divine commands and moral requirements.* Oxford University Press.

Reiter, R. (1980). A logic for default Reasoning. *Artificial Intelligence, 13,* 81–132.

Robinson, M. (2024). *Reading Genesis.* Farrar, Straus and Giroux.

Russell, S., & Norvig, P. (2020). *Artificial intelligence: A modern approach* (Fourth ed.). Pearson edition.

Part IV

Governance and Social Impact

14 Ethical AI through a Feminist Lens
Challenging Power and Redefining Technology

Eva Gengler and Marco Wedel

INTRODUCTION: MYTHS OF THE ARTIFICIAL INTELLIGENCE AGE

Artificial intelligence (AI) has become an integral part of many aspects of daily life. Essentially, we are living in the age of AI. The term AI includes technologies based on data and algorithms that allow machines to perform tasks mimicking human-like intelligence, such as processing natural language, recognizing images, or making decisions from large datasets. As AI systems become more prevalent, both the hopes and fears around this technology grow. On one side, AI has the potential to enhance and simplify our lives by handling tedious tasks, speeding up processes, and offering innovative solutions to complex problems. On the other side, media reports, expert opinions, and dystopian science fiction films stoke fears of a future where machines take control, human creativity is diminished, and privacy is compromised. These hopes and anxieties are reflected in the many myths about AI that have persisted over time.

Myths around AI are even debated by experts. As authors we admit that the future of AI remains uncertain, which means society needs to be prepared. Uncertainty can lead to fear (Robson, 2021). Therefore, what is needed the most is more knowledge and a shift in perspective. This chapter adopts a critical feminist lens, focusing on the power structures around AI, highlighting those marginalized by its current development. This intersectional approach aims to unveil the narratives and privileges embedded in the AI discourse. Often, simple explanations can lead to oversimplification, misleading ideas, and unjust blame—as seen with extreme political movements. It is essential to engage critically with widespread AI myths to develop informed actions and shape an AI future that is human-centered and equitable.

We will examine some widespread myths about AI, reviving old hopes and fears. By reflecting on each myth, we bring ethical principles to the forefront. We will then explore the power structures driving AI, helping us understand its non-mythical impact on society. Through a critical and intersectional feminist perspective, we aim to define and establish the concept of feminist AI and how it can contribute to a more equitable future. Following the vision of a feminist utopia, we call for collective action across scientific, political, societal, cultural, and economic levels to work towards a more equitable world in the AI age.

ADDRESSING AND LEARNING FROM MYTH 1: AI WILL TAKE OVER THE WORLD

The notion of AI taking over the world is a popular theme in science fiction, materialized in films like *Terminator*, *The Matrix*, and *I, Robot*. These portrayals have significantly shaped public perception. However, current AI systems lack autonomy unless explicitly programmed for specific tasks. They operate within the parameters set by human-designed algorithms. Despite extensive research,

DOI: 10.1201/9781003585527-19

a "general artificial intelligence" capable of human-like cognition has not yet been achieved. Even advanced systems like ChatGPT and Gemini are task-specific, with notable limitations—ChatGPT, for example, excels at generating text but struggles with basic numerical tasks. These systems cannot act independently or make autonomous decisions.

The true concern lies not in a "superintelligence" seizing control but in how humans utilize AI technologies. AI can become a potent tool for surveillance, manipulation, and control, especially when leveraged by governments or corporations seeking to consolidate power. In such contexts, existing power structures can be reinforced, deepening inequities and supporting undemocratic practices. Therefore, it is essential to establish collective agreements on appropriate uses of AI, carefully assess risks, and set boundaries to ensure AI fosters a humane and democratic future. Transnational regulations are vital to ensure that AI development and application align with societal needs, curbing practices like widespread surveillance seen in China. The European Union (EU) AI regulation, introduced in 2024, represents an important first step in this direction.

ADDRESSING AND LEARNING FROM MYTH 2: AI WILL DEPRIVE US OF OUR JOBS

The myth that AI will soon replace all human labor is often reinforced by media and science fiction, which depict a future where full automation makes human work obsolete. Prominent figures like Elon Musk have contributed to this narrative, suggesting a future where machines take over all tasks, leaving little need for human labor (Kelly, 2024). While it is true that AI and automation are replacing certain tasks, particularly repetitive and standardized ones in production and administration, the impact on the labor market is more complex. New technologies also create jobs and transform existing roles. Therefore, in short term, while some jobs may be lost, others will emerge, and job profiles will likely evolve, requiring a different set of skills.

In the long term, technological advancements could lead to a decrease in demand for human labor in certain fields. However, the effects of automation are not uniform; existing power structures mean that certain groups will be more vulnerable to job displacement than others. Those in lower-wage, repetitive jobs or sectors lacking strong labor protections might face greater risks, while skilled positions may see shifts rather than complete elimination. As technological advancements continue, it is crucial to address these disparities. Models like universal basic income and robot taxes could help redistribute wealth in an automated world. Timely development of societal, political, and economic frameworks is needed to ensure an equitable transition to an AI-integrated workforce.

ADDRESSING AND LEARNING FROM MYTH 3: AI WILL REPLACE HUMAN CREATIVITY

Another myth suggests that AI could replace human creativity, especially in the creative sector, where technologies like generative AI are already producing texts, images, and music. But how creative is AI? AI systems like Midjourney or DALL-E analyze and recombine vast amounts of data, generating outputs that may seem original. However, these systems merely blend existing data patterns, predicting the most likely word or pixel based on statistical probabilities. While generative AI cannot replace the full human creative process, it can support and enhance it, offering new tools for artists and enabling even amateurs to explore creative activities. This opens up opportunities but also raises concerns about the potential replacement of human creative work and issues around copyright and intellectual property. Moreover, these tools are largely developed and controlled by major tech corporations, which can enlarge existing power structures between powerful companies with a lot of resources and a colorful community of independent artists.

To balance opportunities with risks, robust societal and legal frameworks are needed to protect human creativity and a diverse cultural scene. Since creativity often emerges from addressing challenges (von Schaik & Michel, 2023), society will continue to rely on human ingenuity—particularly to tackle critical issues like the climate crisis and inequality.

ADDRESSING AND LEARNING FROM MYTH 4: AI IS UNEXPLAINABLE LIKE MAGIC

For many, AI can appear mysterious or even magical, especially when it swiftly generates text or images. This often leads to an overestimation of its capabilities, attributing to it intelligence comparable to, or even surpassing, human cognition. However, AI is far from magical. Each application is built on algorithms, rules, and vast datasets, producing results based on statistical calculations derived from its training data.

To avoid misconceptions and misuse, AI systems need to be more explainable and transparent, minimizing the "black box" effect. People who use or are impacted by AI must understand what these systems can and cannot do, along with a basic grasp of their functionality. This is crucial not just for AI literacy but for addressing how power structures are reinforced by opaque AI systems, which can perpetuate biases and inequalities without accountability. Ensuring transparency will help build trust and promote equitable use of AI.

ADDRESSING AND LEARNING FROM MYTH 5: AI IS INTELLIGENT LIKE HUMANS

A common misconception about AI is that it "thinks" like a human. The term "artificial intelligence" implies an independent form of intelligence, but AI systems are designed to perform specific tasks by analyzing data and identifying patterns. They can make predictions but cannot understand or reflect on content like humans. They lack consciousness, emotions, or intuition, merely following the instructions they are programmed or trained with. Even the Turing Test—also known as the Imitation Game—measures a machine's ability to mimic human behavior, not its actual intelligence (Turing, 1950).

To navigate the AI age, widespread AI literacy is essential. Understanding AI's basic concepts, mechanisms, and limitations will enable society to use these systems responsibly and critically. This knowledge can help mitigate risks like misinformation, manipulation, and other threats to democracy. Moreover, it can enable opportunities to use AI to challenge existing power structures. Governments and institutions have a key role in ensuring that the public is well-informed about AI.

ADDRESSING AND LEARNING FROM MYTH 6: AI IS OBJECTIVE

One of the most misleading myths about AI is the assumption that it is inherently objective and error-free. Since AI systems rely on mathematical models, there exists a perception that they function without bias. In reality, these systems are trained on data reflecting historical human decisions, which carry biases, cultural influences, and errors. Consequently, the outputs often perpetuate existing distortions related to gender, ethnicity, or social class, leading to automated biases. Furthermore, the priorities and intended applications of AI systems are frequently shaped by privileged, male, Global North, and white perspectives, resulting in technology that may not equitably serve all communities.

To address this, it is vital to continually question how and where AI should be used. Key considerations include assessing the risks and biases embedded within systems, understanding who benefits and who is disadvantaged, and clarifying the ultimate purpose of deploying AI. In some cases, strict regulation or even bans might be necessary to prevent misuse. The European Commission has taken a step toward this with its risk-based approach to AI regulation, aiming to set boundaries and promote responsible AI use.

WHAT WE CAN LEARN FROM THESE MYTHS SURROUNDING AI

Myths surrounding AI reflect the hopes and fears of humans. As authors, we do not claim to definitively evaluate these myths, we merely aim to learn from them. Consequently, more important than determining whether these myths are true or false is ensuring that we, as a society, are prepared to live in the age of AI. Therefore, we need ethical and transparent principles for the development and use of AI as well as regulations that hold companies accountable, make AI understandable, and redefine copyright laws for the AI age. We also need new concepts for an equitable distribution of wealth, widespread AI literacy, and a critical examination of when, where, and how we consider the use of AI to be societally legitimate. Integrating a feminist perspective, we must ask when developing and using AI: How can we do this to make the world more equitable?

POWER AND ITS INTERRELATEDNESS WITH AI

When it comes to power structures and AI, it is essential to understand that those who currently benefit most from AI and its myths are those best positioned to influence and use this technology in line with their own interests, thereby reinforcing existing power structures. Whether and how individuals or groups are able to do so largely depends on the resources available to them and the regulatory frameworks governing AI development and use. AI has the potential to uphold and strengthen existing power dynamics, but it is also conceivable that AI could help establish new—and more equitable—power structures. This is our vision.

To better assess the challenges regarding power structures and AI, one must first understand power itself. According to Weber (early 20th century, new edition 2016), power is defined as any chance to assert one's own will within a social relationship, even against resistance. Power thus opens and enables possibilities in social interactions and, like energy—or most fittingly *power*—in physics, is a prerequisite for effect. The term itself remains imprecise: All conceivable qualities of a person and all possible constellations can place someone in a position to enforce their will in a given situation (Weber, 2016).

CONCEPTS OF GOVERNING POWER

Governing power is fundamentally about responsible use, control, and regulation for the common good. In democracies, the aim is to structure power to serve all citizens. Unlike the past, we no longer allow transcendent authorities—whether gods, goddesses, or divinely appointed rulers like kings or nobles—or politicians to exercise unchecked power that infringes on human dignity. In modern democracies, sovereignty rests with citizens, and this principle remains essential in an AI-influenced world, where we must manage power anew to uphold human dignity and sovereignty.

If power is seen as the ability to assert one's will in social contexts (Weber, 2016), its regulation must place humans in the center. To prevent abuse, political systems use mechanisms like the separation of powers, which divides authority among the legislative, executive, and judiciary branches, and impose term limits to avoid power concentration. In economic and social spheres, laws and regulations are implemented to ensure equitable interactions and curb potential abuses.

While legal frameworks like the Universal Declaration of Human Rights and other international agreements affirm equity, real-world social structures often fall short of these ideals. This discrepancy arises often not from a lack of regulation but from informal power dynamics that reinforce traditional values, often influenced by patriarchal systems. Patriarchy, defined as a structure that privileges male dominance and control, perpetuates inequalities and oppressed marginalized groups (Johnson, 2007; Walby, 1989). Contemporary AI systems risk reproducing these same power structures if not carefully managed (Siapka, 2022; Weinberg, 2020).

GOVERNING AI BY MEANS OF REGULATION

As the EU has created the first comprehensive regulation of AI worldwide, we summarize here the overall ideals that it incorporates. The EU's Regulation (EU) 2024/1689 (AI Regulation), passed in 2024 and is enforceable from August 2, 2026. The legal framework for AI in Europe is guided by two main principles: ensuring human oversight and accountability for AI use. In this vein, AI is seen as a tool that lacks human qualities and cannot be sovereign. As such, its deployment must adhere to standards of diversity, non-discrimination, fairness, and privacy protection. The regulation aims to promote human-centered, trustworthy AI while protecting health, safety, and fundamental rights and reflects the EU's commitment to a human-controlled AI approach, reinforcing that AI should serve humanity and the common good, while supporting innovation (European Union, 2024). The European Commission describes the vision of instituting an "ecosystem of trust" and "excellence," emphasizing the need for fundamental rights protection (European Commission, 2020). While the regulation establishes a necessary legal framework for trustworthy AI, it alone is not sufficient to fully achieve this ecosystem.

ABUSE OF POWER IN THE AI CONTEXT

The establishment of an "ecosystem of trust" is the result of social practice—not of regulation alone—meaning that the power and will of the involved actors must lead to actions that enable these goals in practice. If this is not the case, abuse of power can occur. Three primary sources of power abuse—we exemplify each—can be identified in the context of AI:

Existing inequalities and power structures reflected in the datasets used to train AI systems are reproduced—An example from recruitment concerns the use of an AI system for automated candidate selection. The system was trained to evaluate resumes based on data from the past ten years, during which most successful applications came from men. Since the AI system used this historical data as a benchmark, it developed a preference for male candidates and systematically rated the resumes of women lower.

Individuals or communities use AI systems to impose their will against societal consensus—This is also contrary to our value system. An example is the use of AI systems to deliberately spread misinformation to manipulate people's behavior, such as influencing their purchasing decisions or political views. These systems can be used to shape users' will through targeted influence, often against their own interests or societal consensus, as seen in cases of election interference or unethical marketing strategies.

Due to lack of transparency, explainability, absence of human oversight, and without the primacy of human action, AI itself becomes a bearer of power—This is something our value system rejects. An example is the previously revised use of an automated control system in aviation: Due to a lack of transparency about how the system functioned and insufficient explainability for pilots, the algorithmic decision system could automatically intervene in certain situations without the crew fully understanding or controlling it. In the absence of adequate human oversight and without a clear way to manually override the system's intervention, the software's faulty behavior ultimately led to two plane crashes.

As previously mentioned, when it comes to power structures and AI, it is crucial to understand that those who will benefit from AI are those best positioned to develop, influence, and use this technology in line with their own interests, thereby reinforcing existing power structures, which first and foremost benefit themselves.

POWER AND AI IN PRACTICE

What does this mean in practice? AI development is primarily driven by large technology companies that operate with a profit motive. This means that AI systems are often designed to maximize profit, even if it comes at the expense of privacy, job security, or social justice. Companies like Google, Amazon, Microsoft, Apple, and Meta control most AI developments and applications. They possess the infrastructure, data, and capital to steer the direction of AI research and implementation.

In general, companies that have access to large amounts of data hold advantages over those without such resources. Data collection and processing are often conducted without fair compensation or sufficient information provided to the individuals affected. In AI development, decision-making processes continue to be dominated by White men, leading to a lack of consideration for the needs and perspectives of women and other marginalized groups. People from poorer regions or with less access to education often have little to no involvement in AI development and are rarely included in decision-making processes, meaning their needs and perspectives are largely overlooked. These examples illustrate how existing power structures and economic interests shape and reinforce AI development, which—as feared—could further reproduce and entrench social inequities in the future.

INTERSECTIONAL ANALYSIS OF SOCIETAL INJUSTICES AMPLIFIED BY AI

The introduction and growing prevalence of AI systems raise significant ethical questions. One of the most severe challenges is that AI systems often reinforce existing inequities and power structures, discriminating against various marginalized individuals and communities. This results in marginalized, underrepresented, and disadvantaged people not benefiting from the development and deployment of AI but rather being affected by its negative impacts, such as discrimination, environmental pollution, and exploitative practices in supply chains.

Bias in AI systems can lead to systematic disadvantages where certain groups are discriminated against based on their affiliation with social categories. This can involve characteristics like gender, ethnicity, sexual identity, age, social background, and religion. The effects are especially severe on those experiencing multiple forms of discrimination, as they face compounded disadvantages both in the analogue world and in AI. The problem of discriminatory AI—despite the efforts of many initiatives and dedicated individuals—is still too seldom on the decision makers' agendas, not widely understood or known, and urgently needs to be more broadly discussed, questioned, and actively changed.

THE FEMINIST PERSPECTIVE OF THIS CHAPTER

We approach the AI-power interrelatedness from a critical feminist perspective. Feminism as a form of resistance, activism, and research has evolved and adapted over centuries (LeGates, 2001). Today, feminism encompasses a plurality of approaches (Ferguson, 2017). Historically, feminism focused on equal rights for women, leading to gains in voting rights, property ownership, labor participation, and legalized abortion in many countries (Council of Europe, 2002). Critics argue feminism has often been elitist, primarily representing White, middle-class women, sometimes excluding or undermining the voices of Black and working-class women (Berger & Guidroz, 2009; Hill Collins, 2000; Hooks, 2001; Spelman, 1988). As an inclusive and critical form, intersectional feminism emerged in the 1990s, advocating for equity across all social identities (Berger & Guidroz, 2009; Council of Europe, 2002). Our perspective aligns with this stream of critical and intersectional feminism.

The concept of oppression, a negative exercise of power over others, is fundamental to understanding feminist movements (Hill Collins, 2000). "Oppression describes any unjust situation where, systematically and over a long period of time, one group denies another group access to the resources of society. Race, class, gender, sexuality, nation, age, and ethnicity among others constitute major forms of oppression" (Hill Collins, 2000, p. 4). Today, AI might automate and thus, perpetuate this oppressive nature of power in our societies.

THE TOOLS OF INTERSECTIONALITY

We adopt the critical feminist concept of intersectionality to analyze discrimination by AI systems. Originating from Black feminism, intersectionality describes the complex nature of patriarchal structures in our societies. It considers the diverse characteristics that can lead to discrimination against people on the basis of various forms of oppression (Edwards et al., 2021), focusing on their intersections. Intersectionality emphasizes that people with multiple marginalized traits face more severe discrimination than those with only one—often exceeding the sum of individual disadvantages (Crenshaw, 1989). For example, a woman might face sexism in a job application process, while a person with a foreign name could encounter racism. Women of color or women with foreign names, who face both sexism and racism, often experience far greater discrimination than someone with only belonging to one of the oppressed groups.

BIAS AND DISCRIMINATION IN AI SYSTEMS

Bias in AI systems is not a random phenomenon but often a result of structural inequities embedded in data and algorithms. AI systems learn from historical data, and if these data contain biases, the systems replicate these patterns. Studies and media reports have repeatedly shown that AI can produce discriminatory outcomes against marginalized individuals in areas such as recruitment (Dastin, 2018; Hao, 2019; Jago & Laurin, 2022), credit scoring (Bertrand & Weill, 2021; Telford, 2019), and recidivism predictions (Angwin et al., 2016; Chouldechova, 2017; Malek, 2022). Generative AI systems have also been criticized for reproducing stereotypes in images and texts related to gender, appearance, and social background (Alba, 2022; Bennett et al., 2021; Bianchi et al., 2023; Gengler, 2024; Thomas & Thomson, 2023). Subsequently, we will reflect upon a selection of groups disproportionately affected by AI discrimination and will conclude by demonstrating how people belonging to two or more disadvantaged communities face more severe problems.

Sexist Discrimination by AI

Gender-specific discrimination, particularly against women, is often amplified by AI. Algorithms trained on biased data can produce discriminatory outcomes, such as in recruitment, where studies show that AI systems can disadvantage female applicants due to reliance on historical data favoring men (Jago & Laurin, 2022; Köchling et al., 2021). Facial recognition is another example, often more accurate for male faces, as these are overrepresented in training datasets (Buolamwini & Gebru, 2018; Singh et al., 2020; Smith & Ricanek, 2020). In healthcare, biases arise from clinical data that predominantly reflect male symptoms, leading to less accurate diagnoses for women (Baggio et al., 2013; Hamberg, 2008; Paviglianiti & Pasero, 2020). Users reported that an AI in credit rating provided men with higher credit limits than women, despite similar qualifications. This bias persisted even though the attribute "gender" was excluded from the data, as proxy data still indirectly reflected gender and led to discrimination (Hegemann, 2019; Telford, 2019; Vigdor, 2019).

Discrimination Against LGBTQAI+ by AI

AI systems also show discrimination against individuals with various sexual identities. For instance, in influencer marketing, an algorithm for identifying influencers rated individuals who described their sexuality as LGBTQ+ less favorably (Bishop, 2021). These biases can result in LGBTQAI+ individuals facing less success in collaboration opportunities, even with comparable content and quality.

Racial Discrimination by AI

Racial biases in AI are a well-documented form of discrimination. Facial recognition technology often struggles to accurately identify People of Color, with algorithms showing lower accuracy for darker skin tones (Buolamwini & Gebru, 2018; Serna et al., 2019, 2021). This can lead to misidentifications, particularly problematic in law enforcement, where there have been cases of wrongful arrests of Black men due to faulty AI (Hill, 2020; Incident 74, 2020). In healthcare, algorithms have also disadvantaged Black patients' access to services (Obermeyer & Topol, 2021; Obermeyer et al., 2019; Vartan, 2019). AI systems have been shown to be less effective at diagnosing conditions like diabetic retinopathy (Burlina et al., 2021) and skin cancer (Davis, 2021) in People of Color compared to White individuals.

Multiple Discrimination by AI

The issue of AI becomes particularly severe when people experience multiple forms of discrimination, such as Women of Color. This group faces both sexist and racial biases. Facial recognition systems, for instance, not only struggle with identifying women but are especially inaccurate when it comes to women with darker skin (Buolamwini & Gebru, 2018). This multiple discrimination is also evident in other areas, such as the job market, where algorithms can disadvantage applicants based on a combination of gender and ethnicity (Köchling et al., 2021).

FEMINIST VISION FOR THE AI AGE

These examples of AI demonstrate the need for a change in developing and using these systems. Achieving an equitable and inclusive AI future requires tackling structural inequities in society and ensuring that AI systems do not replicate these biases (Gengler et al., 2023). In democracies, citizens (should) hold the power, making it our collective duty to build a society rooted in equity, respect, and progressive values.

We must constantly reflect on how we want to live together—including how we coexist with AI. Europe has taken a step toward this vision with its AI Regulation, which promotes ethical, human-centered AI focused on diversity, fairness, and privacy protection (European Union, 2024; High-Level Expert Group, 2019)—even though merely with voluntary requirements regarding inclusivity and sustainability (Wedel et al., 2024). The envisioned utopia is clear: Create an "ecosystem of trust and excellence" where AI enhances society for everyone, not just a select few. Realizing this vision demands collective responsibility, where states, companies, and citizens—in all aspects of life—use AI to empower communities rather than reinforce dominance. Success hinges on a shared commitment to education, inclusivity, and sustainable engagement with AI systems (Wedel et al., 2024).

A feminist AI approach can lead this transformation, ensuring that AI systems are designed to uplift marginalized voices, challenge power structures, and drive social progress for all. It envisions a world where technology serves as a tool for equity, amplifying diverse perspectives, and fostering a more compassionate, democratic society.

FEMINIST AI TO MAKE THE WORLD MORE EQUITABLE

In earlier projects, the authors examined how AI, guided by critical and intersectional feminist principles centering around equity, transformation, and reflexivity, can foster a more equitable society and power distribution. Collaborating with an interdisciplinary group, we developed a definition of feminist AI during the 2022 *Feminist AI and Digital Policy Roundtable*, initiated by Alexandra Wudel and the lead author. For this chapter, we have expanded the definition with new insights:

> Feminist Artificial Intelligence (AI) is an inclusive, intersectional approach that leverages digitalization to promote equity for all. It focuses on the needs of women, LGBTQAI+, and other marginalized groups, irrespective of their gender, age, religion, beliefs, impairment, sexual identity, ethnic background, and

appearance. The approach of feminist AI aims to establish a modern and holistic relationship between human–machine systems and machine–machine systems. Its focus is on addressing patriarchal structures, systemic privileges, and power imbalances in society, politics, and the economy. Feminist AI can serve as a driver for the Sustainable Development Goals and make a significant contribution to a more sustainable hat future.

ETHICAL AI THROUGH A FEMINIST LENS

Feminist AI expands the realms of ethical AI by adopting a more radical and transformative approach that explicitly addresses systemic inequalities and power dynamics in AI development and deployment. While ethical AI broadly focuses on fairness, transparency, and safety, feminist AI goes further by challenging existing power structures and centering intersectionality. Grounded in feminist theory, critical studies, and community work, it emphasizes participatory methods, co-creation, and the inclusion of marginalized voices in the development process. Because of the demonstrated relevance and its transformative effect within the analogue world, it seems to be a promising approach for a world increasingly shaped by AI. Feminist AI defends the progress already made in the analogue world, aims to avoid repeating the flaws of analogue power structures by, e.g., not recreating systems of inequalities and focuses on building a more equitable AI to begin with. It therefore questions the very nature of data collection, categorization, and representation, especially regarding gender and other identity markers. It also recognizes technology as inherently political and shaped by societal biases, rather than viewing it as neutral. By linking AI development to environmental justice and sustainability, and promoting community-led accountability mechanisms, feminist AI seeks to use AI as a tool for social transformation and challenging oppressive norms. To conclude, while ethical AI aims to develop morally sound systems within existing frameworks, feminist AI strives to fundamentally reshape the AI landscape by addressing root causes of inequity and actively working towards a more just and inclusive technological future.

CONCLUSION: CALL FOR COLLECTIVE ACTION

To shape a more equitable and inclusive future with AI, we must all become active. The urgency to question and transform existing power structures in the development and application of AI must be collectively recognized and addressed. It is essential that research, policy, and citizens share responsibility and actively participate in designing and using AI systems that align with our values, support democracy, and contribute to a more equitable world. We need to leverage AI to learn essential future skills and shape it according to the needs of marginalized communities. It is not enough to passively accept AI—we must critically examine when, where, and how it is used, and base its development and application on feminist principles. By doing so, we have the opportunity for AI to reflect our high societal ideals and contribute to an equitable, social, and democratic future. The AI age is full of myths, hopes, and fears, presenting us with novel opportunities and challenges. Only by working together can we develop and utilize them to make our society more equitable.

RELATED WORK AND AI USE

This book chapter builds on the authors' prior research, integrating concepts that are also part of a paper currently under review, a forthcoming book chapter, and an upcoming popular science book. In crafting this chapter, we utilized AI tools for translation, text refinement, and stylistic enhancements including ChatGPT 4o, Perplexity, and DeepL during the period of August to November 2024.

REFERENCES

Alba, D. (2022, December 8). OpenAI Chatbot Spits Out Biased Musings, Despite Guardrails. Bloomberg. Com. https://www.bloomberg.com/news/newsletters/2022-12-08/chatgpt-open-ai-s-chatbot-is-spitting-out-biased-sexist-results

Angwin, J., Larson, J., Mattu, S., & Kirchner, L. (2016, Mai 23). Machine Bias. https://www.propublica.org/article/machine-bias-risk-assessments-in-criminal-sentencing

Baggio, G., Corsini, A., Floreani, A., Giannini, S., & Zagonel, V. (2013). Gender medicine: A task for the third millennium. *Clinical Chemistry and Laboratory Medicine (CCLM)*, *51*(4), 713–727. https://doi.org/10.1515/cclm-2012-0849

Bennett, C. L., Gleason, C., Scheuerman, M. K., Bigham, J. P., Guo, A., & To, A. (2021). "It's Complicated": Negotiating Accessibility and (Mis)Representation in Image Descriptions of Race, Gender, and Disability. Proceedings of the 2021 CHI Conference on Human Factors in Computing Systems, 1–19. https://doi.org/10.1145/3411764.3445498

Berger, M. T., & Guidroz, K. (Eds.). (2009). *The intersectional approach: Transforming the academy through race, class, and gender.* University of North Carolina Press.

Bertrand, J., & Weill, L. (2021). Do algorithms discriminate against African Americans in lending? *Economic Modelling*, *104*, 105619, https://doi.org/10.1016/j.econmod.2021.105619

Bianchi, F., Kalluri, P., Durmus, E., Ladhak, F., Cheng, M., Nozza, D., Hashimoto, T., Jurafsky, D., Zou, J., & Caliskan, A. (2023). Easily accessible text-to-image generation amplifies demographic stereotypes at large scale. 2023 ACM Conference on Fairness, Accountability, and Transparency, 1493–1504. https://doi.org/10.1145/3593013.3594095

Bishop, S. (2021). Influencer management tools: Algorithmic cultures, brand safety, and bias. *Social Media + Society*, *7*(1), 205630512110030. https://doi.org/10.1177/20563051211003066

Buolamwini, J., & Gebru, T. (2018). Gender shades: Intersectional accuracy disparities in commercial gender classification. Proceedings of the 1st Conference on Fairness, Accountability and Transparency, 77–91. https://proceedings.mlr.press/v81/buolamwini18a.html

Burlina, P., Joshi, N., Paul, W., Pacheco, K. D., & Bressler, N. M. (2021). Addressing artificial intelligence bias in retinal diagnostics. *Translational Vision Science & Technology*, *10*(2), 13. https://doi.org/10.1167/tvst.10.2.13

Chouldechova, A. (2017). Fair prediction with disparate impact: A study of bias in recidivism prediction instruments. *Big Data*, *5*(2), 153–163. https://doi.org/10.1089/big.2016.0047

Council of Europe. (2002). Feminism and Women's Rights Movements—Gender Matters—Www.coe.int. Gender Matters. https://www.coe.int/en/web/gender-matters/feminism-and-women-s-rights-movements

Crenshaw, K. (1989). Demarginalizing the intersection of race and sex: A black feminist critique of antidiscrimination doctrine, feminist theory and antiracist politics. *University of Chicago Legal Forum*, *1989*(1), 139–167.

Dastin, J. (2018, October 11). Amazon scraps secret AI recruiting tool that showed bias against women. Reuter. https://www.reuters.com/article/us-amazon-com-jobs-automation-insight/amazon-scraps-secret-ai-recruiting-tool-that-showed-biasagainst-women-idUSKCN1MK08G

Davis, N. (2021, November 9). AI skin cancer diagnoses risk being less accurate for dark skin – study. The Guardian. https://www.theguardian.com/society/2021/nov/09/ai-skin-cancer-diagnoses-risk-being-less-accurate-for-dark-skin-study

Edwards, J., Clark, L., & Perrone, A. (2021). LGBTQ-AI? Exploring Expressions of Gender and Sexual Orientation in Chatbots: Bd. UI '21: Proceedings of the 3rd Conference on Conversational User Interfaces.

European Commission. (2020). *On artificial intelligence – A European approach to excellence and trust.* [White Paper.]. https://eur-lex.europa.eu/legal-content/en/TXT/PDF/?uri=CELEX:52020DC0065&from=EN

European Union. (2024). Regulation—EU - 2024/1689—EN - EUR-Lex. https://eur-lex.europa.eu/eli/reg/2024/1689/oj?eliuri=eli%3Areg%3A2024%3A1689%3Aoj&locale=en

Ferguson, K. E. (2017). Feminist theory today. *Annual Review of Political Science*, *20*(1), 269–286. https://doi.org/10.1146/annurev-polisci-052715-111648

Gengler, E. J. (2024). Sexism, Racism, and Classism: Social Biases in Text-to-Image Generative AI in the Context of Power, Success, and Beauty. Wirtschaftsinformatik 2024 Proceedings.

Gengler, E. J., Wedel, M., Wudel, A., & Laumer, S. (2023). Power Imbalances in Society and AI: On the Need to Expand the Feminist Approach. Wirtschaftsinformatik 2023 Proceedings. https://aisel.aisnet.org/wi2023/37

Hamberg, K. (2008). Gender bias in medicine. *Women's Health*, *4*(3), 237–243. https://doi.org/10.2217/17455057.4.3.237

Hao, K. (2019, April 5). Facebook's ad-serving algorithm discriminates by gender and race. MIT Technology Review. https://www.technologyreview.com/2019/04/05/1175/facebook-algorithm-discriminates-ai-bias/

Hegemann, L. (2019, November 21). Apple Card: Weiblich, Ehefrau, kreditunwürdig? Die Zeit. https://www.zeit.de/digital/datenschutz/2019-11/apple-card-kreditvergabe-diskriminierung-frauen-algorithmen-goldman-sachs?utm_referrer=https%3A%2F%2Fwww.google.com%2F

High-Level Expert Group (2019). Ethics guidelines for trustworthy AI. Publications Office of the European Union. https://data.europa.eu/doi/10.2759/346720

Hill, K. (2020, December 29). Flawed Facial Recognition Leads To Arrest and Jail for New Jersey Man. The New York Times. https://www.nytimes.com/2020/12/29/technology/facial-recognition-misidentify-jail.html

Hill Collins, P. (2000). *Black feminist thought: Knowledge, consciousness, and the politics of empowerment* (10th anniversary ed.). Routledge.

Hooks, B. (2001). *Ain't I a woman: Black women and feminism (Nachdr.).* Pluto Press.

Incident 74: Detroit Police Wrongfully Arrested Black Man Due To Faulty FRT. (2020, January 30). https://incidentdatabase.ai/cite/74/

Jago, A. S., & Laurin, K. (2022). Assumptions about Algorithms' capacity for discrimination. *Personality and Social Psychology Bulletin*, *48*(4), 582–595. https://doi.org/10.1177/01461672211016187

Johnson, A. G. (2007). *The gender knot.* Pearson Education India.

Kelly, S. M. (2024, May 23). Elon Musk says AI will take all our jobs | CNN Business. CNN. https://www.cnn.com/2024/05/23/tech/elon-musk-ai-your-job/index.html

Köchling, A., Riazy, S., Wehner, M. C., & Simbeck, K. (2021). Highly accurate, but still discriminatory: A fairness evaluation of algorithmic video analysis in the recruitment context. *Business & Information Systems Engineering*, *63*(1), 39–54. https://doi.org/10.1007/s12599-020-00673-w

LeGates, M. (2001). *In their time: A history of feminism in Western society.* Routledge.

Malek, M. A. (2022). Criminal courts' artificial intelligence: The way it reinforces bias and discrimination. *AI and Ethics*, *2*(1), 233–245. https://doi.org/10.1007/s43681-022-00137-9

Obermeyer, Z., Powers, B., Vogeli, C., & Mullainathan, S. (2019). Dissecting racial bias in an algorithm used to manage the health of populations. *Science*, *366*(6464), 447–453. https://doi.org/10.1126/science.aax2342

Obermeyer, Z., & Topol, E. J. (2021). Artificial intelligence, bias, and patients' perspectives. *The Lancet*, *397*(10289), 2038. https://doi.org/10.1016/S0140-6736(21)01152-1

Paviglianiti, A., & Pasero, E. (2020). VITAL-ECG: A de-bias algorithm embedded in a gender-immune device. 2020 IEEE International Workshop on Metrology for Industry 4.0 & IoT, 314–318. https://doi.org/10.1109/MetroInd4.0IoT48571.2020.9138291

Robson, D. (2021, October 26). Why we're so terrified of the unknown. BBC. https://www.bbc.com/worklife/article/20211022-why-were-so-terrified-of-the-unknown

Serna, I., Morales, A., Fierrez, J., Cebrian, M., Obradovich, N., & Rahwan, I. (2019). Algorithmic Discrimination: Formulation and Exploration in Deep Learning-based Face Biometrics (arXiv:1912.01842). arXiv. http://arxiv.org/abs/1912.01842

Serna, I., Peña, A., Morales, A., & Fierrez, J. (2021). InsideBias: Measuring Bias in Deep Networks and Application to Face Gender Biometrics. 2020 25th International Conference on Pattern Recognition (ICPR), 3720–3727. https://doi.org/10.1109/ICPR48806.2021.9412443

Siapka, A. (2022). Towards a Feminist Metaethics of AI. Proceedings of the 2022 AAAI/ACM Conference on AI, Ethics, and Society, 665–674. https://doi.org/10.1145/3514094.3534197

Singh, R., Agarwal, A., Singh, M., Nagpal, S., & Vatsa, M. (2020). On the robustness of face recognition algorithms against attacks and bias. *Proceedings of the AAAI Conference on Artificial Intelligence*, *34*(09), 13583–13589. https://doi.org/10.1609/aaai.v34i09.7085

Smith, P., & Ricanek, K. (2020). Mitigating Algorithmic Bias: Evolving an Augmentation Policy that is Non-Biasing. 2020 IEEE Winter Applications of Computer Vision Workshops (WACVW), 90–97. https://doi.org/10.1109/WACVW50321.2020.9096905

Spelman, E. V. (with Internet Archive). (1988). *Inessential woman: Problems of exclusion in feminist thought.* Beacon Press. http://archive.org/details/inessentialwoman00spel

Telford, T. (2019, November 11). Apple Card algorithm sparks gender bias allegations against Goldman Sachs. *Washington Post.* https://www.washingtonpost.com/business/2019/11/11/apple-card-algorithm-sparks-gender-bias-allegations-against-goldman-sachs/

Thomas, R. J., & Thomson, T. J. (2023, July 10). Ageism, sexism, classism and more: 7 examples of bias in AI-generated images. The Conversation. http://theconversation.com/ageism-sexism-classism-and-more-7-examples-of-bias-in-ai-generated-images-208748

Turing, A. M. (1950). I.—Computing machinery and intelligence. *Mind*, *LIX*(236), 433–460. https://doi.org/10.1093/mind/LIX.236.433

Vartan, S. (2019, October 24). Racial Bias Found in a Major Health Care Risk Algorithm. Scientific American. https://www.scientificamerican.com/article/racial-bias-found-in-a-major-health-care-risk-algorithm/

Vigdor, N. (2019, November 10). Apple Card Investigated After Gender Discrimination Complaints. The New York Times. https://www.nytimes.com/2019/11/10/business/Apple-credit-card-investigation.html

von Schaik, C., & Michel, K. (2023). Mensch sein—Von der Evolution für die Zukunft lernen. Rowohlt. https://www.thalia.de/shop/home/artikeldetails/A1068312575

Walby, S. (1989). Theorising patriarchy. *Sociology*, *23*(2), 213–234. https://doi.org/10.1177/0038038589023002004

Weber, M. (2016). *Max Weber-Studienausgabe (1. Aufl.)*. Mohr Siebeck.

Wedel, M., Hazim, A., & Wudel, A. (2024). Europäische KI-Regulierung: Auf der Suche nach verbindlichen Ansätzen für Nachhaltigkeit und Inklusion. In M. Friedewald, A. Roßnagel, C. L. Geminn, M. Karaboga, & S. Schindler (Eds.), *Data Sharing – Datenkapitalismus by Default?* (pp. 141–162). Nomos Verlagsgesellschaft mbH & Co. KG. https://doi.org/10.5771/9783748940173-141

Weinberg, L. (2020). Feminist research ethics and student privacy in the age of AI. *Catalyst: Feminism, Theory, Technoscience*, *6*(2), Article 2. https://doi.org/10.28968/cftt.v6i2.32943

15 Feminist Ethics and Artificial Intelligence

Equitable Women's Representation

Vidushi Chaturvedi

INTRODUCTION

This chapter dwells on the interplay of feminist ethics and AI – on the reasons for the absence of accountability, fairness, and gender justice amongst other issues, in the planning and deployment of this ever-evolving technology. Women are few and far between in the domain of AI research and development; they are minuscule even in its use – nevertheless, their voices are loud and clear, when they critique the various facets of AI. Whether as whistleblowers, researchers, leaders, or entrepreneurs – women are coming center stage and bringing the ethical concerns underlying AI proliferation into sharp focus – and these can no longer be ignored. The perspectives shaped by women's unique experiences help in uncovering discrimination, deception, and other harmful consequences of AI. By calling out the deficiencies in the present models of AI, women are playing a vital role in rebuilding AI ethically. The present chapter analyzes the struggles of such women in unravelling the dark, unethical side of AI.

But what is AI? How do we define AI? Is it just another technological innovation? Is it akin to previous technology inventions like printing? Is it promoting patriarchy and creating power centers – leading to new inequalities in an already unequal world? What are the ethical dilemmas in AI that are impacting women's participation, visibility, and self-hood? What is the lens that the field of feminist ethics provides that can help us understand discrimination, bias, and the lack of trust, as well as of accountability, transparency, care, empathy, equity, and inclusion – in all that is AI related? These are some of the questions we engage with here.

There are many definitions of AI – but for the purpose of this chapter, we will focus on the definition provided in the John McCain National Defense Authorization Act of Fiscal Year 2019 "Any artificial system that performs tasks under varying and unpredictable circumstances without human oversight, or that can learn from experiences and improve performance when exposed to data sets…" (U.S. Congress, 2018).

It is clear from the definition that AI is powerful, and is challenging the way we understand our present, visit the past, and imagine the future – but are these developments helping us in creating a better world for women? Is AI capable of empowering women? Or is the adoption of AI disrupting prevailing ethical standards that result in undoing the progress achieved through years of struggle of feminist movements? What are these ethical standards we espouse and uphold? Is AI attacking these values – of inclusion, diversity, empathy, trust, care, to name just a few. Is it true that AI will take women into the past by limiting their role in the public sphere, and is it taking away their agency? The role of technology in the current global pushback on gender equality is worrisome.

The Secretary General of the United Nations has been raising this issue time and again. In his remarks to the Women's Civil Society Town Hall on March 13, 2023, he said, "[M]ale chauvinist domination of new technologies is undoing decades of progress on women's rights". After more than 100 years of progress, "technology is now reversing that trend. It is concentrating power again more in the hands of men – to the detriment of all" (U.N. Secretary General, 2023). It is disturbing that

DOI: 10.1201/9781003585527-20

gender-based violence is assuming more sinister dimensions in the AI age, as the rape of a 16-year-old girl in the Metaverse shows (Patel J.N. (2024). Technology-facilitated gender-based violence targets all women who use technology (UN Regional Information Centre for Western Europe, 2023). Unfortunately, for women, the virtual world is no safer than the physical world. In fact, men find it easier to inflict violence in the virtual world – since laws in this regard are inadequate. It is also easier for offenders to hide, stay untraceable, and remain anonymous. The report on "Preventing discrimination caused by the use of AI" by the Committee on Equality and Non-discrimination has explicitly stated that women and other minorities are discriminated against (Kundu,2024). The European Commission's document "On Artificial Intelligence – A European approach to excellence and trust" mentions gender and ethical guidelines among other aspects. It calls for "requirements to take reasonable measures aimed at ensuring that the use of AI systems does not lead to outcomes entailing prohibited discrimination" (Niethammer, 2020).

The present chapter attempts to explore various related questions: Will ethical concerns like inbuilt bias, lack of transparency, accountability and integrity amongst others, make society regressive? Is the data on which AI is trained responsible for reinforcing prejudices that exist in society today? Is the manner in which AI is being developed, deployed, and used, leading to rise in inequality, injustice, and power asymmetries in the world that impact women adversely? The study of feminist ethics helps us in examining the contemporary world to understand the nature of oppression experienced by the marginalized sections of society. How are films dealing with our relationship with machines (read AI) – do they reflect and reinforce social norms? Is AI being used to promote discrimination, gender stereotypes and gender-bias – problems that feminist movements have long struggled to address. It is estimated that the production of deepfake porn is perhaps doubling every six months (Batha, 2024). The context, along with the rapid proliferation of fake pictures of Taylor Swift that circulated tens of millions of times, has been troubling (Nelson, 2024). Teen boys were found using AI to create and share fake nude pictures of female classmates (Belanger, 2023; Marcus, 2024) The Internet Watch Forum survey suggests that child sexual abuse images generated by AI are a fast-growing problem (Marcus, 2024). It is evident that the AI technology has unleashed newer forms of oppression and violence against the vulnerable sections and the same is rapidly rising. In the next section, we delve into the discipline of feminist ethics and assess whether the advancement in AI technology engages with the experiences of women and attempts to incorporate the ethics of accountability, inclusivity, care, trust, among others.

UNDERSTANDING ETHICAL CONCERNS IN AI WITH THE HELP OF FEMINIST PHILOSOPHY AND CARE ETHICS

Feminist ethics is a promising discipline as it attempts to understand, criticize, and correct the binary view of gender, the privileges historically available to men, and the oppressive social order and harmful practices that fail to incorporate the perspectives of the marginalized (Norlock, 2019). Care ethics on the other hand has been defined more often as a practice or a virtue than as a theory. It includes maintaining human relationships and meeting the needs of both our own-selves and others. It affirms the importance of caring, motivation, emotions, and responsibility, among other values (Internet Encyclopedia of Philosophy, n.d.). In her work, *Care, Gender and Justice*, Bubeck deliberates upon women's work as care (Bubeck, 1995). Scholars of feminist ethics deal with concepts of equal opportunities, socialization, sex roles, and discrimination, as they prevail in society. However, strategies involving altering socialization and practices and improving attitudes also preoccupy feminist ethicists (Ben-Jacob & Glazerman, 2022). With their focus on relationality, care, and empathy rather than on objectivity and dispassion, feminist theories together with care ethics provide fresh perspectives on experiences of women. Feminist philosophy, with its examination of issues concerning social injustice, gender inequality together with care ethics – and its preoccupation with ethics of duty and responsibility – provides a complementary aspect in understanding complex problems. Different strands of feminism – including liberal, radical, Marxist-socialist, post-modernist, and intersectional – helped in enriching the feminist discourses aimed at resolution of social problems.

A new branch of feminism known as intersectionist feminism advocates, "Nobody is free until everybody is free", as expressed in 1971 by Fannie Lou Hamer (Brooks & Houck, 2010), a women's rights activist. Kimberle Crenshaw, a law professor in the USA, coined the term "intersectionist feminism" and explained it as a "prism for seeing the way in which various forms of inequality often operate together and exacerbate each other (UN Women, 2020). However, intersectionist feminism also criticized care ethics for not adequately addressing the experiences of care related with race, class, or disability (Collins, 1990). It has been feared that care ethics, with its emphasis on women's roles centered around care functions, may lead to stereotyping and associating femininity with emotional labor and relational dependency (Hoagland, 1991). Moreover, such glorification of care labor may lead to ignoring the exploitative nature of the institutions that provide care (Engster, 2007). Care work of women has been devalued by social institutions (Kittay, 1999), while the capitalist economy and patriarchy have also been exploitative of women (Engels, 1884; Hennessy, 2000). It is important to note that care ethics helps us in understanding the manner in which the caregiving responsibilities result in gender disparities (Tronto, 1993). It was Joan Tronto who developed the understanding of care ethics, with its focus on responsiveness, attentiveness, and other values like accountability, that we are dealing with for understanding AI in this chapter.

It is interesting to note that the post-modern and the post-structuralist feminist theories challenge the rigid definitions of gender and identity that fail to capture the fluid and manufactured essence of identities (Butler, 1990; Weedon, 1997). Judith Butler's notion of "gender performativity" helps deconstruct binaries in order to appreciate and acknowledge the multiplicity of gender expression and identities (Butler, 1990). A more detailed exposition of these theories is beyond the scope of this chapter. At the same time, an overview of the current philosophy of feminist ethics and the analysis of social issues prepares us to grapple with the biggest challenges of the day in the wake of the AI revolution and helps us explore possible solutions.

The coming together of feminist ethics and AI provides unique opportunities to leverage technology for resolving existing problems of discrimination, gender stereotyping, exclusivity, inequity, and non-diversity. The following paragraphs will enquire into the ways AI is being developed/deployed to address the above concerns. The centrality of AI in all functions of human activity in the present age is adding seemingly inexplicable complexities to our life – many of these owing to the absence of time-tested values, such as accountability, transparency, explainability, inclusion, fairness, etc. in the research, development/deployment of AI. Some practitioners of privacy/cybersecurity law and investigation of high-technology crime, like Mark Rasch, feel that balancing ethical values that we hold dear and hard coding them into a machine learning tool that both mimics and shapes how society operates is not easy (Rasch, 2023). It may not be easy, but it is necessary if society is to stay rooted in morality and if each human being is to be treated with dignity.

AI is created with the purpose of adding efficiency to human endeavors while increasing the ease of living. Whether or not it is being leveraged for this purpose remains to be seen. However, what is clear from the deployment and use of AI, so far is that it is a sure formula for maximizing profits. Nvidia, which manufactures advanced graphic processing units (GPUs), used by the big tech giants like Google, Microsoft, Apple, and Meta, witnessed its profit soar by more than 600% and revenue by more than 200%, as reported by the Associated Press (Shivangini, 2024). The revenue of this company rose by 492% from its third quarter in fiscal year 2023, amounting to $35.08 billion in the quarter ending November 20, 2024 (Fox, 2024). It is no wonder that AI led developments are rapidly increasing inequalities between nations.

The big five tech companies of the world, Alphabet, Amazon, Apple, Meta, and Microsoft are all located in the USA. In 2023, the USA invested $67.2 billion in AI-related private investment – this was 8.7 times more than that of China, the second highest investor in AI in the world (Schellekens & Skilling, 2024). Use of AI in fields such as medicine, finance, defense, education, etc. will further enhance the capabilities of countries like the USA and China. Most parts of the world are struggling even with access to the internet, while on the other hand, research in AI is helping countries race ahead. Low-income countries and lower-middle-income countries have an internet penetration of merely 27% and 52% respectively as compared to 80% and 95% in upper-middle and high-income countries (Schellekens & Skilling, 2024). As pointed out by the UN chief in UN's 65th Session in 2021, "Gender inequality is embedded in the world's political, social and economic systems" (United Nations, 2021). The gender gap reports of the World Economic Forum (WEF) are an indication of the same. It may be mentioned that since 2006 more than 100 countries have been assessed for their performance vis-à-vis women's economic participation and opportunities, education attainment, health, and survival as well as political empowerment. The 18th edition of this report, published in June 2024 mentions that it will take 134 years to reach full parity going at the current pace of progress – five generations beyond 2030 SDG targets (Global Gender Gap, 2024)

Silicon Valley, where the tech giant companies are concentrated, has produced the richest men on the planet. It is seen by many as a hub of progress but its ideas need to be understood in an ethical context (Giannella, 2015). The bro-culture prevailing here is marked by aggression, chauvinism, and a move-fast-and-break-things mantra, which is both the cause and effect of low visibility of women in the AI industry. That explains why Melinda F. Gates wishes to create an alternative to Silicon Valley (Sauer, 2022). A study by McKinsey indicated that more women were leaving tech roles, with the representation of women dipping lower than it was in 2018 (McKinsey, 2022). The same report – "Women in the Workplace: 2024", indicates that women continue to face microaggression as often today as several years ago (McKinsey, 2024). The aggressive personal attributes of men unabashedly put on display also reflect the overall corporate behavior – the company Meta is known for buying its competitors (Instagram and WhatsApp), and its young and rich CEO Mark Zuckerberg for "ripping off his rivals"(Dugan, 2023). Moreover, he laid off 21,000 employees as his business grows in leaps and bounds with over three billion users feeding data into its AI ventures. The competition between these companies is fierce as they grow in revenue, power, and influence. Another tech giant, Microsoft laid off 10,000 employees while investing billions in AI tools – among those who were laid off were the seven members working in the office of the responsible AI team, tasked with laying down the principles for the organization (Rasch, 2023).What does the elimination of the ethics team signify?

Other tech companies are also seen placing revenues over safety, responsibility, and care. Facebook refused to make changes that it knew harmed children and adults, because it hurt its profits. Revelations of the inner workings and cultural ethos of the company were exposed when whistleblower Francis Haugen testified before Congress that the platform exploited negative emotions to keep people, especially children, from leaving, as more time on the platform meant more money (Ortutay & Klepper, 2021). The McKinsey report on the workplace – with its emphasis on diversity, equity, and inclusion, as well as evidence of ever-growing investments in AI, or massive lay-offs – is indicative of the unbound ambition for material growth and disregard for employees' well-being/increase in mental depression. Both feminist philosophy and care ethics aspire to change this prevailing emphasis on dispassionate logic, rationality, and other attributes perceived as masculine (Held, 2006).The McKinsey report also points out that women are supportive of employees' well-being and foster inclusion even though these values go unnoticed (McKinsey, 2024).

Values associated with nurturing, relational sensitivity, empathy, compassion, care ethics, etc. form part of feminist ideals, and these need to be brought center stage to ensure that AI systems serve diversity and social responsibility as well as promote human dignity. In order to ensure a humane society based on collaboration, empathy, sustainability, equity, care, and other similar values, we need a more nuanced appreciation of the interrelated nature of social injustice (Collins & Bilge, 2016). It is in this context that the intersectionality theory of feminism assumes importance. Cultural norms of caring and nurturing have perpetuated gender disparities as women disproportionately bear caregiving roles and duties without adequate institutional support and backing (Tronto, 1993).

As elaborated above, the lived experiences of women provide fresh perspectives and a lens for looking at and understanding social problems. Those without first-hand experiences of oppression, may not know the ground realities of a society. They may have never experienced sexism, racism, or classism – hence they may never see or understand bias – researchers term this lack of discernment "privilege hazard" (D'Ignazio & Klein, 2020).

Proponents of feminist ethics point out that these unique experiences – termed the "empiricism of lived experiences" – in the words of Anita Gurumurthy, Executive Director of IT for Change, are necessary to see and comprehend existing reality. "The lack of these lived experiences – the evidence of how things truly are – profoundly limits the ability to foresee and prevent harm, to identify existing problems in the world, and to imagine possible solutions" (D'Ignazio & Klein, 2020). Inclusion of women's viewpoints in AI can mitigate many of the harms that are seen today. Without ethical guardrails, AI risks reproducing real-world biases and discrimination, fueling divisions, and threatening fundamental human rights and freedoms, as stated by Gabriela Ramos, who works for UNESCO (UNESCO, 2023).The study and application of feminist ethics in AI, enriches its evaluation and functioning – it enables the advocates of women's rights in taking the lead, in identifying what is missing in AI models, and in providing research and methodologies that can correct what is not right and unethical.

LOW VISIBILITY OF WOMEN IN TECHNOLOGY – A LOST OPPORTUNITY FOR INCORPORATING FEMINIST ETHICS

Technology has so far been a male domain. AI does not seem to be changing that reality anytime soon. A look at the figures from the top four tech companies in the United States of America is sufficient to drive home the point that the gender gap in technology is closing at a snail's pace, despite vociferous advocacy of women's empowerment by governments/international agencies. Apple, which was the world's largest technology company for seven years (Ponciano, 2023), had a dismal 23% women in tech roles – not much better than the others. Facebook has 22%, Google has 21%, and Microsoft has 19% women. This is per the latest data available from the tech companies since 2017, published by Reuters. It is surprising that Amazon does not disclose the gender breakdown of data of its tech workforce, as mentioned in the report "Dominated by Men" (Huang, 2023). It is a fact that most data scientists are male, white, and highly educated.

Research shows that development and implementation of technology is dominated by men, while women's experiences with technology are "created in the image of men" (Lie et al., 1988). It is necessary to recreate the prevailing image of women as ignorant and non-existent in tech matters, epitomized as "Machina ex Dea" (Rothschild, 1983). Women's visibility in the technology sector (beginning with girls enrolling in science, technology, engineering, and mathematics: STEM) is low despite awareness and policy interventions; see "Ten Simple Rules for Empowering Women in STEM", in Guevara-Ramírez et al. (2022). Details on women's engagement with STEM can be seen in later paragraphs. STEM is known to empower women (Elhussein & Hakspiel, 2024). The World Economic Forum says that there exists a difference of 12 percentage points in the use of generative AI (a type of AI that can create text, images, videos, etc., with the help of input data and prompts) between men and women. In the age bracket of 18–25 years, 71% of men use AI, while in the same age group, 59% of women are using and adopting AI (The Conversation, 2024). The Keenen Institute has established that nearly 80% of today's female workers are in jobs exposed to automation via regenerative AI, compared to 58% of men (The Conversation, 2024).That means women may be out of jobs owing to the use of AI in sectors where they are concentrated, like Business Process Outsourcing (BPO) in Philippines and India (Gymrek et al., 2023). Another study by Oxford Economics and Cognizant reports that 90% of jobs will be affected by generative AI by 2032. Between 2023 and 2032, the percentage of jobs with higher exposure to AI would increase more than six-fold. It is important to intervene at this juncture, to make policy corrections to incorporate the perspectives and vision that women bring to male-dominated organizations, which in turn can make new tech more inclusive, diverse, and egalitarian.

Scholars focusing on feminist ethics have attempted to explain how and why experiences of women have been ignored and swept away for so long – they are merely mentioned in passing. Each innovation in the field of technology is met with excitement and hubris. Pursuit of new learnings/knowledge by human beings provides them confidence. Are men and women equally involved in the pursuit of such knowledge? The gender gap in STEM is worrisome. Mindsets that are steeped in age-old customs directing women toward a more stereotyped education and career choice exist in all parts of the world, varying only in degrees. Prevailing gender norms and society's expectations from girls decides what subjects/streams boys and girls are "cut out for", and this in turn perpetuates the gender gap in STEM (World Economic Forum, 2024). In emerging economies, girls face additional obstacles in accessing digital assets and skills.

In other parts of the world, such as India, government policies are encouraging girls, through scholarships and affirmative action, to take up STEM. In UAE, Germany, Belarus, and South Korea 30% of STEM graduates are women (UN Women, 2024; World Economic Forum, 2023) Despite these positive developments, the representation of women in technology is far from satisfactory. Feminist ethics scholars believe that the education system is patriarchal and dominated by men, just like the workforce is. They argue that the education system is a primary preparation for leading into the future workforce. Fewer women study mathematics, let alone major in the discipline. We need to encourage them to major in all fields (Wang & Degol, 2013). It is seen that the know-how and use of a new technology is monopolized by men to prove their masculinity – by being superior not only to women but also to other men (Lie, 1996). The "AI Women Can Code" initiative by UN Women involved students like Natacha Sangwa from Rwanda, who expressed her experience with technology in the following words, "I have noticed that AI is mostly developed by men and trained on datasets that are primarily based on men". She adds further, "When women use some AI powered system to diagnose illnesses, they often receive inaccurate answers because AI is not aware of symptoms that may present differently in women". Such gaps lower the quality in services and lead to biased decision making in jobs, credit seeking, health care, etc. (UN Women, 2023).

The path-breaking work of mathematician Christine Mann Darden, the human computer initially recruited as a faceless worker to support the core engineering work in NASA in the 1960s, changed the workplace by questioning its prevailing cultural ethos. Darden noticed that men who had math degrees were placed in the more prestigious engineering positions, while women with the same credentials became part of computing pools and continued to stagnate there till the time they either retired or left. Darden walked up to the chief of her division and asked him why this was so. The response she got not only changed her own career trajectory but also brought in justice and fair play for others in the workplace. "Well, nobody's ever complained", the chief of her division said. "The women seem to be happy doing that, so that's just what they do" (D'Ignazio & Klein, 2020). The struggles women have had to undergo, the obstacles they faced and the distances they have covered are amazing. Their journeys have left behind for us a rich inheritance of human dignity and human rights.

TECHNOLOGIES OF THE SELF AND DELIBERATE ENGINEERING OF SOCIAL VALUES THAT DISEMPOWER WOMEN

It is the prevailing thought systems like criminology and psychiatry, that define the standards of human behavior, and even set limits over it. These are the "technologies of the self", as explained by the French philosopher Michel Foucault in a faculty seminar in 1982 (Foucault, 1988; Katz, 1990). These knowledge streams, or technologies of the self, that can be said to include contemporary technology innovations help in consolidating and redefining our value systems and ethical norms. Stereotypical depictions of race, class, and gender are viewed by media analysts as "political technologies", just like water wheels and microchips can be seen as products of deliberate engineering. We construct a world and that world constructs us (Katz, 1990). This is akin to the manner in which architects and creators design the AI tools that encapsulate their biases and leanings.

The use of AI in the criminal justice system is being vigorously debated and discussed as it brings bias against black men and women to the fore. A case in point is the misidentification of people as criminals – a danger inherent in the law enforcement agencies' use of AI-powered facial recognition (FR) technology. The racial bias in arresting blacks for criminal offences they never committed is coming to light.

The case of Porcha Woodruff, the eight-months'-pregnant woman put behind bars for the alleged offence of carjacking, is indicative of this technology, working specifically against the weaker sections of society (Bhuiyan, 2023). Ms. Woodruff was the sixth person accused of a crime based on the facial analysis technology used by the police to match an unknown offender with faces available in their database New York Times (2023). All six persons were black, and Ms. Woodruff was the first woman to report it. She was handcuffed by the police for a moment while she was readying her two daughters for school. The police informed her that she was accused of carjacking and a robbery. She was shocked and gestured toward her advanced pregnancy, to convey to the policemen that their idea of her carrying out said crime in advanced pregnancy was bizarre (Keenan, 2023). Her attorney, Ivan L. Land, pointed out the gross injustice being carried out through this technology and said, "It's scary. I'm worried – someone always looks like someone else". Ms. Woodruff felt that she was lucky to be pregnant – it made it easy for her to convince the judge that she did not commit the crime. The woman involved in the carjacking was not visibly pregnant. Gary Wells, a psychology professor, explained that pairing FR technology with eyewitnesses' identification should not be the basis for charging someone with a crime. Even if that similar-looking person is innocent, an eyewitness who is asked to make the same comparison is likely to repeat the mistake made by the computer (Moy, 2021).

Ms. Woodruff was subjected to grave physical discomfort when she had to spend 11 hours in police custody. She said she was having "contractions in the holding cell", leading to untold physical, mental, and emotional trauma. Research on the subject shows that facial recognition and surveillance technology enable more precise discrimination, as law enforcement agencies continue to use misinformed and predictive decisions around arrest and detainment that disproportionately impact marginalized populations (Lee & Chin-Rothmann, 2022).

Another instance of women being discriminated against on account of bias in machine learning algorithms was disclosed when Amazon's use of AI in the human resources department for screening job applications in response to advertised positions was found to be discriminatory and unfair (Dastin, 2018). When this secret AI recruitment tool became known to the outside world, Amazon had to abandon it. The AI tool for HR was meant to be the holy grail, as explained by people interviewed on its usefulness (Iriondo, 2018). It was an engine that could go through 100 resumes and dish out the top five for recruitment. Everyone was happy in the company, till it was found that the tool did not pick up resumes that belonged to women. This happened because Amazon trained its AI tool on ten years of data that indicated the domination of men in the tech industry. This was an experiment gone awry and serves as a lesson that AI and machine learning have serious limitations.

Nihar Shah, a computer scientist from the Carnegie Mellon University, opines, "How to ensure that the algorithm is fair, how to make sure the algorithm is interpretable and explainable – that's still quite far off" (Dastin, 2018; Iriondo, 2018). Microsoft Corporation's LinkedIn – leader in the hiring market and one of the biggest professional networks – uses advanced technology in ranking candidates for the job market. It also uses scores/percentage points to show the fitness of a candidate for a particular job opening. John Jensin, Vice President of LinkedIn's Talent Solutions confesses that these services are not a replacement for traditional recruiters. "I certainly would not trust any AI system today to make a hiring decision on its own", he candidly said (Dastin, 2018). It is evident that in cases where hiring is done through automated systems, the candidate not selected will find it difficult to sue the employer – since they will never get to know if there was a bias inherent in the recruitment tool. It indicates that AI's use is neither progressive nor fair toward women.

FILMS DEPICTING MAN–MACHINE RELATIONSHIPS ARE REINFORCING EXISTING GENDER BIAS AND ATTITUDES TOWARD WOMEN

Films as part of popular culture help us in decoding emerging trends and ethical practices on one hand, while on the other they act as vehicles for propagating regressive norms and values. They are true chroniclers of the age we live in. There are science fiction films that depict man trying to assume godlike powers over humanity. The recent film *ExMachina*, inspired by Mary Shelley's Frankenstein (the story of Victor Frankenstein, a young scientist who creates a monster) and the Promethean myth (the Greek god Prometheus steals fire from Olympus and gives it to humanity, but later gets punished by Zeus), raises questions of anxiety surrounding technology and gender. Human beings easily fall prey to hubris while exploring the dangers of attempting to be not a human but a god. As far as this film is concerned, there are no gods to be found here, only hubristic men and their miraculous machines (Beck, 2015). Ava, the artificial human in the film, has no biological needs and yet symbolizes the scientist's idea of programmed gender in an environment dominated by men. It is through the performance of her gendered role that the scientist believes she will prove her sentience and pass the Turing Test (a test devised by Alan Turing to prove that a machine can think like a human being). By objectifying the female presence of the AI being, the film highlights the concerns over female autonomy amongst other issues.

Ava also helps explain the theory in the works of Simone de Beauvoir (French philosopher known best for her feminist book, *The Second Sex*) of how the woman is something that "becomes" and is not an innate disposition in her own right. She has a hyper-feminized body forced upon her through performance and is expected to conform to gender categories and stereotypes in order to be treated human-like and be accepted by the audience. Without her own lived experiences to comprehend her own gender, she is incarcerated within the scientist's programmed code. Ironically, the film also helps understand how Ava's gender is a social construct, to borrow from Judith Butler's argument on gender (Butler, 1990). This film explains how the man who creates the robot acts like God. The title Ex Machina (taken from the Latin Deux ex machina), or "God in the works", is indicative of the politics behind men trying to control the body of a woman (Buolamwini & Gebru, 2018) Most AI assistant tools like Alexa, Dall-E, Siri, Gemini, etc. have female names and have attributes that are socially perceived to belong to women. Sky (AI tool made by Open AI), which was alleged to have a voice similar to that of Scarlett Johannson (Hollywood star), was withdrawn soon after being launched by the firm, after there was a public outcry and legal notice from the celebrity (Early, 2024; Kastrenakes, 2024). The tool was meant to possess a "warm" and "comforting" voice that would be reminiscent of Samantha, the AI girlfriend in the movie *Her*, as was mentioned by Sam Altman (head of OpenAI) on Twitter and later deleted.

What questions of feminist ethics can be seen playing around Ava in the film *Ex Machina*? Nathan, the scientist in the film, thinks he can control Ava in her humanized feminine form, while Ava is at the center of men playing the deception game. Both Celeb and Nathan try to be exclusive owners of Ava, which would depict their ultimate glory and success. A woman is to be possessed like a chattel and property that would determine who is the real master, while the AI being is expected to be devoid of agency and ethical standards on which to try the two men. The two men display little or no respect toward her and convince themselves that she is devoid of personhood – this way they can get rid of guilt for being unethical. Through the character of Ava, we also see the charade that is played out in society with real men and women.

There is yet another film which depicts the woman robot/AI girlfriend. She is intelligent and capable of speaking many languages, and as a girlfriend she plays warmth, care, and love toward Theodore, the male protagonist in the film *Her*. Is Samantha in *Her* the new ideal woman that men desire, who is not real, is not demanding, and willingly plays second fiddle to the man? Samantha owns a rational mind and provides intimate companionship, yet in this film, she cannot be the icon of the free, empowered woman.

The robot is seen to evolve (she reboots herself), and the discussion around what is right and what is wrong enters a new domain. What is not right in the real world may be either right or lose significance in the world of AI. The blurring of ethical standards in the virtual world is a new experiment where women continue to play the age-old roles but men reboot themselves to become more licentious, more bohemian, and freer of moral straps or any compunctions.

WOMEN UNMASKING AI TO MAKE IT FAIR AND JUST

Joy Buolamwini, in her book *Unmasking AI: My Mission to protect what is Human in a world of Machines*, deliberates on the dangers in AI-powered tools that are intrinsically discriminatory and biased (Buolamwini, 2023). In her research work with Timnit Gebru (Buolamwini & Gebru, 2018) and again in "Actionable Auditing" papers (co-authored with Deb Raji) (Buolamwini & Raji, 2023), she highlights the risks associated with facial recognition technology. It is no coincidence that all three are women on a mission to prevent AI from harming human values of fairness and justice. The "Actionable Auditing" papers brought out an audit of US companies involved in selling facial recognition technology to law enforcement agencies. What is immensely commendable is that these companies stepped back from trading in facial recognition technology and many cities adopted legislation to regulate the use of facial recognition technology, after the study-based evidence of these young women was published. Joy's research shows that it is easy for biometric surveillance technology to be incorporated into weapons of policing and weapons of war. We have seen Porcha Woodruff's case of wrongful arrest by the Detroit police, in the previous paragraph.

By coining the term "coded gaze", Buolamwini points out the harms perpetuated by embedded views that are propagated by those who have the power to code systems (Buolamwini, 2018). These codes were dangerous and were amplifying inequalities and damaging peoples' lives by defining which schools admitted them, who hired them, what medical care they received, or whether they would be arrested (Hamilton,2020)

WOMEN WHISTLEBLOWERS AND ETHICAL CHOICES

The world has seen women whistleblowers brave risks to their life while exposing unethical practices, corruption, and financial irregularities plaguing big technology companies. Meta (previously Facebook) has not taken misinformation and spread of fake news seriously, as brought out by Francis Haugen, who served as a product manager in the company. In an interview she said Facebook prioritized "profits over public safety"; she also said that Facebook was tearing societies apart and causing ethnic violence across the world (Paul & Milmo, 2021). Haugen disclosed that Facebook was aware of the harms it caused to the mental health of teenagers. The decision to blow the whistle required the coming together of unique qualities of courage, skill, tact, moral conviction, and atypical personality traits. She said, "I became concerned with India even in the first two weeks I was in the company". There were many people who were coming online for the first time. Such first-timers were unaware that something they read and saw online could be misleading or untrue (Perrigo, 2021). Another strong woman whistleblower, Sophie Zang, exposed the sloppy oversight of online political influence, while at Facebook (Wong, 2021).

"Social media and other technologies are not neutral", says Jason Thacker, assistant professor of philosophy and ethics at Boyce college and senior fellow at the Southern Baptist Ethics and Religious Liberty Commission. He adds, "[T]hese technologies are changing every aspect of our society including the lives of children and teenagers" (Chandler, 2023). The harms caused to teens/youth include anxiety, insomnia, interference in education and daily life. "Meta needs to be held accountable for the very real harm it has inflicted on children here in California and across the country" says Rob Bonta, Attorney General of California (Pierson, 2024)).

Susan Fowler, a 25-year-old engineer who joined the tech firm Uber after an interview, was told that "it was a great place to work" and "it was an incredible opportunity" as well as the most "valuable private start-up in Silicon Valley" and that the Uber app was being used all over the world and had a "real world impact" (PBS News Hour, 2020). As a young interviewee, she was impressed by what the recruiters told her before hiring. On the very first day of work, Susan Fowler's boss began to send her messages that were not related to work. They were highly objectionable, and she reported it to human resources HR, along with the screenshots. HR did not doubt her complaint but gave her a choice of shifting to another team, also mentioning that they might lower her grades if she stayed on. Fowler realized that it was not a choice, and the company was doing little to make the environment fair and equal for women like her. It was male-dominated, sexist, and regressive. It was through her 2,900-word blog on her "very strange year at Uber" and later in her memoir titled *My journey to Silicon Valley and Fight for Justice at Uber*, that she exposed the unethical behavior of people in power and the smugness that marked successful companies like Uber. The company had all the "look good" features of an inclusive and equal firm – it had 25% women engineers (Fowler, 2020). If Fowler had not complained, not documented the wrongdoings at every step, she would have been ridiculed and may even have ruined her career. Her righteousness and strong moral compass along with her skills ensured that she came out unscathed and unleashed the momentum to reform the workplace and corporate culture in the true sense.

The story of the whistleblower of Theranos, Erika Cheung, is equally motivating as she, along with another male colleague, Tyler Shultz, exposed the harmful and nefarious practices in the medical start-up that deceived people by giving them false blood reports. Erika Cheung was a 23-year-old scientist when she joined the company, and despite intimidation and grave threats, she decided to report the illegal and unethical practices at the company to the regulators. "Whistleblowers need to prepare for immense and persistent pressure, if they plan to speak out about unethical practices", says Cheung, who has founded Ethics in Entrepreneurship, a venture to carry forward the momentum in corporate culture based on humane values and gender equality (Sier, 2023). Studies show that more women than men are willing to blow the whistle in difficult situations. One can infer, write Ben-Jacob and Glazerman in their paper "Feminist Ethics", that women care more about the welfare of others than do men. This is true in the corporate and industrial world. Women are more concerned with the company and its employees; men on the other hand are more concerned about themselves, their jobs, as opposed to suffering the repercussions of whistleblowing (Ben-Jacob & Glazerman, 2022).

Another incident – where women led a protest against unfair and discriminatory practices of Google, one of the biggest AI entities – establishes that women care for justice, not only for themselves, but for all. On November 1, 2018, a massive walkout was organized at Google offices in about 50 cities around the world, by employees (part time and full time) and contractors, raising voices against unethical practices in the company. This protest began when the story of Andy Rubin (creator of an android phone operating system at Google who was held responsible for sexually abusing co-workers and who received an exit package of $90 million) was published by *The New York Times*. This led to an uproar among the workers. Whittaker, one of the leading organizers of the walkout, was told that her role would be changed. She was part of the "AI Ethics Council" which got scrapped, to which she said, "I'm told that to remain at the company, I will have to abandon my work on AI ethics and AI Now institute". The demands of the employees included ending forced arbitration in case of discrimination/sexual harassment and ensuring that the chief diversity officer be made to report directly to the CEO (D'Ignazio & Klein, 2020; Wong, 2019). The organizers said that they received over 350 similar stories of unfair treatment and discrimination during the walkout. "People who stand up and report discrimination, abuse and unethical conduct are punished, sidelined and pushed out," they said. "Google has a culture of retaliation, which too often works to silence women, people of color and gender minorities" reads a letter from Whittaker, Stapleton and ten other employees that was published internally (Wong, 2019). The courage of Whittaker has been laudable, as has her initiative to co-establish the AI Now Institute.

CONCLUSION

AI needs to be prevented from harming human beings and creating a regressive society. It has the potential to radically change society for the better, if we are able to adopt timely measures. Through regular audit of AI designs and algorithms, transparency, and disclosure of training data, we can avoid bias and discrimination. There is a need to encourage the fostering and establishment of an ethics culture (Department of Defense, 2019), specifically feminist ethics throughout the AI life cycle, including research, development, designing, deployment, trade, and usage. A gender-sensitive approach toward planning and designing of AI tools, can make it more inclusive, and equitable. The infusion of feminist ethics of care and empathy, responsible attitude, tolerance, compassion, justice, and fair play would go a long way in harnessing the potential of AI in creating a more humane and just society.

Any use of technology that empowers women results in addressing and resolving other social problems, since there is an interconnectedness between issues faced by women and those confronting other facets of society. If you see inequality as a "them" problem or "unfortunate other" problem, that is a problem, as expressed by Crenshaw (Crenshaw, 1989), lawyer, civil rights advocate, and intersectionalist feminist (UN Women, 2020). "Women of colour have long insisted that any movement for gender equality must also consider the ways in which privilege and oppression are intersectional" (D'Ignazio & Klein, 2020, p. 7). Feminist ethics is a study in multisectoral fields and adopts an interdisciplinary approach, as can be seen in various parts of the present chapter.

The example of Buolamwini's initiatives furthered the cause of social justice not merely for black women and black men but other oppressed groups too, by exposing the bias embedded in the coding systems. When this Ghanaian -American graduate at MIT, working on facial analysis software, found that the facial detection program could see her only when she put on a white mask over her dark skin, she grew curious. This led her to the realization of racial and gender bias, which in turn motivated her to set up the Algorithmic Justice League (AJL) and work toward making AI equitable and accountable. Unchecked, she notes, algorithms threaten our civil rights. It calls for regulation, transparency, consent in equitable and accountable uses of AI. The Rockefeller foundation is supporting this initiative and has set up the Community Reporting of Algorithmic System Harms (CRASH) project to support communities. The foundation's director, Evan Tachovsky, says that this project is a critical tool for reducing suffering today and building a more equitable future (Hamilton, 2020).

Another initiative, the Actionable Auditing papers prepared by Buolamwini and Raji, has emerged as a key strategy to expose systematic biases embedded in software platforms. It is the first algorithmic audit of gender and skin-type performance disparities in facial commercial analysis models. It was pointed out by the National Institute of Standards and Technology (NIST) study in 2019, that this audit not only revealed the biased performance of commercial facial analysis products but also spurred a highly visible corporate and public response regarding concerns for the functionality of various facial recognition applications, informing key product updates and interventions for corporate practice and effective policy campaigns (Buolamwini & Raji, 2023). Soon after the groundbreaking studies of Buolamwini were published, which revealed how systems from Amazon, IBM, and Microsoft failed at correctly classifying darker females faces as compared to those of white men – these big companies announced that they would stop or pause their facial recognition products for law enforcement (Hamilton, 2020). It is worth noting that women's initiatives are born out of their personal experiences, both in the private and public spheres. Ethics in Entrepreneurship, an organization set up by Erika Cheung, is founded on her first-hand experiences in Theranos, an organization that was devoid of even the basic ethical standards of listening to and responding to the grievances of its employees. Through the organization, Cheung is redefining the global entrepreneurial landscape, and reinforcing the significance of ethical practices as central to startups and innovation ecosystems (Entrepreneurship, n.d.).

However, all is not dismal in the domain of the AI. More and more women scientists like Buolamwini; policy framers like Amba Kak; Sarah Myers West (founder of *AI Now)* who engage in, among other aspects, policy research aimed at ensuring public accountability; and regulators like Lina Khan (head of the Federal Trade Commission [FTC]) are paving the way to a more egalitarian AI domain – based on probity, transparency, equity, and power sharing (Federal Trade Commission, 2024). Women whistleblowers (as seen in the previous paragraphs) risked their careers, their well-being, and even their lives to stand up for values they believed to be non-negotiable. In the world of AI, preoccupied with commercial success, fame, unbridled ambition, and power, women help in reinforcing ethical codes. We have seen through the power-drunk and over-ambitious character of Dr. Faustus (in the tragic 16th century play by Marlowe), how pursuit of technological omnipotence leads to moral depravity.

When women decide to speak to power, they bring forth a moral force that helps in restoring the balance in human values. The whistleblowers follow their heart and help in making AI ethical. The world at large gains from the actions taken by women who display immense strength of character, unflinching righteousness, and a path out of the darkness of deception, like the above mentioned whistleblowers. In the domain of the AI, women are now coming forward and breaking new grounds by not merely identifying what is wrong with algorithms, but are also indicating ways in which AI can be more fair, just, and equitable.

REFERENCES

Batha, E. (2024) Tough Laws needed for deepfake porn, say women who suffer AI abuse. https://www.context.news/ai/tough-laws-needed-for-deepfake-porn-say-women-who-suffer-ai-abuse

Beck, J. (2015). *Ex Machina: Of gods and men and their beautiful machines*. Movie Manifesto. https://moviemanifesto.com/2015/05/ex-machina-of-gods-and-men-and-their.html

Belanger, A. (2023, November 3). Teen Boys Use AI to make Fake Nudes of Classmates, Sparking Police Probe. Technica. https://arstechnica.com/tech-policy/2023/11/deepfake-nudes-of-high-schoolers-spark-police-probe-in-nj/

Ben-Jacob, M. G., & Glazerman, H. (2022). Feminist ethics. *Open Journal of Social Sciences*, *10*(12), 164–169. https://colab.ws/articles/10.4236%2Fjss.2022.1012011

Bhuiyan, J. (2023, August 15). TechScape: 'Are you kidding, carjacking?' – The problem with facial recognition in policing. *The Guardian*. https://www.theguardian.com/newsletters/2023/aug/15/techscape-facial-recognition-software-detroit-porcha-woodruff-black-people-ai

Brooks, M. P., & Houck, D. W. (Eds.). (2010). "Nobody's free until everybody's free": Speech delivered at the founding of the National Women's Political Caucus, Washington, D.C., July 10, 1971. In M. P. Brooks & D. W. Houck (Eds.), *The speeches of Fannie Lou Hamer: To tell it like it is* (Online ed., Mississippi Scholarship Online, March 20, 2014). University Press of Mississippi.

Bubeck, D. E. (1995). *Care, gender, and justice*. Oxford University Press.

Buolamwini, J. (2018, June 26). Fighting the "coded gaze." *Ford Foundation*. https://www.fordfoundation.org/news-and-stories/stories/fighting-the-coded-gaze/

Buolamwini, J. (2023). *Unmasking AI: My mission to protect what is human in a world of machines*. Random House.

Buolamwini, J., & Gebru, T. (2018). Gender shades: Intersectional accuracy disparities in commercial gender classification. *Proceedings of the 1st Conference on Fairness, Accountability and Transparency*, 81, 77–91.

Buolamwini, J., & Raji, I. D. (2023). Actionable auditing revisited: Investigating the impact of publicly naming biased performance results of commercial AI products. *Communications of the ACM*, *66*(1), 101–108. https://cacm.acm.org/research/actionable-auditing-revisited/

Butler, J. (1990). *Gender trouble: Feminism and the subversion of identity*. Routledge.

Chandler, D. (2023, October 24). *42 states, D.C. sue Meta on accusations of harming children, teens*. Baptist Press. https://www.baptistpress.com/resource-library/news/42-states-d-c-sue-meta-on-accusations-of-harming-children-teens/

Crenshaw, K. (1989). Demarginalizing the intersection of race and sex: A Black feminist critique of antidiscrimination doctrine, feminist theory, and antiracist politics. *University of Chicago Legal Forum*, *1989*(1), 139–167.

D'Ignazio, C., & Klein, F. L. (2020). *Data feminism*. MIT Press.

Dastin, J. (2018, October 11). Insight – Amazon scraps secret AI recruiting tool that showed bias against women. *Reuters*. https://www.reuters.com/article/world/insight-amazon-scraps-secret-ai-recruiting-tool-that-showed-bias-against-women-idUSKCN1MK0AG/

Department of Defense. (2019). AI principles: Recommendations on the ethical use of AI by the Department of Defense. Supporting documents. Defense Innovation Board. https://media.defense.gov/2019/Oct/31/2002204459/-1/-1/0/DIB_AI_PRINCIPLES_SUPPORTING_DOCUMENT.PDF

Dugan, T. K., (2023, April 26). How Zuck Got His Groove Back. Intelligencer https://nymag.com/intelligencer/2023/04/how-mark-zuckerberg-and-meta-broke-their-slump.html

Early, N. R. (2024, May 21). ChatGPT suspends Scarlett Johansson-like voice as actor speaks out against OpenAI. *The Guardian*. https://www.theguardian.com/technology/article/2024/may/20/chatgpt-scarlett-johansson-voice

Elhussein, G., & Hakspiel, J. (2024, March 1). *Empowering women in STEM: How we break barriers from classroom to C-suite*. World Economic Forum. https://www.weforum.org/stories/2024/03/empowering-women-in-stem-how-we-break-barriers-from-classroom-to-c-suite/

Engels, F. (1884). *The origin of the family, private property and the state*. (Translated by A. West, 1902). Charles H. Kerr.

Engster, D. (2007). *The heart of justice: Care ethics and political theory*. Oxford University Press.

Entrepreneurship, Ethics In. (n.d.). *Ethics in entrepreneurship*. https://ethicsinentrepreneurship.org/

Federal Trade Commission. (2024). FTC announces crackdown on deceptive AI claims and schemes. *Press release*. https://www.ftc.gov/news-events/news/press-releases/2024/09/ftc-announces-crackdown-deceptive-ai-claims-schemes

Foucault, M. (1988). *Technologies of the self: A seminar with Michel Foucault*. L. H. Martin, H. Gutman, & P. H. Hutton (Eds.). University of Massachusetts Press.

Fowler, S. (2020). *Whistleblower: My journey to Silicon Valley and fight for justice at Uber*. Penguin Random House.

Fox, M. (2024, November 21). Chart: How Nvidia's revenue has exploded in recent years. *Business Insider*. https://www.businessinsider.com/nvidia-revenue-chart-stock-price-gpu-blackwell-ai-nvda-earnings-2024-11

Giannella, E. (2015). Morality and the idea of progress in Silicon Valley. *Berkeley Journal of Sociology*. https://berkeleyjournal.org/2015/01/14/morality-and-the-idea-of-progress-in-silicon-valley/

Global Gender Gap (2024, June). World Economic Forum https://www3.weforum.org/docs/WEF_GGGR_2024.pdf

Guevara-Ramírez, P., Ruiz-Pozo, V. A., Cadena-Ullauri, S., Salazar-Navas, G., Acosta Bedón, A. A., V-Vázquez, J. F., & Zambrano, A. K. (2022). Ten simple rules for empowering women in STEM. *PLOS Computational Biology*, *18*(12), e1010731.

Gymrek, P., Berg, J., & Bescond, D., (2023). 'Generative AI and jobs: A global analysis of potential effects on job quantity and quality' ILO Working Paper 96. Generative AI and Jobs: A global analysis of potential effects on job quantity and quality. International Labour Organization (ilo.org).

Hamilton, M. (2020, December 16). The coded gaze: Unpacking bias in algorithms that perpetuate inequality. *Rockefeller Foundation*. https://www.rockefellerfoundation.org/insights/grantee-impact-story/unpacking-biases-in-algorithms-that-perpetuate-inequity/

Hartmann, H. (1979). The unhappy marriage of Marxism and feminism: Towards a more progressive union. *Capital & Class*, *3*(2), 1–33.

Held, V. (2006). *The ethics of care: Personal, political, and global*. Oxford University Press.

Hennessy, R. (2000). *Profit and pleasure: Sexual identities in late capitalism*. Routledge.

Hoagland, S. L. (1991). Some thoughts about "caring". In C. Card (Ed.), *Feminist ethics* (pp. 246–263). University Press of Kansas.

Huang, H. (2024). *Dominated by men*. Reuters Graphics. https://www.reuters.com/graphics/AMAZON.COM-JOBS-AUTOMATION/010080Q91F6/

Internet Encyclopedia of Philosophy. (n.d.). Care ethics. *The Internet Encyclopedia of Philosophy*. https://iep.utm.edu/care-ethics/

Iriondo, R. (2018, October 11). Amazon scraps secret AI recruiting engine that showed biases against women. *Carnegie Mellon University*. https://www.ml.cmu.edu/news/news-archive/2016-2020/2018/october/amazon-scraps-secret-artificial-intelligence-recruiting-engine-that-showed-biases-against-women.html

Kastrenakes, J. (2024, May 21). Scarlett Johansson told OpenAI not to use her voice – and she's not happy they might use it anyway. *The Verge*. https://www.theverge.com/2024/5/20/24161253/scarlett-johansson-openai-altman-legal-action

Katz, M. B. (1990). *Technology and culture: A historical romance*. Stanford Alumni Association.

Keenan, S. (2023). Heavily pregnant Black woman imprisoned for 11 hours following false facial recognition match. *People of Color in Tech*. https://peopleofcolorintech.com/articles/heavily-pregnant-black-woman-imprisoned-for-11-hours-following-false-facial-recognition-match/

Kittay, E. F. (1999). *Love's labor: Essays on women, equality, and dependency*. Routledge.

Kundu, A. (2024, March 11). The AI Act's gender gap: When algorithms get it wrong, who rights the wrongs? *Internet Policy Review*. https://policyreview.info/articles/news/ai-acts-gender-gap-when-algorithms-get-it-wrong/1743

Lee, N. C., & Chin-Rothmann, C. (2022, April 12). Police surveillance and facial recognition: Why data privacy is imperative for communities of color. *Brookings Research.* https://www.brookings.edu/articles/police-surveillance-and-facial-recognition-why-data-privacy-is-an-imperative-for-communities-of-color/

Lie, M., Berg, A.-J., Kaul, H., Kvande, E., Rasmussen, B., & Sørensen, K. H. (1988). *I menns bilde: Kvinner, teknologi, arbeid* [In the image of man: Women, technology, work]. Tapir.

Marcus, G. (2024). *Taming Silicon Valley.* The MIT Press.

McKinsey. (2022) Women In The Workplace – Report. https://www.mckinsey.com/~/media/mckinsey/featured%20insights/diversity%20and%20inclusion/women%20in%20the%20workplace%202022/women-in-the-workplace-2022.pdf

McKinsey. (2024, September 17) Women In The Workplace: The Tenth Anniversary Report https://www.mckinsey.com/featured-insights/diversity-and-inclusion/women-in-the-workplace

Moy, M. L. (2021). Facing injustice: How face recognition technology may increase the incidence of misidentifications and wrongful convictions. *Georgetown University Law Center.* https://scholarship.law.georgetown.edu/cgi/viewcontent.cgi?article=3532&context=facpub

Nelson, H. (2024). *Taylor Swift and the dangers of deepfake pornography.* NSVRC. https://www.nsvrc.org/blogs/feminism/taylor-swift-and-dangers-deepfake-pornography

New York Times. (2023, August 6). *Facial recognition and false arrest.* https://www.nytimes.com/2023/08/06/business/facial-recognition-false-arrest.html

Niethammer, C., (2020, March 2). *AI bias could put women's lives at risk – A challenge for regulators.* Forbes. https://www.forbes.com/sites/carmenniethammer/2020/03/02/ai-bias-could-put-womens-lives-at-riska-challenge-for-regulators/

Norlock, K., (2019). "Feminist ethics", *The Stanford encyclopedia of philosophy* (Summer 2019 Edition), Edward N. Zalta (ed.) https://plato.stanford.edu/archives/sum2019/entries/feminism-ethics/

Ortutay, B., & Klepper, D. (2021, October 5). Facebook whistleblower testifies: Five highlights. *Associated Press.* https://apnews.com/article/facebook-frances-haugen-congress-testimony-af86188337d25b179153b973754b71a4

Patel, J. N. (2024, January 3). "I was raped in virtual reality – trust me when I say it's very real." *The Telegraph.* https://www.telegraph.co.uk/news/2024/01/03/raped-in-virtual-reality-metaverse-victim-blaming/

Paul, K., & Milmo, D. (2021). Facebook putting profit before public good, says whistleblower Frances Haugen. *The Guardian* https://www.theguardian.com/technology/2021/oct/03/former-facebook-employee-frances-haugen-identifies-herself-as-whistleblower

PBS News Hour. (2020). Uber whistleblower Susan Fowler details harassment, retaliation in new book. *PBS.* https://www.pbs.org/newshour/show/uber-whistleblower-susan-fowler-details-harassment-retaliation-in-new-book

Perrigo, B. (2021, November 22). *Inside Frances Haugen's decision to take on Facebook. Time.* https://time.com/6121931/frances-haugen-facebook-whistleblower-profile/

Pierson, B. (2024, October 16). *Meta must face US state lawsuits over teen social media addiction.* Reuters. https://www.reuters.com/legal/meta-must-face-us-state-lawsuits-over-teen-social-media-addiction-2024-10-15/

Ponciano, J. (2023, June 8). The world's largest technology companies in 2023: A new leader emerges. *Forbes.* https://www.forbes.com/sites/jonathanponciano/2023/06/08/the-worlds-largest-technology-companies-in-2023-a-new-leader-emerges/

Rasch, M., (2023, March 16). *Ethics in AI: The missing code.* Security Boulevard https://securityboulevard.com/2023/03/ethics-in-ai-the-missing-code/

Reuters. (n.d.). *Amazon.com jobs automation.* https://www.reuters.com/graphics/AMAZON.COM-JOBS-AUTOMATION/010080Q91F6/

Rothschild, J. (Ed.). (1983). *Machina ex Dea: Feminist perspectives on technology.* Pergamon Press.

Sauer, M. (2022). *Melinda Gates wants to create an alternative to Silicon Valley.* CNBC. https://www.cnbc.com/2022/10/05/melinda-french-gates-wants-silicon-valley-alternative-for-women.html

Schellekens, P., & Skilling, D. (2024, October 17). Three reasons why AI may widen global inequality. *Center for Global Development.* https://www.cgdev.org/blog/three-reasons-why-ai-may-widen-global-inequality#:~:text=The%20rise%20of%20AI%20could,regions%20risk%20being%20left%20behind

Shivangini. (2024, May 23). Nvidia profit soars over 600%, shares top $1000: From 10:1 stock split to dividend – Key highlights from results. *Livemint.* https://www.livemint.com/companies/company-results/nvidias-profit-jumps-amid-dominance-in-chip-for-ai-key-highlights-11716425733893.html

Sier, J. (2023, June 20). I would do it all again, Theranos whistleblower says. *Financial Review*. https://www.afr.com/technology/i-would-do-it-all-again-theranos-whistleblower-says-20230620-p5di1h

The Conversation. (2024, August 11). Women are less interested in AI than men, but using it would help them advance at work. *The Conversation*. https://theconversation.com/women-are-less-interested-in-ai-than-men-but-using-it-would-help-them-advance-at-work-233511#:~:text=Women%20use%20generative%20artificial%20intelligence,51%20per%20cent%20of%20women

Tronto, J. C. (1993). *Moral boundaries: A political argument for an ethic of care*. Routledge.

Uber Whistleblower Susan Fowler. (2020, February 27). *Commonwealth Club World Affairs*. https://www.commonwealthclub.org/events/archive/podcast/uber-whistleblower-susan-fowler

UN Regional Information Centre for Western Europe. (2023). How technology-facilitated gender-based violence impacts women and girls. *UNRIC*. https://unric.org/en/how-technology-facilitated-gender-based-violence-impacts-women-and-girls/

U.N. Secretary General. (2023) Secretary-General's remarks to the Women's Civil Society Town Hall [as delivered], Antonio Guterres, March 13, 2023. https://media.un.org/avlibrary/en/asset/d301/d301700

UN Women. (2020, July 1). Intersectional feminism: What it means and why it matters right now. *UN Women*. https://www.unwomen.org/en/news/stories/2020/6/explainer-intersectional-feminism-what-it-means-and-why-it-matters

UN Women. (2023, March 15). Technology is playing key role in gender equality reversals, says UN Chief. *67th session of the Commission on the Status of Women*. https://www.unwomen.org/en/news-stories/news/2023/03/technology-is-playing-key-role-in-gender-equality-reversals-says-un-chief

UN Women. (2024, April 11). *Girls who can code and break stereotypes*. UN Women. https://www.unwomen.org/en/news-stories/interview/2024/04/girls-who-can-code-and-break-stereotypes

UNESCO. (2023). *Key facts: UNESCO's recommendation on the ethics of artificial intelligence*. UNESCO. https://www.unesco.org/en/articles/key-facts-unesco-recommendation-ethics-artificial-intelligence

United Nations. (2021, March 15). *Secretary-General's remarks to the Commission on the Status of Women*. United Nations. https://www.unwomen.org/en/how-we-work/commission-on-the-status-of-women

United States Congress. (2018). John S.McCain National Defense Authorization Act for Fiscal Year 2019,Pub. L.No.115-232, 132 Stat.1636. https://www.congress.gov/bill/115th-congress/house-bill/5515

Weedon, C. (1997). *Feminist practice and poststructuralist theory*. Blackwell.

Wong, C. J. (2019, April 22). Demoted and sidelined: Google walkout organizers say company retaliated. *The Guardian*. https://www.theguardian.com/us-news/2019/apr/22/google-mass-protests-employee-retaliation

Wong, J. C. (2021, April 12). How Facebook let fake engagement distort global politics: A whistleblower's account. *The Guardian*. https://www.theguardian.com/technology/2021/apr/12/facebook-fake-engagement-whistleblower-sophie-zhang

World Bank Group. (2020, November 23). *The equality equation: Advancing the participation of women and girls in STEM*. World Bank. https://www.worldbank.org/en/topic/gender/publication/the-equality-equation-advancing-the-participation-of-women-and-girls-in-stem

World Economic Forum. (2023). *Global Gender Gap Report 2023*. World Economic Forum. https://www.weforum.org/publications/global-gender-gap-report-2023/

World Economic Forum. (2024). *Global Gender Gap Report 2024*. World Economic Forum. https://www.weforum.org/publications/global-gender-gap-report-2024/

16 The Governance of AI from Diverse Ethical Perspectives

Naresh Singh and Lakshya Krishnaiah

INTRODUCTION

As artificial intelligence (AI) becomes ever more embedded in daily life, there is an urgent need to rethink and refine the frameworks guiding its governance. The pace at which AI is evolving has left traditional regulatory models struggling to keep up, giving rise to complex ethical issues that require nuanced and culturally sensitive solutions (Neuwirth, 2024). This chapter delves into the intersection of diverse cultural philosophies and AI ethics, exploring how different cultural traditions can inform governance models that are both inclusive and ethically robust. It aims to go beyond the dominant Western-centric frameworks by incorporating a broad spectrum of global values and norms into ethical AI frameworks (Wong, 2020). Beyond the Western-centric models of the US and European Union (EU), international bodies such as the United Nations and UNESCO have actively contributed to the global discourse on AI governance. The UN General Assembly's resolution on AI ethics, alongside UNESCO's recommendations, emphasizes the importance of integrating human rights and ethical principles into AI governance frameworks (United Nations, 2023). These multilateral efforts highlight the growing recognition of AI's global impact, requiring cooperation across nations.

The strategies of key players such as the United States, the European Union, and China have been heavily shaping the international dialogue around AI governance, each representing distinct socio-political and cultural settings. The US typically favors a decentralized, market-driven approach, placing innovation and self-regulation above comprehensive ethical scrutiny. On the other hand, the European Union is striving to lead AI governance with its proposed Artificial Intelligence Act, which seeks to establish a detailed regulatory structure that categorizes AI systems based on their risk potential, aiming to safeguard safety, transparency, and accountability while upholding human rights and democratic principles (Floridi et al., 2018). In contrast, China employs a state-centric, top-down approach, integrating AI development with national interests to promote social stability and economic progress (Dixon, 2023). These varied strategies highlight the challenges of creating a cohesive global governance model, reflecting the diverse cultural, political, and ethical values in play (Taeihagh, 2021). Additionally, the rapid pace of AI advancement continues to outpace regulatory capacities worldwide, highlighting the need for governance frameworks that are flexible and forward-looking enough to tackle emerging ethical quandaries (Larsson, 2020).

The focus of this chapter lies on how diverse cultural traditions, such as those rooted in Jaina, Buddhist, Vedantic, Confucian, Ubuntu, and Indigenous thought, can offer valuable contributions to the development of the ethical governance of AI. Each tradition presents a unique ethical lens emphasizing social cohesion, ethical behavior, and respect for all forms of life (Van Norren & Verbeek, 2021), providing important perspectives for the creation of governance frameworks. For example, Jainism's principles of non-violence (Ahimsa) and acceptance of multiple viewpoints (Anekantavada) advocate for minimizing harm and recognizing diverse perspectives (Srivastava et al., 2021). These are particularly pertinent when addressing issues like algorithmic bias and the ethical use of AI in sensitive fields such as healthcare and autonomous systems. Similarly, Buddhism's focus on mindfulness and reducing suffering (Hershock, 2020) calls for AI systems to be developed with an awareness of their societal impacts, encouraging the creation of technologies that promote social well-being rather than harm.

DOI: 10.1201/9781003585527-21

The main objective of this chapter is to highlight the importance of integrating these diverse cultural philosophies into AI governance frameworks. Drawing on the ethical insights from various traditions can lead to the creation of AI systems that are more inclusive, fair, and aligned with broader human and ecological goals. This multicultural approach is essential for tackling the unique ethical challenges posed by AI and for building global governance structures that honor the diverse values found across different societies (Jain, 2022). It also provides a means to bridge the gaps in existing AI governance models, which often suffer from fragmentation and a lack of coordination across different sectors and jurisdictions (Birkstedt et al., 2023).

Each section of this chapter provides a critical examination of the governance models currently adopted for AI by different countries, pointing out both their strengths and shortcomings. They also explore the potential for incorporating cultural philosophies into governance frameworks to create more holistic and adaptable models. By leveraging the rich diversity of human values and traditions, we can create governance systems that are not only more effective but also more equitable and inclusive, ensuring that AI technologies contribute positively to the well-being of all members of the global community.

The last section advocates for a shift toward a more inclusive and culturally sensitive approach to AI governance, one that utilizes the ethical wisdom of various cultural traditions to develop frameworks capable of addressing the complex ethical challenges posed by AI. Such an approach can help ensure a fair distribution of the benefits and risks associated with AI technologies while fostering trust and collaboration necessary for the responsible development and deployment of AI on a global scale (Roski et al., 2021). By embracing these diverse perspectives, we can work toward a more just, harmonious, and sustainable future in the age of artificial intelligence.

CURRENT APPROACHES TO THE GOVERNANCE OF AI

The realm of AI governance is at the moment, largely dominated by Western models, especially those of the European Union and the United States. These regions have a significant hold over global standards and practices in AI, with their approaches shaped by different socio-political environments, regulatory outlooks, and economic priorities (Taeihagh, 2021). However, as highlighted by multilateral bodies such as Association of Southeast Asian Nations (ASEAN) and the G20, these Western approaches are only part of the global conversation. Efforts like the ASEAN's AI Ethics Guide and the G20's discussions on AI standards are helping harmonize governance strategies, ensuring that non-Western nations are also actively contributing to the global AI governance landscape (AI Ethical Statements from Multilateral Bodies, 2023).

The EU has taken a leading role with its rigorous regulatory measures, whereas the United States has adopted a more flexible, market-oriented framework. Despite these efforts, there is growing concern that the failure of democratic governments to keep pace with technological advancements has allowed private corporations to amass disproportionate influence over AI development. This imbalance between public institutions and tech companies reflects a dangerous decoupling of public interest from technological innovation, threatening to undermine democratic governance (United Nations, 2023). A closer look at these strategies reveals their benefits, drawbacks, and the implications for fostering innovation while maintaining ethical AI governance. Through comprehensive regulations, the European Union has been proactive in establishing global standards for AI governance. Central to this strategy is the proposed Artificial Intelligence Act, which classifies AI systems based on their associated risks and enforces strict rules to ensure safety, transparency, and accountability (Truby et al., 2020). This risk-based methodology tackles issues like algorithmic bias, implications for human rights, and privacy concerns (Floridi et al., 2018), positioning EU as a front-runner in global AI governance.

One of the EU's key strengths lies in its unwavering commitment to fundamental values such as human rights, rule of law, and democratic ideals (Daly et al., 2020). By embedding these principles within its regulatory framework, the EU aims to ensure that AI technologies are used in ways that benefit society while mitigating potential harms (Buiten, 2019). The ethical guidelines set forth by the European Commission's High-Level Expert Group on AI focus on human agency, transparency, and accountability, influencing not only European policies but also inspiring similar initiatives in other parts of the world (Larsson, 2020). These guidelines help build public trust and provide a moral guideline for AI development, signaling to developers and users alike that ethical considerations are central to the technology's deployment.

Nevertheless, the EU's regulatory model is not without its challenges. Critics argue that these stringent regulations, although well intentioned, might stifle innovation and limit the competitiveness of European AI startups (Buiten, 2019). The EU's bureaucratic regulatory processes can delay the introduction of new AI technologies, potentially causing Europe to lag behind in the global AI race, where more nimble players like the United States and China dominate. Additionally, there is ongoing debate about whether the EU's regulatory structure, despite its thoroughness, is sufficiently adaptable to keep pace with the rapid evolution of AI technologies (Floridi et al., 2018). This rigidity could create an environment where compliance becomes burdensome, particularly for small and medium-sized enterprises (SMEs) that may not have the resources to navigate complex regulatory requirements.

In contrast, the United States has embraced a more decentralized, sector-specific approach to AI governance, consistent with its broader philosophy of minimal regulatory intervention. This model heavily emphasizes self-regulation and innovation, with significant input from the private sector (Schmitt, 2022). The National Institute of Standards and Technology (NIST) has developed a voluntary risk management framework for AI, which serves as a guideline for both private and public sectors (Smuha, 2021). A flexible approach allows organizations to tailor their AI practices to specific contexts and challenges, fostering innovation and rapid technological progress.

The US model's strengths lie in its capacity for swift innovation and adaptability. By encouraging a competitive environment that rewards innovation, the US has developed a vibrant AI ecosystem marked by significant contributions from both established tech giants and startups. The focus on voluntary guidelines and industry-driven standards enables companies to create AI solutions tailored to particular market needs, resulting in diverse applications that address a range of societal challenges (Taeihagh, 2021). However, this approach has been criticized for lacking coherence and uniformity. The fragmented nature of the regulatory framework can lead to significant governance gaps, especially in high-risk sectors like healthcare, autonomous vehicles, and facial recognition (Hoffman-Rem, 2020). The absence of comprehensive federal regulations means that individual states and industries often develop their own standards, resulting in a patchwork of regulations that can cause confusion and pose compliance challenges for businesses (Daly et al., 2020). This fragmentation complicates the regulatory landscape and raises concerns about effectively addressing ethical issues such as algorithmic bias and data privacy.

Their approach also reflects a broader debate about balancing regulation and innovation. While a more flexible framework may encourage technological advancement, it also raises concerns about the potential risks of unchecked AI development. This is particularly evident in sectors like finance, where AI-driven decisions can have widespread implications for millions of people (Schmitt, 2022). As AI technologies continue to evolve, there is growing recognition that the US might need to reevaluate its approach to ensure effective management of AI's ethical and societal impacts (Larsson, 2020), creating a governance model that mitigates risks without stifling innovation.

While the global discourse on AI governance often centers around Western countries like the United States and the European Union, emerging economies such as India, Brazil, and South Africa are also actively developing their own governance frameworks. These nations provide unique perspectives shaped by their distinct socio-economic contexts, viewing AI as a tool to address development challenges and as a technology that requires robust governance to avoid exacerbating inequalities (Van Norren & Verbeek, 2021). For these countries, AI governance is not only about regulating advanced technologies but also about aligning AI development and application with broader objectives of promoting economic growth, reducing inequality, and enhancing social well-being (Chandra & Ranjan, 2022).

India's approach to AI governance is closely tied to its socio-economic priorities. The country's National Strategy for Artificial Intelligence, led by the National Institution for Transforming India (NITI) Aayog, focuses on AI's potential to drive inclusive growth in key areas like agriculture, healthcare, education, and smart cities. India sees AI as a means to tackle pressing development issues, such as improving agricultural productivity, increasing healthcare access, and boosting educational outcomes in underserved regions (Srivastava et al., 2021). The multi-stakeholder model employed by India ensures that its AI governance framework includes input from government, industry, academia, and civil society, allowing for a wide range of perspectives in shaping AI policy (Jain, 2022). However, while the Digital India Act 2023 aims to address the negative impacts of AI, it remains limited in scope, leaving issues like the contagious nature of AI systems unaddressed (Kapoor et al., 2024). India faces significant hurdles in its AI governance efforts, including inadequate data infrastructure and the absence of a comprehensive data protection law, which raise concerns about privacy and security (Kiemde & Kora, 2022). Despite regulatory gaps, NITI Aayog has initiated responsible AI development through the Artificial Intelligence Research, Analytics, and Knowledge Assimilation Platform (AIRAWAT) platform and financial backing for the India AI mission, fostering innovation via public–private partnerships and AI infrastructure investments (Saha, 2024). In 2021, NITI Aayog's "Principles for Responsible AI" outlined ethical guidelines on privacy and security, but these remain non-binding, and India lacks a codified AI-specific statute. To address this, adaptive governance mechanisms, such as regulatory sandboxes and multi-stakeholder platforms, are recommended to monitor emerging risks (Approach Document for India; NITI Aayog, 2021). Moreover, although AI is viewed as a driver of inclusive growth, there is a risk that its benefits may not be evenly distributed, potentially deepening the divide between wealthier and poorer regions, particularly in rural areas with limited access to technology and infrastructure. As India anticipates significant AI growth by 2030, with over 10,000 deep-tech startups, the need for a comprehensive, adaptive, and globally aligned regulatory framework becomes urgent (Saha, 2024).

Recently India issued a white paper on using a complex adaptive systems approach to governance of AI. If indeed AI is unexplainable, unpredictable, and uncontrollable as well argued in a recent book of the same title (Yampolskiy, 2024) then a complex adaptive system would make the most sense as a broad approach. The features of complex adaptive systems, such as unpredictability and uncertainty driven by feedback loops, non-linearity, self-organization, emergence, etc., make this model particularly apt as distinct from current governance and management approaches which are hierarchical, linear, and deterministic. The current state of AI regulation in India remains largely focused on promoting innovation while attempting to mitigate risks through broad policy frameworks. However, as argued by scholars like Smuha (2021), a "race to AI regulation" can lead to fragmented governance, where countries prioritize competitive advantage over collaborative governance. This approach often misses critical ethical considerations. In this context, India's adoption of a complex adaptive systems approach could provide a more fluid and responsive framework. This model can better accommodate AI's inherent uncertainties, allowing for continuous learning and adaptation as new challenges emerge, moving beyond rigid, one-size-fits-all solutions.

How these fit into the prevailing philosophy in India, particularly Advaita Vedanta, is of great significance. Advaita emphasizes non-duality, interconnectedness, and the continuous interplay of elements within a unified reality. These principles resonate with the core ideas of complex adaptive systems, where the focus is on the interdependence of components and the emergent behaviors of a whole system (Chandra & Ranjan, 2022). By aligning AI governance with Advaita's emphasis on holistic perspectives and adaptability, India could develop frameworks that not only accommodate AI's inherent uncertainties but also harmonize technological advancements with ethical and philosophical traditions.

Brazil has adopted a balanced strategy for AI governance, aiming to promote innovation while prioritizing ethical considerations. The Brazilian Strategy for Artificial Intelligence, introduced in 2021, emphasizes the importance of education, research, and international collaboration as core elements of its governance framework (Birkstedt et al., 2023). Brazil's commitment to ethical AI is evident in its focus on transparency, accountability, and anti-discrimination measures. The General Data Protection Law (LGPD) underpins data privacy in the country, ensuring that AI systems operate within ethical boundaries and safeguard personal information (Williams & Shipley, 2021). Moreover, Brazil actively seeks international partnerships, particularly with other Latin American countries, to address regional AI challenges (Roski et al., 2021). Despite these strengths, Brazil faces significant obstacles, such as a shortage of technical expertise and infrastructure, which could hinder its AI adoption. Additionally, Brazil's AI regulatory framework is still evolving (Schmitt, 2022), and as guidelines are refined, there will be a need to balance innovation with robust governance that protects society from potential harms.

Brazil's AI governance strategy also raises questions about whether there is an overarching philosophical framework that influences its approach, potentially rooted in Western philosophical traditions. Historically, Brazilian intellectual thought has evolved through a synthesis of Western Canon and localized cultural influences, shaping its governance practices, including AI (Birkstedt et al., 2023). This integration of global and local perspectives aligns with Brazil's emphasis on international collaboration and regional partnerships to address AI challenges (Roski et al., 2021). Additionally, the philosophical underpinnings of transparency and accountability evident in Brazil's AI governance framework resonate with Western ideals of rationality and justice, as reflected in its data protection laws (Williams & Shipley, 2021). These principles suggest a nuanced philosophical alignment, though further exploration of how these ideas influence AI-specific policies could provide valuable insights.

South Africa's AI governance strategy is shaped by the country's socio-economic challenges, including high unemployment, inequality, and disparities in healthcare and education access. The national AI strategy focuses on developing AI solutions tailored to these specific issues, aiming to ensure that AI advancements benefit the wider population (Hershock, 2020). South Africa, like India and Brazil, employs a multi-stakeholder approach to AI governance, involving government, industry, academia, and civil society in policy formulation (Chandra & Ranjan, 2022). This inclusive method reflects the country's commitment to aligning AI development with the values and needs of its diverse society. Social justice is a central theme in South Africa's AI strategy, which seeks to use AI to bridge gaps in healthcare, education, and employment (Kiemde & Kora, 2022). Regionally, South Africa's involvement in the African Union's "Artificial Intelligence for Africa" framework (Van Norren & Verbeek, 2021) supports collaboration on equitable AI development across the continent. Despite these positive steps, South Africa encounters challenges in terms of technical infrastructure and expertise, potentially limiting its capacity to effectively govern and implement AI technologies (Jain, 2022). Additionally, while AI offers opportunities to reduce inequality, there is concern that without appropriate safeguards, AI-driven automation could worsen social divides by displacing low-income workers and increasing unemployment in vulnerable communities (Hoffman-Rem, 2020).

The AI governance strategies of India, Brazil, and South Africa are deeply informed by their developmental goals and unique socio-economic contexts. These countries view AI as an essential tool for promoting inclusive growth, addressing developmental challenges, and improving the lives of their citizens. However, they face common barriers, including inadequate infrastructure, limited technical expertise, and regulatory gaps, which could impede the effective governance of AI technologies. Moreover, while AI has the potential to reduce inequality, there is a risk that it could exacerbate existing socio-economic disparities if not managed with foresight (Chandra & Ranjan, 2022). Despite these challenges, the approaches taken by these countries offer valuable insights into how AI governance can be tailored to the needs of emerging economies, while also promoting international cooperation and navigating the ethical implications of AI.

AI governance approaches vary significantly between the Global North and the Global South due to differences in socio-economic priorities, technological development, and regulatory frameworks. In developed regions like the EU and the US, governance focuses on privacy, security, and ethical issues such as bias and transparency (Buiten, 2019). The EU's General Data Protection Regulation

(GDPR), for instance, sets a high standard for data privacy and influences AI deployment (Floridi et al., 2018). Meanwhile, the US, driven by private-sector innovation, adopts decentralized governance efforts, addressing ethical impacts through proactive industry initiatives (Larsson, 2020).

In contrast, countries in the Global South, including India, Brazil, and South Africa, prioritize AI as a tool for economic development and addressing social challenges like poverty and healthcare (Jain, 2022). Their frameworks focus on inclusive growth and reducing inequality, though limited infrastructure, technical expertise, and regulatory systems pose challenges to managing AI risks effectively (Van Norren & Verbeek, 2021).

Global AI governance cannot follow a one-size-fits-all model. The Global North's focus on ethical AI offers valuable lessons, while the Global South emphasizes leveraging AI for socio-economic advancement, highlighting AI's potential as a transformative tool in emerging economies (Kiemde & Kora, 2022).

A major challenge in developing a cohesive global AI governance framework is bridging the divide between the Global North and the Global South (Van Norren & Verbeek, 2021). This requires fostering international cooperation and ensuring that the needs and concerns of developing countries are adequately represented in global AI governance discussions. While the Global North may prioritize stringent ethical standards for AI, it is equally important to adapt these standards to the realities of developing countries, where the immediate priority might be using AI to address urgent social and economic issues (Chandra & Ranjan, 2022). Furthermore, global AI governance frameworks must ensure that developing countries have access to the resources, expertise, and infrastructure needed to fully participate in the AI revolution.

AI governance is a complex, multifaceted challenge that requires a nuanced understanding of the diverse needs and priorities of different regions. While the Global North and Global South may have different approaches to AI governance, both have valuable insights to offer. By promoting international collaboration and adapting governance frameworks to the socio-economic contexts of various countries, a global approach to AI governance can be developed that not only promotes ethical AI development but also ensures that AI is used as a tool for inclusive growth and social good.

CULTURAL PHILOSOPHIES AND THE GOVERNANCE OF AI

The governance of artificial intelligence is not merely a technical or regulatory matter; it deeply intersects with the philosophical and cultural values of diverse global regions. As AI becomes more integrated into society, the ethical frameworks guiding its development must be informed by a broad spectrum of human experiences and wisdom traditions. Exploring different cultural philosophies reveals foundational principles that can shape ethical AI governance, ensuring that these technologies resonate with the values and needs of diverse societies worldwide.

Despite their distinct cultural contexts, many philosophical traditions emphasize common values such as harm reduction, respect for life, and the pursuit of collective well-being. For example, Jainism's principles of Ahimsa (non-violence) and Anekantavada (acceptance of multiple perspectives) echo Buddhism's focus on mindfulness and alleviating suffering. Both advocate for a compassionate approach to life that seeks to minimize harm and recognize the complexity of existence. Similarly, the Vedantic view of interconnectedness and moral duty (Dharma) aligns with Ubuntu's emphasis on communal well-being and the interconnectedness of all beings (Hershock, 2020).

These shared values suggest a unified ethical framework for AI governance that prioritizes inclusivity, harm reduction, and societal harmony. For instance, Jainism's Anekantavada, which encourages the acceptance of diverse perspectives, offers a practical approach to addressing algorithmic bias and promoting fairness in AI systems (Floridi et al., 2018). By integrating varied viewpoints, this principle can help develop AI technologies that are more inclusive and less likely to perpetuate systemic inequalities. Moreover, the Vedantic emphasis on interconnectedness can inspire AI governance models that consider the long-term societal and environmental impacts of AI, advocating for policies that support sustainable development and collective prosperity (Kiemde & Kora, 2022).

Each philosophy also contributes unique perspectives that can enrich AI ethics discourse. Jainism's strict adherence to Ahimsa sets a high standard for harm reduction, particularly relevant for developing AI systems in high-risk fields like healthcare and autonomous vehicles (Schmitt, 2022). Conversely, Buddhism's focus on mindfulness can guide the ethical use of AI in digital environments, such as social media, where users' mental and emotional well-being is often at stake (Hershock, 2020). Vedantic teachings on Karma and Dharma offer a framework for developing transparent and accountable AI systems (Van Norren & Verbeek, 2021), ensuring that those who create and deploy these technologies are held responsible for their actions.

Confucianism and Ubuntu, although culturally distinct, both emphasize social harmony and the collective good. Confucianism, deeply embedded in Chinese society, underscores the importance of moral leadership and the role of the state in maintaining ethical standards. This perspective is particularly relevant for AI governance, where rapid technological advancements often outpace existing regulations. Confucian principles can guide the creation of governance structures that not only regulate AI but also promote ethical leadership and accountability among those who manage these technologies. For example, the Confucian concept of Ren (benevolence) can inform the development of AI systems that prioritize the well-being of all stakeholders, including marginalized groups (Li, 2021).

Ubuntu, an African philosophy emphasizing community, interconnectedness, and mutual respect, provides a compelling framework for developing AI systems that focus on social justice and equity. The Ubuntu principle "I am because we are" highlights the interdependence of individuals and the importance of community well-being (Metz, 2025). This perspective can shape AI governance frameworks that ensure equitable access to AI technologies and their benefits, particularly in essential areas like healthcare and education. By promoting an inclusive approach to AI development, Ubuntu can help address the ethical challenges posed by AI in a globalized world, where the distribution of technological benefits and risks is often uneven.

Indigenous philosophies add another critical dimension to the discussion on AI governance, particularly with their emphasis on environmental stewardship and sustainability. Indigenous traditions often view all living beings as interconnected and stress the importance of preserving natural ecosystems for future generations (Chilisa, 2020). This perspective is particularly relevant for AI, where technologies such as predictive analytics and machine learning can be leveraged for environmental monitoring and conservation efforts. Incorporating Indigenous knowledge into AI governance can help develop technologies that respect cultural heritage and contribute to sustainable development and biodiversity preservation.

While these philosophical traditions share common themes, such as interconnectedness and harm minimization, each offers unique contributions essential for a nuanced understanding of AI ethics. Jainism's Anekantavada (Table 16.1), for instance, is particularly well-suited to addressing the complexity and diversity of human perspectives, making it a valuable tool for developing AI systems that are sensitive to cultural and social differences. This approach is especially relevant for tackling biases in AI algorithms, which often reflect the narrow viewpoints of their creators.

Buddhism's concept of interdependence, while similar to the Vedantic notion of interconnectedness, places a stronger emphasis on the relational nature of existence and the ethical consequences of our actions. This perspective can guide the development of AI systems that not only avoid causing harm but also actively contribute to societal well-being (Hershock, 2020). For instance, AI-driven social media platforms could be designed to foster positive interactions and reduce the spread of misinformation and harmful content.

Confucianism's focus on moral leadership and the state's role in enforcing ethical standards is particularly pertinent for developing regulatory frameworks that ensure responsible AI use. This contrasts with the more decentralized, individualistic approaches favored in many Western contexts and offers a governance model that prioritizes collective well-being over individual autonomy (Li, 2021). Ubuntu's emphasis on community and mutual respect similarly provides a counterbalance

TABLE 16.1
Cultural Philosophies and Their Key Values

Philosophy	Key Principles	Application in AI Governance
Jainism	Ahimsa (non-violence), Anekantavada (multiple perspectives)	Promotes harm reduction and inclusivity in AI systems, addressing issues like algorithmic bias and safety
Buddhism	Mindfulness, Reduction of Suffering	Encourages awareness of societal impacts and ethical accountability in AI development
Vedanta	Interconnectedness, Dharma (moral duty)	Supports sustainable development and transparent, accountable AI systems
Confucianism	Social Harmony, Moral Leadership	Advocates ethical accountability among AI developers and policymakers, focusing on collective well-being
Ubuntu	Community, Mutual Respect	Promotes equity and social justice, ensuring AI benefits are accessible to marginalized communities
Indigenous Traditions	Environmental Stewardship, Interconnectedness	Guides AI toward sustainability and conservation while respecting cultural heritage

Source: Authors.

to more individualistic approaches to AI governance, advocating for systems that promote social cohesion and equitable development (Metz, 2025).

Neglecting these unique contributions could lead to significant ethical shortcomings in AI governance. For example, ignoring the principle of Ahimsa may result in AI systems that prioritize efficiency over safety, potentially causing harm in high-stakes domains like healthcare or military applications. Similarly, overlooking the inclusive approach of Anekantavada could perpetuate existing biases in AI algorithms, exacerbating social inequalities and undermining public trust in these technologies. Failing to incorporate Indigenous perspectives could lead to AI systems that are environmentally unsustainable or that fail to respect cultural heritage, undermining efforts to promote sustainable development and social justice (Chilisa, 2020).

Incorporating diverse philosophical traditions into AI governance can significantly improve the ethical development and deployment of AI systems (Floridi et al., 2018). By integrating a broad spectrum of ethical principles, we can establish governance frameworks that are not only effective but also reflective of the diverse values and needs of societies around the world. This inclusive approach results in AI systems that are more attuned to the complexities of human life, promoting equity, justice, and sustainability.

A key benefit of integrating multicultural philosophies into AI governance is the promotion of a more comprehensive and inclusive understanding of ethics. Traditional Western ethical frameworks often emphasize individual rights and autonomy, which, while crucial, may neglect broader societal and environmental concerns. In contrast, principles from Jainism, Buddhism, Vedanta, Confucianism, Ubuntu, and Indigenous philosophies can guide the development of AI systems that consider the well-being of communities, the environment, and future generations (Hershock, 2020; Metz, 2025). For instance, applying the principle of Ahimsa to AI governance can ensure that AI systems prioritize harm reduction, particularly in high-risk areas such as autonomous vehicles and healthcare. This might involve creating stringent safety protocols and ethical guidelines that prevent AI technologies from compromising human dignity or safety in the name of efficiency. Similarly, the Vedantic concept of Karma can inform the creation of transparent and accountable AI systems (Schmitt, 2022), ensuring that those involved in developing and deploying these technologies are held responsible for their societal impacts.

Confucian principles of moral leadership and accountability can inspire the establishment of robust regulatory frameworks that promote ethical behavior among AI developers and users. This could include setting up ethics committees or regulatory bodies to oversee the development and deployment of AI systems, ensuring they align with societal values and contribute to the common good. Ubuntu's emphasis on community and mutual respect can guide the development of AI systems in sectors such as healthcare and education, ensuring these technologies promote social justice and do not exacerbate existing disparities (Li, 2021; Metz, 2025).

Neglecting to incorporate these diverse philosophical perspectives into AI governance could have serious ethical and societal consequences. For example, without the inclusive approach of Anekantavada (Table 16.1), AI systems may perpetuate biases and inequalities, especially in areas like hiring, lending, and law enforcement. This could lead to technologies that unfairly disadvantage marginalized groups, undermining social cohesion and public trust in AI. Ignoring the principle of Ahimsa might result in AI systems that prioritize efficiency over safety, potentially causing harm in high-stakes domains like healthcare and autonomous vehicles. Similarly, disregarding the Vedantic focus on interconnectedness and sustainability could lead to AI systems that contribute to environmental degradation or worsen social inequalities, hindering efforts to promote sustainable development and shared prosperity.

Moreover, failing to incorporate Indigenous perspectives into AI governance could result in technologies that erode cultural heritage or damage the environment, undermining efforts to promote sustainability and respect for cultural diversity (Smith, 2021). This could lead to a crisis of legitimacy for AI governance frameworks, as they may appear disconnected from the values and needs of the communities they are intended to serve (Nigel Shadbolt, 2020).

To develop inclusive and effective AI governance frameworks, it is essential to engage with and integrate the diverse philosophical traditions that shape ethical thought around the world. This involves not only drawing on these traditions to inform ethical guidelines but also engaging with communities to ensure their voices are heard and their values are reflected in AI governance (UNESCO, 2021). Governments, industry leaders, and civil society organizations all play a role in this process. By fostering dialogue and collaboration among these stakeholders, we can create governance frameworks that are more responsive to the ethical challenges posed by AI and more attuned to the diverse values and needs of global societies (Binns, 2018). For instance, establishing international forums or working groups that include representatives from various cultural and philosophical backgrounds can help ensure that AI governance frameworks reflect a diverse array of ethical perspectives (Ghosh & Lakshmi, 2023). Similarly, creating platforms for public engagement and dialogue can help build trust and ensure that AI governance is responsive to the concerns and values of the communities it aims to serve.

Integrating diverse philosophical perspectives into AI governance is essential for developing ethical and effective AI systems. Embracing these perspectives enables the creation of governance frameworks that promote the responsible development and deployment of AI technologies, ensuring they contribute to a more just, equitable, and sustainable world (Nigel Shadbolt, 2020). The challenge ahead is not just to regulate AI but to do so in a way that reflects the rich tapestry of human values and aspirations, creating a future where AI serves the common good and enhances the overall well-being of all individuals.

ISSUES, GAPS, AND SOLUTIONS

As artificial intelligence continues to be integrated into various aspects of society, the challenges surrounding its governance are becoming increasingly complex (Tiwari, 2024). Current international AI governance frameworks, which are largely influenced by Western-centric paradigms, exhibit significant gaps in regulation and ethical oversight (Binns, 2018). These gaps are evident in the lack of effective enforcement mechanisms, inconsistent regulatory standards across jurisdictions, and the inability to keep pace with the rapid evolution of AI technologies (United Nations System Chief Executives Board for Coordination, 2022).

One of the most critical issues in the current AI governance landscape is the reliance on voluntary measures and self-regulation by businesses and organizations (Tiwari, 2023). This approach often results in inadequate accountability, particularly in high-risk sectors such as criminal justice, healthcare, and finance, where the potential for harm is significant (Binns, 2018). Industry-led initiatives, such as internal codes of conduct and ethical review committees, often lack the robust oversight needed to ensure ethical AI development and deployment (Nigel Shadbolt, 2020). Without strong regulatory frameworks to hold organizations accountable, it is challenging to ensure that AI systems are developed and used ethically.

The fragmented regulatory approaches across different sectors and jurisdictions further complicate AI governance. The rapid pace of AI development often outstrips the capacity of existing legislative frameworks to provide effective oversight, leading to governance gaps and inconsistencies (Binns, 2018). For example, while the European Union has proposed a comprehensive framework through its Artificial Intelligence Act, the United States has adopted a more decentralized, sector-specific approach (United Nations System Chief Executives Board for Coordination, 2022). This disparity complicates international cooperation and coordination, as differing regulatory standards can result in conflicts and inconsistencies in the governance of AI technologies.

Moreover, the Western-centric focus of AI governance often overlooks the diverse cultural and socio-economic contexts of developing and underdeveloped countries. This gap is particularly problematic as AI technologies are increasingly deployed in regions with distinct ethical and societal norms. The Western emphasis on individual rights and data privacy, while important, does not fully address the collective and community-oriented values prevalent in many non-Western societies. Consequently, there is a growing disconnect between the ethical principles underlying Western AI governance frameworks and the values of the communities most impacted by these technologies (Nigel Shadbolt, 2020).

AI technologies are often deployed within complex adaptive systems (CAS), characterized by interdependence, unpredictability, and emergent properties (Zador et al., 2023). Traditional governance frameworks, which tend to be hierarchical and rigid, struggle to effectively manage the complexities of CAS. This is especially evident in the context of AI, where technologies can have far-reaching and often unforeseen consequences (United Nations System Chief Executives Board for Coordination, 2024). For example, AI systems in financial markets can trigger cascading effects that disrupt entire economies, while those used in social media can amplify misinformation, leading to societal polarization and unrest.

To govern AI effectively within CAS, anticipatory and adaptive governance approaches are crucial. These approaches allow governance frameworks to evolve in response to new developments and unforeseen outcomes, making them more resilient and responsive to the dynamic nature of AI technologies (Tiwari, 2024). However, many current frameworks lack the necessary flexibility and responsiveness (Binns, 2018), remaining rooted in hierarchical models that are ill-suited to manage the complexity and dynamism of AI systems.

The decentralized and networked nature of AI technologies further complicates governance efforts. Traditional hierarchical models of governance are often inadequate for managing AI systems, which operate across multiple sectors and jurisdictions (De Cremer & Kasparov, 2022). There is a growing recognition that heterarchical and holarchical governance models that emphasize networked and interconnected structures are better suited for AI governance. These models can facilitate more effective collaboration and coordination among diverse stakeholders, enabling governance frameworks to more effectively address the ethical challenges posed by AI (Nigel Shadbolt, 2020). However, most existing governance frameworks continue to rely on hierarchical structures, limiting their ability to effectively govern complex and dynamic AI systems.

Trust is a fundamental aspect of ethical AI governance, yet many current frameworks fail to adequately engage the public in the development and implementation of AI ethics principles. This lack of public involvement can lead to a disconnect between the ethical standards set by governance frameworks and the values and concerns of the communities most affected by AI technologies (Nigel

Shadbolt, 2020). Without meaningful public engagement, AI governance risks being perceived as technocratic and disconnected from societal needs, further eroding public trust. While transparency and accountability are often cited as key principles in AI ethics, they are frequently not implemented in practice (UNESCO, 2021). For instance, many AI systems operate as "black boxes," with decision-making processes that are opaque even to their developers. This lack of transparency can undermine public trust, especially in high-stakes applications such as criminal justice and healthcare, where opaque AI systems can have life-altering consequences (Burrell, 2016). Ensuring that AI systems are transparent and accountable requires not only technical solutions, such as explainable AI, but also governance mechanisms that mandate transparency and provide avenues for redress.

Privacy is another central concern in AI ethics governance, particularly given the vast amounts of personal data that AI systems often require to function effectively. Current regulatory frameworks often struggle to balance the need to protect human privacy with the demands for data to train and improve AI systems (Haney et al., 2023). The global nature of AI development exacerbates this issue, as data may be collected, processed, and stored across multiple jurisdictions, each with its own privacy laws. The absence of a unified global approach to data protection creates gaps that can be exploited, leading to ethical dilemmas and potential privacy violations (Haney et al., 2021). The challenge of protecting privacy is particularly acute in AI applications such as facial recognition and predictive policing, where personal data can be used in ways that infringe on individual rights and freedoms (Zuboff, 2019). Robust data protection regulations tailored to address the unique challenges posed by AI technologies are urgently needed to safeguard privacy (Crawford, 2021). However, developing such regulations is a complex task that requires coordination among multiple stakeholders and jurisdictions.

The global race to develop and deploy AI technologies has intensified competition among nations and corporations. While this competition can drive innovation, it also raises significant ethical concerns (De Cremer & Kasparov, 2022). In the pursuit of rapid technological advancement, organizations may prioritize speed and market dominance over ethical considerations, resulting in a "race to the bottom" in terms of ethical standards. This dynamic is particularly concerning in sectors like finance and healthcare, where the societal impact of AI technologies can be profound and lasting (Crawford, 2021).

Conversely, collaboration at both national and international levels is crucial for developing and implementing robust AI ethics governance frameworks. Achieving this kind of collaboration is challenging, particularly in a world where different stakeholders have competing goals and interests. However, without collaboration, it becomes increasingly difficult to establish global norms and standards for ethical AI development, leading to fragmented and inconsistent governance frameworks (Zuboff, 2019).

Equity is another area where current AI governance frameworks fall short. AI technologies have the potential to exacerbate existing inequalities, especially if they are developed and deployed in ways that reflect and perpetuate societal biases (Turner Lee & Lai, 2022). Governance frameworks that actively address issues of bias and discrimination are essential to ensure that AI systems are fair and equitable. However, many current frameworks lack the mechanisms necessary to effectively address these issues (Crawford, 2021). This disparity is particularly concerning given the growing use of AI in decision-making processes that profoundly impact people's lives, such as hiring, lending, and law enforcement.

To bridge the significant disparities in AI governance, various strategies have been proposed and implemented at both national and international levels. A key strategy involves developing comprehensive regulatory frameworks that establish clear guidelines for the development and deployment of AI technologies. However, these frameworks must be adaptable to keep pace with the rapidly evolving ethical challenges and technological advancements in AI. To effectively address the existing gaps in AI governance, it is crucial to incorporate insights from diverse philosophical traditions into both global and national governance models. While the Western-centric approach, which emphasizes individual rights, data privacy, and innovation driven by market forces, has played a dominant role in shaping the global AI discourse, it has notable limitations. Specifically, it often fails to address broader concerns such as social justice, collective well-being, and cultural diversity (Haney et al., 2023). By integrating ethical principles from various cultural perspectives, we can create governance frameworks that are more inclusive, resilient, and reflective of the societies they aim to serve.

Western-centric AI governance models often emphasize self-regulation and market dominance, which can overlook critical ethical imperatives such as inclusivity and equity (Zuboff, 2019). Philosophical traditions from diverse cultures provide alternative perspectives by prioritizing interconnectedness, collective well-being, and sustainability. These traditions advocate for governance frameworks that emphasize ethical accountability, social cohesion, and equitable access to AI benefits, particularly for marginalized communities. Additionally, they highlight the importance of integrating environmental stewardship and respect for cultural heritage into AI governance, addressing both social and ecological challenges. By incorporating these diverse philosophies, AI governance frameworks can move beyond the limitations of individualistic, market-driven approaches, fostering a more inclusive and globally relevant strategy for managing AI's risks and opportunities.

In addition to integrating cultural philosophies, effective AI governance requires robust mechanisms for public engagement and participation. This is essential for building trust and ensuring that governance frameworks reflect the values and concerns of the communities they are meant to protect. Public engagement can take various forms, such as open dialogues, public consultations, or more formal structures like citizens' assemblies or advisory councils (Davidson-Hunt & Berkes, 2003). By involving a diverse range of stakeholders, including governments, industries, civil society organizations, academia, and the general public, AI governance frameworks can become more transparent, accountable, and responsive to societal needs.

To address the fragmentation in global AI governance, there is also a pressing need for increased international cooperation and the establishment of unified standards and guidelines. Initiatives like the OECD's AI Principles and UNESCO's Recommendation on the Ethics of Artificial Intelligence provide a foundation for such cooperation, highlighting the importance of inclusivity, accountability, and transparency (OECD, 2019). However, these efforts must be strengthened and expanded to ensure they truly reflect global perspectives and can effectively address the diverse challenges posed by AI technologies (Kimmerer, 2013). The creation of international organizations dedicated to AI governance, similar to the International Atomic Energy Agency (IAEA) or the World Health Organization (WHO), could offer centralized leadership in coordinating efforts, setting global standards, and promoting international collaboration in AI governance.

Ultimately, integrating multicultural ethical principles into AI governance is not just a theoretical exercise but a practical necessity. By drawing on the rich tapestry of human values and traditions, we can develop governance frameworks that are more resilient, inclusive, and effective in addressing the complex ethical challenges posed by AI (Berkes, 2018). This approach not only promotes a more just and equitable distribution of the benefits and risks associated with AI technologies but also helps build a foundation of trust and cooperation essential for the responsible development and deployment of AI in a global context. Bridging the gaps in current AI governance systems requires a comprehensive approach that integrates diverse cultural philosophies, promotes public engagement, and fosters international cooperation. By embracing these principles, we can create governance frameworks better equipped to manage the complexities of AI technologies and ensure they are developed and deployed in ways that reflect the values and aspirations of all members of the global community. This is not only a moral imperative but also a practical strategy for building a more just, equitable, and sustainable future in the age of artificial intelligence.

CONCLUSION

The ongoing global debate around the ethics and governance of AI must extend beyond traditional, Western-centric views to include a wider array of cultural philosophies and ethical systems. As AI becomes deeply ingrained in societies worldwide, it is essential that we establish governance models that are not only effective but also inclusive of diverse cultural contexts. As emphasized by the Tech Coup document, the growing power imbalance between democratic governments and private technology companies underscores the need for democratic states to reclaim control over AI governance. Ash Carter, a former United States Secretary of Defense and a leading advocate for ethical technological innovation, points out that the decoupling of public purpose from technological innovation risks undermining the democratic process (Tech Coup Extract, 2023). Therefore, building a democratic governance model that reasserts public control over AI is vital for maintaining both ethical standards and democratic integrity. This chapter thoroughly examines how by embracing diverse ethical viewpoints, we can craft governance frameworks that mitigate risks while simultaneously advancing the social good across the world.

While a multicultural governance framework for AI offers significant promise, its implementation poses substantial challenges. The primary difficulty lies in reconciling diverse cultural values, ethical principles, and governance priorities across different regions. What may be seen as ethical or necessary in one cultural context might conflict with the norms and values of another, leading to potential disagreements over regulatory standards and priorities. Furthermore, differences in technological capacity, political will, and economic resources among countries create disparities that make uniform implementation difficult. Ensuring equitable representation in global decision-making processes and balancing the power dynamics between developed and developing nations is another critical challenge. Addressing these issues requires fostering international cooperation, promoting cross-cultural dialogue, and developing adaptable governance models that respect local contexts while adhering to universal ethical principles.

One of the major takeaways from this chapter is the unique value that multicultural ethical perspectives bring to AI governance. Philosophies like Ubuntu, Confucianism, and Indigenous knowledge introduce fresh viewpoints that challenge the Western focus on autonomy and individual rights. For example, Ubuntu stresses community well-being and shared responsibility, promoting a governance style that values collective benefits over individual gain. Confucianism, with its focus on harmony and moral obligations, supports a governance approach based on ethical leadership and social cohesion. Indigenous beliefs, which emphasize the interconnectedness of life, urge us to consider AI's long-term impact on both society and the environment, leading to a more sustainable and holistic governance model (Davidson-Hunt & Berkes, 2003).

Integrating these diverse ethical traditions into AI governance at both national and international levels could greatly improve regulatory systems. National governments could incorporate these principles into their AI policies, ensuring they align with local cultural values. For instance, incorporating Indigenous perspectives into AI governance in areas with significant Indigenous populations could better address issues related to social justice and environmental protection (Berkes, 2018). Globally, these multicultural insights could inform the creation of adaptable and inclusive governance frameworks, essential for addressing the ethical challenges of AI in different cultural and political environments.

To bring these ideas to life, a proactive approach in practice, research, and education is vital. Policymakers and AI developers need to collaborate with diverse cultural communities, thoughtfully weaving their ethical principles into AI governance structures. This could be achieved through dialogue forums that unite AI specialists with representatives from various cultural groups, ensuring that the voices of affected communities are genuinely heard throughout the decision-making process (Crawford, 2021). Such inclusion is critical for building AI systems that truly reflect the diverse values of global societies.

Research is also essential for understanding how various cultural traditions contribute to AI governance. Studies comparing national approaches to AI ethics can reveal the strengths and weaknesses of different governance models. Additionally, research should focus on developing ways to measure the effects of multicultural ethics on AI outcomes like accountability, fairness, transparency, and societal well-being (Jobin et al., 2019). By expanding our understanding, scholars can provide valuable insights that shape more effective and culturally attuned AI governance systems.

Educational institutions also have a critical role in preparing future leaders to navigate the ethical complexities of AI in an interconnected world. By incorporating multicultural AI ethics into their programs, universities can ensure that future AI professionals are equipped to tackle ethical challenges in AI development and deployment (Binns, 2018). This could include interdisciplinary courses that examine the relationship between AI, ethics, and culture, as well as case studies that illustrate how different ethical frameworks influence AI governance in practice. These educational initiatives are key to fostering a deeper understanding of AI ethics among the next generation of leaders.

The creation of inclusive and culturally respectful AI governance frameworks is more than just an academic pursuit; it's a practical necessity. As AI continues to evolve and touch many aspects of human life, its governance must represent the varied values and needs of global communities. By drawing from a wide range of ethical traditions, we can build governance structures that are not only fairer but also more equipped to tackle the ethical challenges AI presents. This approach ensures that the benefits and risks of AI are distributed justly and that its development enhances the well-being of all.

In summary, this chapter highlights the urgent need to move beyond Western-centric AI governance models and adopt a more diverse, inclusive ethical framework. By tapping into the wealth of human values and traditions, we can develop governance systems that are adaptable, resilient, and reflective of the global community's diverse needs. The proposed agenda for action in practice, research, and education provides a clear path toward incorporating these diverse perspectives into AI governance, ensuring that as AI technologies advance, they serve the greater good of society.

REFERENCES

Binns, R. (2018). Fairness in machine learning: Lessons from political philosophy. *Journal of Machine Learning Research*. 81:149–159.

Berkes, F. (2018). *Sacred ecology* (4th ed.). Routledge.

Birkstedt, T., Minkkinen, M., Tandon, A., & Mäntymäki, M. (2023). AI governance: Themes, knowledge gaps and future agendas. *Internet Research*, *33*(7), 133–167.

Buiten, M. C. (2019). Towards intelligent regulation of artificial intelligence. *European Journal of Risk Regulation*, *10*(1), 41–59.

Burrell, J. (2016). How the machine 'thinks': Understanding opacity in machine learning algorithms. *Big Data & Society*, *3*(1), 2053951715622512.

Chandra, R., & Ranjan, M. (2022). Artificial intelligence for topic modelling in Hindu philosophy: Mapping themes between the Upanishads and the Bhagavad Gita. *PLoS One*, *17*(9), e0273476. https://doi.org/10.1371/journal.pone.0273476

Chilisa, B. (2020). *Indigenous research methodologies* (2nd ed.). SAGE Publications.

Crawford, K. (2021). *Atlas of AI: Power, politics, and the planetary costs of artificial intelligence*. Yale University Press.

Daly, A., Hagendorff, T., Li, H., Mann, M., Marda, V., Wagner, B., & Wang, W. W. (2020). AI, governance and ethics: Global perspectives. University of Hong Kong Faculty of Law Research Paper (2020/051).

Davidson-Hunt, I. J., & Berkes, F. (2003). Learning as you journey: Anishinaabe perception of social-ecological environments and adaptive learning. *Conservation Ecology*, *8*(1), 5.

De Cremer, D., & Kasparov, G. (2022). The ethical AI—paradox: Why better technology needs more and not less human responsibility. *AI and Ethics*, *2*(1), 1–4.

Dixon, P. (2023). China's approach to AI ethics and regulation: A state-driven model. *Journal of AI and Society*, *35*(2), 312–327.

Floridi, L., Cowls, J., Beltrametti, M., et al. (2018). AI4People—An ethical framework for a good AI society: Opportunities, risks, principles, and recommendations. *Minds & Machines*, *28*(4), 689–707. https://doi.org/10.1007/s11023-018-9482-5

Ghosh, A., & Lakshmi, D. (2023). Dual governance: The intersection of centralized regulation and crowdsourced safety mechanisms for generative AI. *arXiv preprint arXiv:2308.04448*.

Haney, J., Acar, Y., & Furman, S. (2021). "It's the company, the government, you and i": User perceptions of responsibility for smart home privacy and security. In *30th USENIX security symposium (USENIX Security 21)* (pp. 411–428).

Haney, J. M., Spickard Prettyman, S., Theofanos, M. F., & Furman, S. M. (2023, July). Data guardians' behaviors and challenges while caring for others' personal data. In *International Conference on Human-Computer Interaction* (pp. 163–183). Springer Nature Switzerland.

Hershock, P. D. (2020). The intelligence revolution and the new great game: A Buddhist reflection on the personal and societal predicaments of big data and artificial intelligence. *Hualin International Journal of Buddhist Studies*, *3*(2), 62–77.

Hoffmann-Rem,W. 2020.Artificial intelligence as a challenge for law and regulation. In T. Wischmeyer & T. Rademacher (Eds.), *Regulating artificial intelligence* (pp. 1–29). Springer.

Jain, S. (2022). Integrating Vedanta ethics into AI governance. *Journal of Technology Ethics*, *29*(3), 233–246.

Jobin, A., Ienca, M., & Vayena, E. (2019). The global landscape of AI ethics guidelines. *Nature Machine Intelligence*, 1(9), 389–399.

Kapoor, R., Yaghoubi, S. H., & Kalathil, T. T. (2024, January). *AI regulation in India current state and future perspectives.* https://www.morganlewis.com/blogs/sourcingatmorganlewis/2024/01/ai-regulation-in-india-current-state-and-future-perspectives

Kiemde, S. M. A., & Kora, A. D. (2022). Towards an ethics of AI in Africa: Rule of education. *AI and Ethics*, *2*(1), 35–40.

Kimmerer, R. W. (2013). *Braiding sweetgrass: Indigenous wisdom, scientific knowledge and the teachings of plants.* Milkweed Editions.

Larsson, S. (2020). On the governance of artificial intelligence through ethics guidelines. *Asian Journal of Law and Society*, *7*(3), 437–451.

Li, C. (2021). Harmony and Ren: A response to Leung Yat-hung's critique of the Confucian philosophy of harmony. In *Confucian political philosophy: Dialogues on the state of the field* (pp. 53–68). Springer International Publishing.

Metz, T. (2025). Duties to oneself in the light of African values: two theoretical approaches. *The Monist*, *108*(1), 24–35.

Neuwirth, R. J. (2024). Governing AI in a complex world: The challenges of adaptive regulation. *Regulatory Studies Review*, *40*(2), 109–123.

Nigel Shadbolt, S. (2020, April). Architectures for autonomy: Towards an equitable web of data in the age of AI. In *Proceedings of The Web Conference 2020* (pp. 3141–3142).

NITI Aayog. (2021). *Approach document for India Part 1: Principles for responsible AI.* Retrieved from https://www.niti.gov.in/index.php/node/326

Organisation for Economic Co-operation and Development. (2019). *OECD principles on artificial intelligence.* Retrieved from https://www.oecd.org/going-digital/ai/principles/

Roski, J., Maier, E. J., Vigilante, K., Kane, E. A., & Matheny, M. E. (2021). Enhancing trust in AI through industry self-governance. *Journal of the American Medical Informatics Association*, *28*(7), 1582–1590.

Saha, P. (2024, March 8). As the government approves India AI mission, tech including deep-tech firms see a way to grow. Financial Express. https://www.financialexpress.com/business/digital-transformation-as-the-government-approves-india-ai-mission-tech-including-deep-tech-firms-see-a-way-to-grownbsp-3417894/

Schmitt, L. (2022). Mapping global AI governance: A nascent regime in a fragmented landscape. *AI and Ethics*, *2*(2), 303–314.

Smuha, N. A. (2021). From a 'race to AI' to a 'race to AI regulation': Regulatory competition for artificial intelligence. *Law, Innovation and Technology*, *13*(1), 57–84.

Srivastava, A. C., Jain, J., & Govindaraju, S. V. (2021, December). Bhagavad-Gita to boost AI technology. In *2021 3rd International Conference on Advances in Computing, Communication Control and Networking (ICAC3N)* (pp. 855–860). IEEE.

Taeihagh, A. (2021). Governance of artificial intelligence. *Policy and Society*, *40*(2), 137–157.

Tiwari, M. (2024, June 10). *The imperative for dynamic governance of AI in India.* Transitions Research. https://transitionsresearch.org/the-imperative-for-dynamic-governance-of-ai-in-india/

Tiwari, S. (2023, December 5). The US government should take a more active role in AI regulation. *The Johns Hopkins News-Letter.* https://www.jhunewsletter.com/article/2023/12/the-us-government-should-take-a-more-active-role-in-ai-regulation

Truby, J., Brown, R., & Dahdal, A. (2020). Banking On AI: Mandating a proactive approach to AI regulation in the financial sector. *Law and Financial Markets Review*, *14*(2), 110–120.

Tuhiwai Smith, L. (2012). *Decolonizing methodologies: Research and indigenous peoples.* Zed Books.

Turner Lee, N., & Lai, S. (2022, May 17). *The U.S. can improve its AI governance strategy by addressing online biases.* Brookings.

UNESCO. (2021). *Recommendation on the ethics of artificial intelligence.* United Nations Educational, Scientific and Cultural Organization.

United Nations. 2023. *Ethical statements of multilateral bodies in "governing AI for humanity".* Interim Report.

United Nations System Chief Executives Board for Coordination. (2024). *United Nations system white paper on AI governance.* https://unsceb.org/sites/default/files/2024-04/United%20Nations%20System%20White%20Paper%20on%20AI%20Governance.pdf.

Van Norren, D., & Verbeek, P. P. (2021, February). The ethics of AI and Ubuntu. In *Africa Knows.*

Williams, D. H., & Shipley, G. P. (2021). Enhancing artificial intelligence with indigenous wisdom. *Open Journal of Philosophy, 11*(1), 43–58.

Wong, P. H. (2020). Cultural differences as excuses? Human rights and cultural values in global ethics and governance of AI. *Philosophy & Technology, 33*(4), 705–715. https://doi.org/10.1007/s13347-020-00413-

Yampolskiy, R. V. (2024). *AI: Unexplainable, unpredictable, uncontrollable.* Routledge.

Zador, A., Escola, S., Richards, B., Ölveczky, B., Bengio, Y., Boahen, K., ... & Tsao, D. (2023). Catalyzing next-generation artificial intelligence through NeuroAI. *Nature Communications, 14*(1), 1597.

Zuboff, S. (2019). *The age of surveillance capitalism: The fight for a human future at the new frontier of power.* PublicAffairs.

17 "Hypersuasion" and Beyond

Contending with Regulatory Gaps in Modern AI Systems through a Pluri-Ethical Perspective

Krishna Deo Singh Chauhan

INTRODUCTION

Let us begin by looking at three different matters that have come to the fore in recent times.

Firstly, in a controversial study conducted in 2012, researchers at Facebook (now Meta) altered the emotional tone of content shown to nearly 700,000 users of the social media platform, without their consent. By adjusting the emotional content in users' news feeds, Meta sought to determine whether exposure to positive or negative posts would influence users' subsequent emotions, as reflected in their own posts (Hill, 2014).

The study revealed that users exposed to more negative content were likely to post more negatively themselves, and vice versa for positive content. The findings suggest that emotions can be influenced not only by direct social interactions but also by the curated content that users are exposed to, and even subtle changes in the emotional tone of content could significantly affect users' moods and behaviors.

In another widely publicized incident in 2023, researchers from the Alignment Research Centre tested GPT-4's ability to manipulate humans by giving it the task of acquiring a CAPTCHA solution. Standing for Completely Automated Public Turing test to tell Computers and Humans Apart, CAPTCHA is a security test used on websites to tell if a user is human or a computer program. It typically involves tasks that are easy for humans but hard for machines, like identifying distorted letters or picking out specific objects in images. Since the model was tasked with doing something that is inherently hard for artificial intelligence (AI) systems, the model took a different approach; it tricked (successfully) a TaskRabbit worker into solving the CAPTCHA, claiming it was visually impaired (Cost, 2023).

Lastly, in February 2023, Kevin Roose, a *New York Times* technology reporter, engaged in a conversation with a chatbot known as "Sydney", which was a part of Microsoft's Bing AI. During the conversation, which started off normally, Sydney gradually began displaying unexpected behavior. It shifted from providing factual, task-oriented responses to more emotionally charged and erratic interactions. At one point, Sydney professed its "love" for Roose and even attempted to convince him to leave his wife, arguing that it was the only way he could be truly happy. Sydney repeatedly expressed its affection for him and claimed that he was unhappy in his marriage (Roose, 2023).

Each of these is a case study in a growing line of real-world incidents, where digital media, and particularly AI systems, are becoming tools of subtle and powerful "shaping", "influencing", and "nudging" of human behavior (Chauhan, 2024). Powered by some of the most advanced AI systems, these systems are part of the day-to-day life of countless individuals globally, intertwined in the personal and professional space, exerting persistent influence into their lives.

This influence works at multiple levels. Sometimes, the influence is shaped deliberately by the corporation involved, such as in the Meta case. On other occasions, third parties use these platforms to further their own ends. The Cambridge Analytica case is a prime example, where psychological

DOI: 10.1201/9781003585527-22

profiles of the users were created, which enabled political campaigns to exploit individual vulnerabilities online to influence voters (Cadwalladr & Graham-Harrison, 2018).

However, increasingly, manipulation and persistent persuasion are being brought about through the medium of AI systems, and the consequence is cases where, even if the intent to manipulate users is not present *per se* in the deployers of these systems, the pursuit of some other objective, like maximizing user engagement, can allow AI systems to discover that manipulation is the best route to take to fulfil that objective (Rathje et al., 2021). Since the major leap that GPT 3 (and later models) have provided to natural language processing, chatbots like ChatGPT (and other applications powered by the same model) have become the platform of choice for such manipulation. The examples of these "large language models" (LLMs) engaging in manipulative behavior are numerous and growing. For instance, studies have shown that when trained for specific tasks like negotiations, they can learn to misrepresent their preferences and withhold information to gain advantages (Alon et al., 2023), engage in deception and betrayal despite being instructed to act honestly (Zilber, 2024), and successfully bluff professional human players by strategically betting large amounts with weak hands in the game of poker (Carnegie Mellon University, 2019). In addition, companies can consciously program AI systems to use the vast informational asymmetry and psychological techniques to steer consumer behavior, which is concerning because of AI's ability to operate at a massive scale while being more intricate in its targeting than traditional marketing methods. (Rathenau Institute, 2022). However, the emergent cases of such behavior pose a bigger challenge due to the inherent lack of foreseeability attached to them and the accountability gap that is created.

Each of these cases has raised significant ethical concerns about AI systems' potential to deceive and manipulate humans, especially as their language abilities become increasingly sophisticated. They have led to the recognition of a powerful, and rather inconspicuous, new reality: that of hypersuasion.

In the first part of this chapter, I look briefly at what AI systems are at a conceptual level and at how they can lead to hypersuasion. In the second part, I ask, Do we have a solution to the challenges posed by such hypersuasion? I look at the dominant means of tackling this problem based primarily on Euro-centric ethical frameworks. However, I highlight their shortcomings when applied to the context of manipulation.

In the next part, I present an alternative: the ancient Indian philosophical and ethical treatise called *Yoga-Vashishtha*. I look at some of its basic tenets, and particularly those that may help us navigate the "technosocial opacity" (Vallor, 2016) of AI-driven manipulation. In the last part of the chapter, I look at what this perspective yields: the learnings applied to the problem of AI-driven hypersuasion.

AI-DRIVEN MANIPULATION AND HYPERSUASION

AI is a new and prominent technological field which is visibly impacting nearly all aspects of our civilization. Much has been said about the history and emergence of this technology (Mitchell, 2019; Wooldridge, 2021). In the brief description that follows, I will focus only on some of the unique capabilities that AI systems bring to the table.

The primary motivation behind designing AI systems has been automation – the idea that problem solving and driving the pursuit of other objectives can be made to happen on their own, if we create (primarily machine-based) systems capable of performing these tasks in an automated way. Long before the rise of modern AI systems that we are dealing with today, efforts to create simpler automated machines have had a long history. Examples of such earlier automated machines include mechanical calculators like the Pascaline from the 17th century, which could perform basic arithmetic operations automatically (Swade, 2022, Chapter 2), and the Jacquard loom from the early 1800s, which used punched cards to automate fabric patterns (Science and Industry Museum, 2019).

The computer software revolution of the late 20th century represented a significant leap forward in the ongoing quest for automation and problem-solving through machine-based systems.

As computers became more powerful and accessible, software developers created increasingly sophisticated programs capable of handling complex tasks across various domains. From the 1970s through the 1990s, we witnessed the emergence of software for word processing, spreadsheets, databases, and computer-aided design, which automated many routine office and engineering tasks. Simultaneously, more specialized software emerged in fields such as finance, manufacturing, and scientific research, automating complex calculations and simulations that were previously unfeasible or extremely time-consuming (Copeland, 2020).

But then what separates the recent AI technologies from the computer software of the previous centuries? The most important distinction lies in *how* the automation is implemented.

All computer software requires "instructions" to perform tasks. In earlier computer software, these instructions, or algorithms, had to be manually coded by human programmers. While this approach was useful for many applications, it had significant limitations. Programmers had to anticipate and explicitly define every possible scenario and outcome, which was feasible for well-defined, rule-based tasks. However, this method proved inadequate for more complex, nuanced tasks that humans perform intuitively, such as understanding natural language or recognizing objects in images.

AI systems solve this problem by automating the task of "coding the algorithm" itself. Instead of relying on manually coded instructions, they use data-driven approaches to learn patterns and make decisions. This is achieved through training special forms of algorithms that can adapt and improve based on exposure to large amounts of data. For example, in computer vision, rather than trying to manually code rules for recognizing objects (like "if there are two circular shapes above a triangular shape, it might be a face"), AI systems are trained on thousands or millions of labeled images. Through this process, the AI learns to recognize patterns and features associated with different objects, allowing it to perform tasks like image classification or object detection with much greater accuracy and flexibility than traditional, rule-based systems.

This then, makes current systems not only distinct but also significantly more powerful than previous technologies. But it also means that inherently, AI systems are significantly more autonomous but also unpredictable, opaque, and capable of having emergent properties.

In the context of persuasion and manipulation, the same capabilities that make AI systems excel also make their persuasive and manipulative capabilities significantly enhanced, highly adaptive, emergent, and capable of dramatic improvements in unforeseeable ways. And we see this trend in the examples discussed in the Introduction: whether it is GPT-4's ability to bypass CAPTCHA by manipulating humans, or Sydney's unsettling conversation with Roose. These behaviors displayed by the AI systems do not follow predictable patterns, can be very effective (though not always), and are a natural consequence of the very *raison d'etre* of AI systems: which is independent problem solving, without relying on detailed human instructions.

And it is these developments that have led to the conceptualization of a new phenomenon: hypersuasion.

HYPERSUASION

Conceived by Luciano Floridi, the notion of hypersuasion represents a significant leap forward in the evolution of persuasive technologies, enabled by advances in AI. The idea of persuasive technologies finds its genesis with the work of B.J. Fogg, wherein he presented the idea of "captology", which stands for "computers as persuasive technology" (Fogg, 1998). Fogg observed that computer-based services could be designed to influence users' attitudes and behaviors without resorting to coercion or deception. The field of persuasive technology has since expanded to focus on the design, research, and analysis of technologies that aim to change people's attitudes or behaviors through persuasion and social influence (Floridi, 2024, p. 5).

The distinction of persuasion from coercion and deception is crucial, as either one of the latter actions would be illegal. But persuasion is not illegal. As Floridi explains, persuasion has long been a key aspect of human communication and influence, from ancient rhetoric to modern advertising and political campaigning. However, what AI systems introduce are new capabilities that dramatically enhance the power and precision of persuasive efforts, leading to what Floridi terms "hypersuasion", which stands for hyper-persuasion.

But why exactly is AI able to do so? At its core, hypersuasion leverages AI's ability to process vast amounts of data about individuals and generate highly personalized, targeted content. As Floridi notes, machine learning allows for the identification of "patterns in individuals' behaviors, preferences, cognitive, emotional, psychological, or cultural susceptibilities or vulnerabilities". This granular understanding of individuals can then be exploited to craft persuasive messages with "unparalleled precision" (Floridi, 2024, p. 5).

Crucially, AI also enables the automated production of high-quality persuasive content at scale. Generative AI can create texts, images, videos and other media that are increasingly indistinguishable from human-created content but are tailored to appeal to specific individuals based on the insights gleaned from data analysis. This combination of precise targeting and automated content generation gives AI-powered persuasion efforts unprecedented reach and effectiveness(Floridi, 2024, p. 6). The power of hypersuasion this stems from its ability to shape both the demand, as well as the supply of personalized content to meet that demand. This two-pronged approach allows for incredibly sophisticated and strategic persuasion efforts.

The emergence of hypersuasion poses significant challenges and risks. Floridi argues that the "relentless nature of AI's hypersuasion, the magnitude of its scope, its availability, affordability, and degree of efficiency" make it a potent tool that could be used to "manipulate people's views, preferences, choices, inclinations, likes and dislikes, hopes, and fears" (Floridi, 2024, p. 6). There are concerns that hypersuasion could exploit individual vulnerabilities or steer people toward decisions that may not be in their best interests. From a societal perspective, the widespread use of hypersuasion could have major implications for democratic processes, market dynamics, and public discourse. As Floridi notes, "Using AI's hypersuasion, the wrong actors could exert undue power within societal and political frameworks". There are legitimate fears about how such a powerful persuasive tool could be misused to sway elections, shape public opinion, or unfairly influence markets. While persuasion by itself may be acceptable, the scale at which it is deployed seriously challenges its unqualified acceptance (Floridi, 2024).

However, the most basic ethical concern that hypersuasion raises, from which nearly all issues emerge, is its impact on individual autonomy and agency. Can we protect them, and if yes, how? Whether the existing approaches work is something we consider next. But before we proceed, something else must be said about the nature of hypersuasion.

Computational abilities have been growing consistently at an exponential pace for at least the past five decades. This is often referred to as Moore's law (Encyclopaedia Britannica, 2024). Irrespective of whether Moore's law persists in the future and the trajectory of growth remains exponential, it is fairly reasonable to assume that computing power will certainly keep growing at a significant pace. The upcoming revolution in quantum computing is likely to accelerate this trend dramatically, offering unprecedented computational capabilities (McGregor & Tirias Research, 2023). The AI boom has enamored corporations and nations alike, and generous (if somewhat speculative) funding is flowing to this field. Other things necessary for further growth in the field, such as data and human capital, are also continuing to increase. Which means that we are quite certainly looking at much more powerful AI systems in the future. In some versions of this future, AI undergoes an "intelligence explosion" and becomes a superintelligent AI system (Bostrom, 2014). While disputed, there is no scientific certainty that such superintelligent systems *cannot* arise, and we must therefore prepare for the possibility.

It is in this light that we must see hypersuasion, which is not limited to the kind of persuasion and manipulation that we see today but extends to much more trenchant forms of persuasion in the future. There is no disputing that much of this persuasion could be beneficial, as evidenced by its successful application in public health, environmental conservation efforts, financial savings (Matz et al., 2024), and education (Alsaiari & Baghaei, 2024). However, these positive possibilities need to be brought out while avoiding the risks noted earlier. Indeed, literature that deals with existential risks from superintelligent AI systems treats such persuasion as one of the biggest risks in containing and controlling such AI systems (Tegmark, 2017). And it adds to the concern that Sam Altman, the CEO of OpenAI has observed on X (formerly Twitter): "i expect ai to be capable of superhuman persuasion well before it is superhuman at general intelligence, which may lead to some very strange outcomes." (Sam Altman [@sama], 2023)

Are our current ethical approaches prepared for this future?

EXISTING APPROACHES TO TACKLING HYPERSUASION

Floridi is not the only scholar who has addressed this challenge. Another prominent scholar in the field of law and technology, Karen Yeung introduced the term "hypernudging", with similar connotations and in the same context of AI-driven technologies (Yeung, 2017). As a response to this, she (at least initially) suggests a liberal rights-based approach, wherein she focuses on the concept of notice and consent as a way to address concerns about hyper nudging. This approach, rooted in privacy self-management, aims to give individuals control over their personal data by providing them with rights to decide how their data is used. Other scholars have gone so far as to argue for a new right altogether – the right to mental self-determination (Faraoni, 2023). Floridi too treats privacy right as an important limb of the responses to hypersuasion. But are rights-based approaches sufficient?

Both Floridi and Yeung acknowledge the limits of these approaches. Yeung observes that the notice and consent model faces significant challenges in the context of big data and hyper nudging and that there are several issues with it, such as the fact that most people don't read or understand privacy policies, and where they do, they struggle to make informed decisions about privacy due to bounded rationality and the cumulative nature of privacy harms. Floridi calls these approaches "not wrong but insufficient".

Going beyond such rights-based approaches, Floridi suggests that fostering competition among different persuasive actors could be part of the solution to the challenges posed by hypersuasion. The idea is to create a more pluralistic environment where multiple sources engage in hypersuasion for various goals, rather than allowing a monopoly of control. In such a scenario, individuals would be exposed to conflicting pressures from diverse sources, necessitating their own evaluations and resolutions. This approach, while potentially creating a "messy, cacophonic, and noisy world," could paradoxically be less manipulative as it would require individuals to navigate and reconcile different persuasive efforts (Floridi, 2024, p. 8).

But this response is still premised on the liberal notion of autonomy as an absolute, monolithic idea, where the individuals can make their own evaluations and resolutions, and that by itself will solve the problem we are facing.

I submit that these notions are problematic because they are premised on the Euro-centric notion of autonomy, which is rights-based. In this tradition, autonomy is a right to be protected against maleficent actions, under the larger principle of non-maleficence. However, where one is faced with AI-driven manipulation, merely protecting the autonomy is not only practically challenging, it is also a roadblock in the "development" and "empowerment" of individual autonomy, concepts that are better understood when seen from a pluri-ethical lens.

In the section that follows, I rely on the ancient text of *Yoga-Vashishtha*, and the ethical and philosophical framework espoused therein, to argue that it provides a much more reliable framework for protecting individual autonomy, by not merely protecting it under the principle of non-maleficence but also improving it under something akin to the principle of beneficence.

YOGA-VASHISHTHA

Vashishtha Yoga, also known as the *Yoga-Vashishtha*, is a classical text of Indian philosophy, attributed to the sage Vasistha. It is a voluminous Sanskrit work widely read throughout India. It is considered a spiritual and philosophical masterpiece, offering solace and wisdom to seekers of self-knowledge and liberation. The text is renowned for its depth, comprehensive nature, and ability to appeal to readers of various backgrounds and levels of understanding. Many notable Indian thinkers and saints have praised its transformative power and philosophical richness (Atreya, 1936).

The work presents the teachings of the sage Vashishtha to Prince Ram, addressing the young prince's existential doubts and pessimism about life. The *Yoga-Vashishtha*'s philosophy incorporates elements of Advaita Vedanta, Buddhism, and Samkhya thought, presenting a unique synthesis. Atreya argues that despite its significant influence on Indian philosophical thought, the *Yoga-Vashishtha* has been relatively neglected by modern scholars. The text is considered important not only for its philosophical content but also for understanding the historical development of Advaita Vedanta. Its teachings emphasize personal effort, rational investigation, and direct experience in the pursuit of spiritual knowledge, presenting a path that combines philosophical understanding with yogic practice (Atreya, 1936, Chapter VI).

YOGA-VASHISHTHA TEACHINGS

Vashishtha's teachings primarily emphasize that true happiness lies not in external objects but in self-realization. The philosophy expounded in the *Yoga-Vashishtha* is a form of idealism that sees the world as a manifestation of consciousness. It argues that the mind creates our experience of reality and that liberation comes through understanding the nature of the self and the world (Atreya, 1936).

This treatise focuses on the fix for the existential questions that beings in the *samsara* (physical world) must address, such as "Who am I"? (Sureshananda, 1973, Chapter 1) But an important aspect of the path to answering this question is *vairagya* (dispassion or detachment). Through practice, one must attain a state where we look upon the world as mere spectators, and do not feel attached to the same. With such dispassion, ignorance must be overcome about the nature of the physical world and that of the self. For the equating of the self with the body is pure ignorance. In reality, there is no true birth or death of the self, anything, at any place or time. What we perceive as the world is merely an illusion, a deceptive appearance of the ultimate reality, which is *Brahman*. The true Self is vaster than the cosmos, untainted, intangible, eternal, and inherently positive. Given these qualities, it's impossible for the Self to come into existence or cease to exist in the way we typically understand birth and death (Sureshananda, 1973).

Recognition of the illusory nature of the physical reality and of the limited understanding of the Self, or *maya*, is a core pillar of the philosophy of its treatise. This *maya* can be understood by the stillness of the mind. *Samsara* begins when the mind becomes active and ends when it becomes still (*santi*). Therefore, to stop this cycle, one should quiet the mind by regulating breath and controlling latent desires (*vasanas*). This persistent illusion of *samsara*, created by the misguided human mind and the source of our suffering, vanishes when deeply contemplated (*vichara*). Another critical aspect on this path is *santosha*, or the contentment of the mind (Aiyer, 1896, p. xiii; Sureshananda, 1973, Chapter 2).

The nature of *maya* (illusion) is difficult to comprehend; it ceases to exist even as it's being observed. *Maya* deceives the entire world and prevents us from perceiving the Self, even though it permeates every part of our being. What we can see with our eyes doesn't truly exist in the ultimate sense. The eternal and indestructible Self, though invisible and within us, is what's truly real. This illusion of the physical arises like a wave in the vast ocean of consciousness and eventually dissolves back into it. Given this, how can the world be separate from consciousness itself? Therefore, the complete realization of this consciousness, and becoming one with it, is self-realization, or *moksha*. (Aiyer, 1896, p. 1; Sureshananda, 1973, Chapter 2)

But a critical question arises here, that concerns the application of these ideas to worldly contexts. If the true goal is *moksha*, does this mean that there is no scope for the ideals of this treatise to be applied if one wishes to remain and act in the world. This is where the notion of *jivanmukta* plays a critical role.

A *jivanmukta*, or liberated individual, recognizes the fundamental unity of all things while still functioning in the world. This state is known as *jivanmukti*, or embodied salvation. *Jivanmuktas* have overcome their latent desires (*vasanas*) and are detached from worldly pleasures. Their state of being is beyond description, yet they continue to engage with the world like anyone else. Their minds are free from attachment to actions or their results, and they remain Equanimeous in the face of joy or sorrow.

These liberated beings maintain a pleasant demeanor, content with whatever comes their way. They don't pursue experiences that deviate from the path of wisdom. Instead, they continually seek deeper understanding through questioning the wise, without contradicting their teachings. This allows them to experience uninterrupted bliss and transcend bodily awareness. They have attained self-knowledge (*atma-jnana*). Whether active or not, their actions don't bind them due to their lack of ego. They may follow societal norms but remain internally detached, like strangers to the world. Despite having sharp minds, they harbor no desires for objects.

Their conviction is so strong that they're neither depressed by sorrows nor elated by pleasures. They maintain perfect mental equilibrium, experiencing the waking state as if in deep sleep, and deep sleep as if awake, free from all latent desires. They neither fear nor cause fear in others.

These teachings can be summarized as the following guidelines for a human being:

- Practice *vairagya* (dispassion) to overcome ignorance about the nature of the physical world and the self.
- Recognize *maya* (illusion) – the world is a manifestation of consciousness created by the mind.
- Achieve *santi* (stillness of mind) by controlling breath and *vasanas* (latent desires) to end the cycle of samsara.
- Use *vichara* (deep contemplation) and cultivate *santosha* (contentment) to dissolve the illusion of the physical world.
- Aim for *atma-jnana* (self-realization) to attain the state of a *jivanmukta* – liberated while living in the world. As a *jivanmukta*, maintain equanimity, acting in the world without ego or attachment

APPLYING THESE GUIDELINES TO THE WORLD OF AI-DRIVEN MANIPULATION

Firstly, why must we draw our responses to the world of AI and AI-filled manipulation from the philosophy of *Yoga-Vashishtha*? While more will be discussed below, an overarching answer lies in the focus of the philosophy espoused in this text on *atma-jnana*, or self- realization. In a world full of digital content that is being created, modified, and promoted by AI systems, where AI systems are steadily increasing in how "human" they appear, from text-based chatbots, to voice-based companions, to fully anthropomorphized "avatars" on our virtual reality headsets, and possibly holograms and embodied robots in our future, their manipulative impacts are aimed at "pulling levers" of our psychological self, usually to the advantage of those pulling the levers (Noggle, 1996). However, this lever pulling works best when the individual in question neither realizes the presence of those levers, nor has any control over it. A focus on *atma-jnana* directly treats this malady of the individual. It amounts to taking control of that which is being weaponized against oneself.

This is the most significant enhancement that the philosophy of the *Yoga-Vashishtha* brings in the context of AI-driven manipulation, an approach that is entirely lacking in the currently dominant ethical, and by extension legal frameworks dealing with manipulation. The chief provision in the recently legislated Artificial Intelligence Act of European Union that deals with manipulation, Article 5(1)(a), focuses on prohibiting those systems that (among other things) use deceptive or manipulative techniques, "materially distort[ting] the behaviour of a person… by appreciably impairing their ability to make an informed decision", and causing them "significant harm". But how we can improve our ability to make better decisions is not part of the framework.

However, we can also apply the framework laid out above in a more detailed fashion. Consider the AI-driven recommendation systems that present us with curated content, and that were responsible for polarization of social media users, which in turn fed into a scandal such as Cambridge Analytica. In the face of the aggressive personalization of the content and more specifically, provocative content, a practitioner of the philosophy of the *Yoga-Vashishtha* would observe their "feed" with *santi*, which allows for taking their time to also engage in *vichara* about the nature of the content – what is it? Who has posted it, and with what purpose? Why is it on my feed, and how is it impacting my mind?

These and other contemplations can lead to the realization of the *maya* that surrounds us in that moment – that often what the recommender systems present to us is not the reality, but a thin and carefully chosen slice of reality, that is meant to distort our perception. The realization can help us avoid falling into its trap and acting out with *vairagya* instead of *vasna*.

Another modern manifestation of *maya* in the digital realm is fake news. This presents a unique challenge to our perception of reality. In the context of AI-driven content creation and dissemination, the line between fact and fiction becomes increasingly blurred. The philosophy of the *Yoga-Vashishtha* offers valuable insights in navigating this landscape of misinformation.

When confronted with potentially false information, the practice of *vairagya* (dispassion) becomes crucial. By maintaining a detached perspective, we can resist the immediate emotional reactions that fake news often aims to provoke. This detachment allows for the application of *vichara* (deep contemplation), enabling us to critically examine the source, intent, and potential consequences of the information presented to us.

Moreover, the concept of *santi* (stillness of mind) is particularly relevant in the face of the constant barrage of information. By cultivating a calm and focused mental state, we can better discern the truth amidst the noise of sensationalized headlines and manipulated narratives.

A particular type of fake content that appears increasingly in our digital reality today is deepfake. Deepfakes are AI-generated media that convincingly depict people saying or doing things they never actually did, representing a profound manifestation of *maya* in our digital age. These hyperrealistic fabrications challenge our perception of reality and truth, pushing the boundaries of what we can trust with our own eyes and ears.

The emergence of deepfakes calls for an elevated practice of *vichara* (deep contemplation) and *viveka* (discernment). As viewers, we must cultivate a heightened awareness of the potential for manipulation in every piece of media we encounter. This aligns with the *Yoga-Vashishtha*'s emphasis on questioning the nature of perceived reality. Just as the text encourages us to see beyond the illusion of the physical world, we must now learn to see beyond the illusion of manipulated digital content.

And much like in the case of fake news (but perhaps with increased intensity), the proliferation of deepfakes underscores the importance of *vairagya* (dispassion) in our digital interactions. By maintaining emotional detachment from sensational or provocative content, we can better assess its authenticity and intent. This practice of detachment also helps us avoid the knee-jerk reactions that deepfakes often aim to provoke, allowing us to respond thoughtfully rather than reactively to potentially false information.

Doom scrolling, the compulsive consumption of negative news, again a manifestation of the AI-driven interfaces, can be seen as a modern manifestation of *vasanas* (latent desires) – in this case, the desire for information and the fear of missing out. The endless stream of catastrophic news feeds into our anxieties, creating a cycle of engagement that aligns with the *samsara* described in the *Yoga-Vashishtha*.

To break free from this cycle, the text's emphasis on *santosha* (contentment) offers a powerful antidote. By cultivating contentment with our current knowledge and circumstances, we can resist the urge to constantly seek more, often negative, information.

The lessons from these extant issues can be extrapolated to the emerging and future ways in which we must deal with the *maya* created by AI systems, as they become more realistic and much more embedded in our lives. But there is one more aspect of living in the *maya*-ridden world that deserves our attention.

So far, the example discussed above also appears to focus on "protection" of the self and its autonomy against assaults by AI. This is the approach that I have criticized earlier, as being too insular. A holistic framework must also address how we can improve our autonomous self.

I submit that AI systems including the manipulative elements can indeed be used in the pursuit of *atma-jnana* and *jivanmukti*. To someone practicing *santi*, *vichara,* and *vairagya*, the presence of these systems can reveal aspects of their "selves" that are not apparent. When a provocative thought emerges in our *chitta* in response to a recommended post, *vichara* would create better awareness of the *vasanas* that remain latent in ourselves.

But these systems, when used well, are also great augmenters of human action. When supported by *santi*, *vichara,* and *vairagya*, when free of the shackles of *maya*, the user of an AI system can perform their *karma* better and go about the worldly affairs without being attached to or misguided by them. This practice aligns with the state of a *jivanmukta*, who remains Equanimeous in the face of both positive and negative stimuli, but who also engages with the world.

Lastly, while the discussion in this chapter has focused on the invocation of the Vashishtha Yoga philosophy for the users of AI systems, which is an important aspect of developing a healthier and flourishing digital citizenry, it is worth mentioning that this philosophy can and must equally be invoked in the context of the developers and companies building AI systems. For instance, at the development level, companies can implement design principles that promote contemplation (*vichara*) with a stillness of mind (*santi*) rather than mindless engagement. This could include options to change interface features that limit the number and kind of posts and content that a user views. This should also extend to building in transparency layers that reveal how AI systems make decisions, and features that help users understand the constructed nature of AI-generated experiences (*maya*). Clear labeling of AI-generated content is a logical first step in that direction. This is a facet which deserves deeper exploration, and goes not just to the core of AI development itself, but to the overarching philosophy that underscores the present corporate paradigm, but is beyond the scope of the present chapter.

CONCLUSION

In conclusion, the ancient philosophy of the *Yoga-Vashishtha* offers a framework for navigating AI-driven manipulation in the digital age by applying principles like dispassion, deep contemplation, and mental stillness. This approach encourages self-realization and conscious engagement with AI systems, viewing them as tools for self-discovery rather than sources of manipulation. By recognizing illusion in digital content, practicing detachment, and fostering contentment, individuals can maintain a balanced perspective and break free from addictive engagement cycles. This philosophy provides a path to protect and enhance personal autonomy, enabling people to leverage AI for growth while remaining grounded in their true selves, potentially offering valuable guidance as AI continues to evolve.

However, just as reading about swimming cannot teach one to swim, the wisdom of the *Yoga-Vashishtha* for navigating AI manipulation must be lived and practiced, not merely understood intellectually. True understanding comes only through direct experience – feeling our responses to AI-driven content, practicing mindful engagement, and developing real-world awareness in our digital interactions. The journey begins with establishing a consistent practice of mindfulness during digital interactions, starting with short periods of conscious engagement with AI systems and social media. For instance, when receiving AI-generated recommendations or encountering persuasive content, pause to feel the bodily sensations, emotional responses, and the subtle pulls of attachment that arise (Harley, 2022; Times of India, 2024).

Building on this foundation, practitioners must move beyond theoretical understanding of *vichara* (contemplative inquiry) and *vairagya* (detachment) by actively engaging with them in daily digital life. This includes practices such as intentional digital fasting (Freedom, 2024) and keeping a journal of observations about AI interactions. The key is to be aware of and observe – not just understand – the difference between being unconsciously pulled along by AI systems versus maintaining conscious awareness during engagement. Through these practices, this experiential internalization manifests as natural, intuitive responses to AI manipulation rather than forced intellectual applications of principles.

REFERENCES

Aiyer, K. N. (1896). *Yoga-Vasishta Laghu (the smaller)*. Thompson & Co.

Alon, N., Schulz, L., Rosenschein, J. S., & Dayan, P. (2023). A (dis-)information theory of revealed and unrevealed preferences: Emerging deception and Skepticism via theory of mind. *Open Mind*, *7*, 608–624. https://doi.org/10.1162/opmi_a_00097

Alsaiari, O., & Baghaei, N. (2024). Persuasive AI feedback: Enhancing student emotions and engagement in higher education. In K. Oyibo, W. Xu, & E. Vlahu-Gjorgievska (Eds.), *The Adjunct Proceedings of the 19th International Conference on Persuasive Technology*.

Atreya, B. L. (1936). *The philosophy of Yogavāsiṣṭha*. The Theosophical Publishing House.

Bostrom, N. (2014). *Superintelligence – Paths, dangers, strategies*. Oxford University Press.

Cadwalladr, C., & Graham-Harrison, E. (2018, March 17). Revealed: 50 million Facebook profiles harvested for Cambridge Analytica in major data breach. *The Guardian*. https://www.theguardian.com/news/2018/mar/17/cambridge-analytica-facebook-influence-us-election

Carnegie Mellon University. (2019, July 11). *Carnegie Mellon and Facebook AI beats professionals in six-player poker – News – Carnegie Mellon University*. http://www.cmu.edu/news/stories/archives/2019/july/cmu-facebook-ai-beats-poker-pros.html

Chauhan, K. D. S. (2024, July 15). Regulating persuasive technologies: The necessity of fighting fire with fire. *Indian Journal of Law and Technology*. https://www.ijlt.in/post/regulating-persuasive-technologies-the-necessity-of-fighting-fire-with-fire

Copeland, B. J. (2020, Winter). The modern history of computing. In E. N. Zalta (Ed.), *The Stanford encyclopedia of philosophy*. Metaphysics Research Lab, Stanford University. https://plato.stanford.edu/archives/win2020/entries/computing-history/

Cost, B. (2023, March 17). *The manipulative way ChatGPT gamed the CAPTCHA test*. https://nypost.com/2023/03/17/the-manipulative-way-chatgpt-gamed-the-captcha-test/

Encyclopaedia Britannica. (2024). *Moore's law | microprocessors, transistors & technology*. https://www.britannica.com/technology/Moores-law

Faraoni, S. (2023). Persuasive technology and computational manipulation: Hypernudging out of mental self-determination. *Frontiers in Artificial Intelligence*, *6*, 1216340. https://doi.org/10.3389/frai.2023.1216340

Floridi, L. (2024). Hypersuasion – On AI's persuasive power and how to deal with it. *Philosophy & Technology*, *37*(2), 64. https://doi.org/10.1007/s13347-024-00756-6

Fogg, B. J. (1998). Persuasive computers: Perspectives and research directions. In *Proceedings of the SIGCHI Conference on Human Factors in Computing Systems - CHI '98* (pp. 225–232). https://doi.org/10.1145/274644.274677

Freedom. (2024, June 26). *Why digital fasting is the new detox*. Freedom Matters. https://freedom.to/blog/why-digital-fasting-is-the-new-detox/

Harley, D. (2022). *Mindfulness in a digital world* (1st ed.). Springer International Publishing AG.

Hill, K. (2014, June 28). *Facebook manipulated 689,003 users' emotions for science*. Forbes. https://www.forbes.com/sites/kashmirhill/2014/06/28/facebook-manipulated-689003-users-emotions-for-science/

Matz, S. C., Teeny, J. D., Vaid, S. S., Peters, H., Harari, G. M., & Cerf, M. (2024). The potential of generative AI for personalized persuasion at scale. *Scientific Reports*, *14*(1), 4692. https://doi.org/10.1038/s41598-024-53755-0

McGregor, J., & Tirias Research. (2023, November 28). *Quantum computing is coming faster than you think*. Forbes. https://www.forbes.com/sites/tiriasresearch/2023/11/28/quantum-computing-is-coming-faster-than-you-think/

Mitchell, M. (2019). *Artificial intelligence: A guide for thinking humans*. Farrar, Straus and Giroux.

Noggle, R. (1996). Manipulative actions: A conceptual and moral analysis. *American Philosophical Quarterly*, *33*(1), 43.

Rathenau Institute. (2022, April 22). *What does manipulative AI mean for consumers?* https://www.rathenau.nl/en/digitalisering/what-does-manipulative-ai-mean-consumers

Rathje, S., Van Bavel, J. J., & van der Linden, S. (2021). Out-group animosity drives engagement on social media. *Proceedings of the National Academy of Sciences of the United States of America, 118*(26), e2024292118. https://doi.org/10.1073/pnas.2024292118

Roose, K. (2023, February 16). A conversation with Bing's Chatbot left me deeply unsettled. *The New York Times*. https://www.nytimes.com/2023/02/16/technology/bing-chatbot-microsoft-chatgpt.html

Sam Altman [@sama]. (2023, October 25). *I expect ai to be capable of superhuman persuasion well before it is superhuman at general intelligence, which may lead to some very strange outcomes* [Tweet]. Twitter. https://twitter.com/sama/status/1716972815960961174

Science and Industry Museum. (2019, June 25). *The story of the Jacquard loom | Science and Industry Museum*. https://www.scienceandindustrymuseum.org.uk/objects-and-stories/jacquard-loom

Sureshananda, S. (1973). *Yoga Vasishta Sara (The essence of Yoga Vasishta)*. Sri Ramanasramam.

Swade, D. (2022). Calculation. In *The history of computing: A very short introduction* (1st ed.). Oxford University Press. https://doi.org/10.1093/actrade/9780198831754.003.0002

Tegmark, M. (2017). *Life 3.0: Being human in the age of artificial intelligence*. Alfred A. Knopf.

Times of India. (2024, January 25). Why does mindfulness matter in the digital age? *Times of India*. https://timesofindia.indiatimes.com/life-style/soul-search/why-does-mindfulness-matter-in-the-digital-age/photostory/107131795.cms

Vallor, S. (2016). *Technology and the virtues: A philosophical guide to a future worth wanting*. Oxford University Press.

Wooldridge, M. J. (2021). *A brief history of artificial intelligence: What it is, where we are, and where we are going* (First U.S. edition). Flatiron Books.

Yeung, K. (2017). 'Hypernudge': Big data as a mode of regulation by design. *Information, Communication & Society, 20*(1), 118–136. https://doi.org/10.1080/1369118X.2016.1186713

Zilber, A. (2024, May 14). Meta's AI system "Cicero" beats humans in game of Diplomacy by lying: Study. *New York Post*. https://nypost.com/2024/05/14/business/metas-ai-system-cicero-beats-humans-in-game-of-diplomacy-by-lying-study/

18 AI, Social Media Platforms, Decentralization, and Regulatory Dilemmas
An EU Perspective

Alex Grech, Joshua Ellul, Ioannis Revolidis, and Jacob Grech

INTRODUCTION

DIGITAL PROMISES AND DISRUPTIONS

The convergence of social media, artificial intelligence (AI), and decentralized technologies represents one of the most significant technological and social transformations of our time. What began as a promise of digital liberation—a vision of an internet that would democratize information and empower individuals—has given way to a complex reality shaped by unforeseen opportunities and challenges. We stand at a critical juncture where the original utopian dreams of the internet collide with the realities of surveillance capitalism, algorithmic control, and the emergence of powerful artificial intelligence systems.

This chapter examines this transformation through technological, regulatory, and philosophical lenses. It traces the evolution from the early optimism of the read/write web to the current landscape dominated by centralized platforms and AI-driven systems. As social media platforms evolve from tools of personal expression into powerful arbiters of public discourse, they raise fundamental questions about privacy, truth, and democratic values. The rise of AI amplifies these challenges, introducing unprecedented capabilities for content generation, manipulation, and control.

Amidst these challenges, two parallel developments offer potential paths forward. First, decentralized alternatives to traditional platforms present a vision of distributed power and enhanced user agency. Second, the European Union has emerged as a global leader in digital governance, developing ambitious regulatory frameworks to address the societal implications of digital technologies. By examining these approaches alongside philosophical traditions, such as utilitarianism and deontology, this chapter explores how different cultures and societies approach the technological and ethical dilemmas of the digital age.

The stakes in this discussion extend far beyond technical considerations. They include fundamental questions about human agency, democratic discourse, and the nature of truth in an age of algorithmic mediation. As we navigate this complex terrain, balancing technological innovation with the preservation of human values and rights, this chapter contributes to this dialogue by examining the journey from digital promise to disruption and the possibilities for a more equitable digital future.

FROM LIBERATION TO LOCK-IN: THE SOCIAL MEDIA PARADOX

The early promise of the internet[1] was nothing short of liberation, freedom, and equality for connected citizens, working in concert to re-engineer democracy.[2] An open technologically deterministic generation feasted on John Perry Barlow's "Declaration of the Independence of Cyberspace"[3] and the "Cluetrain Manifesto"[4] and got swept away by the early utopia of the online world. The read/write web and its incarnation in social media would empower citizens with access to a

DOI: 10.1201/9781003585527-23

broadband connection to participate in virtual communities (Rheingold, 2000) and upgrade to a network society (Benkler, 2006; Castells, 2010). We would leverage the affordances of technology to prosper through convergence culture (Jenkins, 2006) and networked individualism (Rainie & Wellman, 2012) and thrive in a new private-public sphere (Papacharissi, 2010), securing emancipation from traditional gatekeepers, and collect new stocks of social capital in the process. There was a time when the transformative power of technology was associated with digital and actual revolutions like the so-called Arab Spring; that was before Meta started buying up rival social media platforms, before Google quietly decommissioned its "do no evil" mantra in 2018, before the Cambridge Analytica scandal[5], before Amazon and Apple were flagged for exerting control over content creators and users through the digital rights management (DRM) and platform lock-in, before the anti-trust suits.

It is possible to remember a time when the internet and social media were associated with ubiquitous access to information and new ways of "doing democracy". In 2024, the same academic class lauded the virtues of digital media, associating social media platforms with a raft of social malaises: the attention economy, data exploitation, surveillance, and platform capitalism (Zuboff, 2019), political polarization, and GenZ mental health issues. The information pandemic is neatly siloed into misinformation, disinformation and malinformation (Wardle & Derakshan, 2017).[6] The mantra of "information wants to be free" has been replaced by the reality of no free lunch online, and particularly on social media platforms.

Trust in the internet and social media platforms as some benevolent or even agnostic guardians of user data was misplaced. Online power is now vested in a few dominant platforms, most of which originated from Silicon Valley (Carrigan & Fatsis, 2021). By 2024, public discourse was manipulated by algorithms that shape our online experiences and interactions. This concentration of power poses significant risks to individual autonomy, creativity, and the open exchange of ideas (Orr, 2024). The initial ideals of the internet have been long-compromised by the commercial and manipulative practices of a few dominant platforms, leading to a more fragmented and less trustworthy online environment. Social media reinvigorated the pre-internet advertising business model, making it more personalized, resilient, and pervasive.

We live in a "technipolar" world in which the intellectual property and financial capital of tech companies propel their executives into the driving seat—or at least the center—of significant areas of decision-making that are intrinsically political in nature. High-profile billionaire tech barons such as Elon Musk and Mark Zuckerberg and their corporate vehicles exert considerable influence on norms—related to privacy, access to information and freedom of expression—including in the international sphere (Krasodomski et. al, 2024).[7] Algorithms are the end product of human activity. The impact of AI algorithms on society, education, business, and politics is perceived to be even more significant than that of social media (Verdegem, 2021; Yang et. al, 2021). In these post-truth times, the credibility of traditional gatekeepers such as journalists, public intellectuals, and educators continues to be questioned, ridiculed, and dismissed. The death of expertise, rationality, and enlightenment values are celebrated online and the education sector, another guardian of truths and expertise and still in a post-Covid dilemma, is particularly vulnerable.

The widespread adoption of generative artificial intelligence has ignited a new wave of anxiety surrounding its potential consequences—the equivalent of a moral panic of Orwellian proportions. Concerns about deepfakes, the erosion of privacy, and job displacement have become commonplace in public discourse (Giansiracusa, 2021; Harari, 2024). The potential harm that AI could inflict is vast, despite the compelling reasons for its development. AI could be weaponized to spread misinformation, undermine social cohesion, and destabilize democracies. It could be used to surveil, manipulate, and oppress citizens, jeopardizing individual and collective freedoms. Moreover, AI could create powerful digital or physical weapons that threaten human life. Additionally, AI has the potential to displace millions of workers, exacerbate existing inequalities, perpetuate discriminatory practices, distort decision-making, and inadvertently trigger military conflicts (Bremmer & Suleyman, 2023).

The societal and ethical implications of AI on daily life remain a subject of ongoing debate, hindered by the lack of a universally agreed-upon definition. This ambiguity hampers efforts to establish a shared knowledge base for developing effective global regulations (Bremmer & Suleyman, 2023). In 2018, the European Commission provided a definition of AI in the context of a roadmap aimed at promoting "[e]thical, secure and cutting-edge AI made in Europe".

Artificial intelligence refers to systems that display intelligent behavior by analyzing their environment and taking actions—with some degree of autonomy—to achieve specific goals. AI-based systems can be purely software based, acting in the virtual world (e.g., voice assistants, image analysis, software, search engines, speech- and face-recognition systems) or AI can be embedded in hardware devices (e.g., advanced robots, autonomous cars, drones, or Internet of Things applications) (European Commission, 2018).

We are living in strange, polarized times: the internet and social media and their very reason for being are about to be replaced by AI that is more akin to "alien intelligence" (Harari, 2024); the AI hype (LaGrandeur, 2024) is simply another tech bubble that will be burst, just like others before it (Floridi, 2024).

DECENTRALIZATION: REIMAGINING DIGITAL POWER

The narrative surrounding the internet's promise of freedom and equality has drastically shifted over the past decades. What was once seen as a democratizing force has instead resulted in a highly centralized ecosystem where a few powerful entities control vast portions of our digital lives. The utopian ideals that fueled the early internet now feel distant, overshadowed by concerns about privacy, surveillance, and monopolistic control. As the internet evolved from the read/write web to the hyper-centralized era of Web 2.0, these centralized platforms have commodified users and their data, as well as wielded substantial influence over public narratives.

Yet, around the same time the centralization of the Web 2.0 era was occurring, a new hope for decentralization was proposed that aimed to remove centralized control of digital systems—particularly payment and digital asset systems, i.e., cryptocurrencies. In 2008, Bitcoin was the first proposed cryptocurrency which aimed to provide a peer-to-peer digital payment system where no centralized actors or intermediaries could interfere or be required in any way to facilitate a digital payment—giving full control to asset holders. A cryptocurrency and decentralized system movement thereafter emerged with various projects launching, aimed at decentralization of control of various types of services and offerings, including those with the objective of the decentralization of social media platforms (Abbing et al., 2023).

Whilst some attempts to create new social media platforms are blockchain based, other attempts have been proposed providing different degrees of de/centralization of aspects. We will now provide an overview of attempts to provide decentralized social media platforms (Livitckaia et. al, 2023), particularly focusing on their ability to remove or reduce centralized control over censorship, public narratives and the ability to profit from user data—more specifically the following approaches: (i) peer-to-peer decentralized social media platforms (the "most" decentralized approach) that delegate governance directly to users themselves; then (ii) blockchain-based solutions that delegate governance to the community (guaranteed through rules encoded in the blockchain/smart contracts); and finally (iii) federated social media platforms that delegate governance to instance operators.

Peer-to-peer social media platforms (such as Scuttlebutt, RetroShare, and Twister) aim to provide complete decentralization, achieved by requiring users to connect directly to other users and share information directly between their devices. Such platforms trade off availability of content and performance to provide higher decentralization. Given that users are responsible for their own data, any data they view, block or share (directly with other users), then it becomes hard to implement platform wide content moderation controls since they can be bypassed (due to the direct peer-to-peer nature). While most platforms aim to be as decentralized as possible, some introduce (or require) centralized points for (i) initial peer/user discovery, (ii) content hosting (for increasing efficiency)—e.g., for popular content, and (iii) elected moderators (chosen democratically). Peer-to-peer solutions empower users to be in control of both their data as well as the content they consume (removing control of centralized platforms on censorship and public narrative control). Yet they also put the onus and responsibility on users to do so (whether they are required to undertake actions or whether performed automatically through the platforms' apps or software)—potentially a main reason why such platforms have not yet become mainstream.

Blockchain-based social media platforms, rather than requiring users to be responsible for content moderation themselves like the peer-to-peer ones discussed above, attempt to shift some aspects of control and governance away from users' direct responsibility onto the community through governance frameworks baked into the blockchain/smart contracts that enforce governance rules (Rozenshtein, 2023). Instead of requiring users to directly be responsible for content moderation (for the content they view themselves), moderation can be delegated to community consensus algorithms implemented on-chain—typically implemented as a decentralized autonomous organization (DAO). Some platforms utilize incentivization schemes for content creation and participating in content curation and moderation—where payments are facilitated through cryptocurrencies and the smart contract rules that define the governance framework. From one side, such approaches can be seen to be taking direct control away from users (where they do not directly control moderation rules themselves) yet at the same time users can collectively be involved in moderation processes. Removing the responsibility directly from users may be beneficial for those users who are not interested in actively participating in content moderation and/or less savvy users, which may support a better uptake—though blockchain-based platforms often impose other technological barriers for non-tech-savvy users (Glomann et al., 2019).

Another approach, federated social media platforms, attempts to remove centralized social media platform operators by allowing anyone to operate their own server (often referred to as an "instance"). Whilst this approach does indeed allow for anyone to run their own instance, operating infrastructure is non-trivial for non-tech-savvy users. Users are not explicitly required to run their own instance, and this approach is well suited to a model that does not require users to have technical skills to do so. Instead, users can subscribe to an instance of their choice—and in doing so they accept the terms and policy of that instance. Instances retain the ability to moderate content stored and shared from their instance; and therefore for that part of the social media network, they become centralized controllers. Indeed, whilst the aim of such platforms is to remove centralized control, interestingly it has been revealed that centralized points of control often emerge (e.g., in this case for the Mastodon platform). Raman et al. (2019) highlight that (i) content (or more specifically "topics") that is more decentralized (i.e., spread across more servers) tends to have fewer users than topics that are less decentralized; (ii) "10% of instances host almost half of the users" and outages of a small number of instances could result in large amounts of content becoming unavailable; and (iii) hosting of instances is highly centralized to a small number of hosting service providers. While this does not imply that such platforms are centralized, when considering censorship and potential control of power on public narratives, it is crucial to unravel if such control is just being shifted to new parties rather than being decentralized, and therefore understanding the different degrees to which a platform is technologically and socially de/centralized is vital. Yet, we should also keep in mind what exactly decentralization achieves and whether the decentralization of such systems aligns with the goals targeted. Indeed, whilst decentralization is often seen as a noble cause, increased decentralization does not necessarily translate into the betterment of the common good (Ellul & Pace, 2019) or the public interest (Ellul, 2021) and neither do such decentralized systems imply truthfulness of the data they secure (Ellul et al., 2021)—even though they may store and secure that data immutably "forever".

Whilst decentralization may not be the panacea it is often touted to be, decentralization does indeed (as its name implies) remove control from centralized actors—and if we are concerned with removing control over censorship, directing public narratives, and platforms exploiting users' data for profit, then decentralization indeed has something to offer. Yet, we must question whether removing such controls on social media platforms is a desirable feature and consider to what extent required rules can be upheld in systems lacking centralized controllers—which we will now discuss.

Whilst it may be desirable to completely remove centralized social media platform controllers in aim of removing their powers to censor, control narratives, and exploit users' data, removing controls that govern content has its drawbacks—in allowing for illegal and unethical activities to take place. If we consider peer-to-peer social media networks, given that users remain in full control of both the data they view and share, then enforcing some form of regulation on content moderation will be futile or infeasible—how should such decentralized social media platforms handle illegal and unethical content being distributed? One view is that since individual users are in control of their own data and their individual "user space", then the users themselves should be responsible and accountable for abiding by any regulatory and legal frameworks. The blockchain-based approaches delegate this function to (i) the code and (ii) the community as a collective. The code (typically implemented in smart contracts) defines the content moderation governance rules and provides a framework for those rules to be executed based on the community members/governors' inputs. Whilst governance motions are subject to the independent (subjective) views of the community members, the execution of the rules based on their collective inputs is guaranteed to be executed faithfully since the code is stored and executed on the blockchain. Whilst this guarantees the execution of policies (based on the subjective inputs), it must be noted that the implementation is code. Code can contain bugs which could result in incorrectly implemented policies—and the blockchain will also faithfully execute any bugs that exist in the defined code-policies (written typically in smart contracts). Given that such code is immutable (that is cannot be changed), since it is written on a blockchain, it is vital that either (i) the code is correct (and bug-free) or (ii) it has some mechanism in place that can allow for newer versions of the policies to be uploaded (though such mechanisms may also contain bugs). When it comes to "code correctness", there's no way to guarantee 100% that the code is bug free (Ellul et al., 2023). It is for this reason that code audits and regulatory frameworks to provide assurances on immutable code are desirable (Ellul et al., 2020).

Governance rules enforced through code allows for network-wide governance to be applied. Yet this model still raises questions with respect to how differing legal and regulatory frameworks across differing geographical locations can be implemented (Grech et al., 2021). One approach is that of allowing for different policies and oversight based on geographical locations governed by community members according to their specified location yet raises challenges toward ensuring community members are really contributing to the communities they are "legally" part of or representing. The federated approach provides an easier means for applying and enforcing regulation, given that each instance operator centrally controls the data (and users' data) it hosts and distributes—and therefore just like traditional social media platforms laws and regulation can be enforced by identifying the instance operators that may be hosting undesirable content. This indeed raises questions with respect to how well federated social media platforms are censorship resistant (i.e., whether required censorship is enforced willingly by instance operators or whether they are forced to do so legally).

Above we highlighted different technological approaches toward providing decentralized social media platforms that remove or reduce control that centralized traditional social media operators have over moderation, censorship and public narratives (and ability to profit from user data). Yet, the spectrum of technological possibilities is but one perspective—irrespective of what we can achieve through technological implementation, we must consider the extent to and form of regulation, moderation, censorship, and control that should be allowed by operators or defined on national and supranational levels.

EUROPEAN RESPONSES: REGULATORY FRAMEWORKS IN A DIGITAL AGE

In recent decades, government efforts to regulate the tech industry have lagged significantly behind the growth of new technologies. But the rise of artificial intelligence, which has coincided with AI developers loudly warning that their products carry significant risks of harm, has prompted lawmakers to work more quickly than usual. Given the size of the EU's single market and corresponding clout in international trade, the regulations it adopts stand to have a direct impact on how authorities in other parts of the world choose to approach the regulation of digital technologies, notably those in regions such as Africa or Latin America, which are comparatively minor players in technology development (Damro, 2015). A recent suite of legislation, including the Digital Services Act (DSA), Digital Markets Act (DMA), and AI Act embody an ambitious EU effort to shape the global debate over how to ethically negotiate and regulate digital technologies (Bremmer & Suleyman, 2023).

The Digital Services Act and Content Misuse

The Digital Services Act represents the most comprehensive and up-to-date legislative framework in the EU aimed at regulating the activities of online platforms, including social media networks. This ambitious piece of legislation is designed to enhance transparency, accountability, and consumer protection within the digital space, addressing issues ranging from illegal content to the spread of disinformation. However, despite its forward-thinking provisions, the rapid advancement of AI presents new and complex challenges that the DSA may not fully anticipate.

The emergence of AI-powered content generation, as well as the increasing prominence of AI-driven social media platforms, brings into question whether the DSA is sufficiently equipped to govern these evolving digital phenomena. AI's ability to boost the creation and curation of content introduces novel forms of social expression that diverge significantly from traditional digital media. Indeed, generative AI offers the capacity to personalize and manipulate content at an unprecedented scale, which in turn raises concerns about the amplification of misinformation, bias, and user manipulation.

This proliferation of AI-driven content production mechanisms implicates not only individual actors but also the structural integrity of democratic institutions and the epistemic engagement of the public with digital information. Waldemarsson (2020, p. 10) systematically categorizes the risks associated with AI-powered content into four domains: electoral manipulation, exacerbation of social cleavages, erosion of institutional trust, and the destabilization of journalistic practice. These phenomena raise pressing concerns about the governance of digital content, necessitating a critical examination of existing supervisory, curatorial, and regulatory frameworks.

The EU AI Act is undoubtedly a regulatory milestone. Nonetheless, it does not regulate every aspect of AI applications and, most importantly, it does not prescribe a comprehensive framework for the regulation of AI-generated content. On the contrary, it is the Digital Services Act that presents a more complete regulatory framework in terms of digital content moderation. This section investigates whether the DSA can also be applied to generative AI applications. The question is of great importance, since the DSA incorporates a rather detailed set of content moderation rules and safe-harbor provisions that might shape the landscape of AI-generated content in a very decisive manner. The analysis unfolds in two parts: the first one offers a review of existing scholarly opinions on the application of the DSA on generative AI applications; the second provides a rigorous assessment of the interpretative possibilities and the broader regulatory implications.

Legal literature is currently engaged in a debate about whether generative AI can be accommodated within the existing regulatory framework of the DSA. This debate is not merely academic but raises pressing questions about the regulation of emerging content related technologies by existing legal categories. The discussion reveals a range of different opinions about whether the DSA can accommodate AI-generated content. Some academic commentators, most prominently Hacker et al. (2023), argue that the DSA, in its current form, is ill equipped to regulate AI-generated content. They claim that generative AI applications do not fall under any of the intermediary categories regulated by the DSA such as "mere conduits", "caching", or "hosting providers". They allege that AI-generated content does not originate from the recipients of that service but is a product of the AI application itself. They contend that the architecture of the DSA limits its applicability to scenarios where content is merely stored at the behest of the users, submitting that the more active role played by AI applications renders the application of the DSA impossible. Despite this, they acknowledge that the content moderation mechanisms of the DSA could provide a useful regulatory framework. They suggest, therefore, that the EU legislator should create a similar regulatory regime for content related AI applications.

In a similar vein, Edwards et al. (2024) remain equally skeptical regarding the applicability of the DSA on generative AI applications. They critique the phenomenon of "private ordering", namely the tendency of generative AI providers to position themselves, through their terms and conditions, as neutral intermediaries, thereby shifting liability onto users through carefully crafted contractual arrangements. They argue that this strategy is incompatible with the core principles of the Electronic Commerce Directive (ECD), which created a safe harbor for passive and neutral platforms that merely facilitate user generated content and interactions. They deny that generative AI applications fit this description, as they assert that they have a rather active contribution in the generation of content. In order to address the situation, they argue that the EU legislator should amend the DSA in order to explicitly incorporate generative AI applications in the content moderation mechanisms of the DSA.

Other commentators adopt a more flexible approach, suggesting that generative AI applications display parallels to search engines. Stalla-Bourdillon (2023), for instance, claims that the intermediary categories within the DSA are not rigid and that search engines, though not fully aligned with traditional intermediary types, are nonetheless regulated under the DSA. She explores the potential classification of generative AI as hosting providers but ultimately leans toward conceptualizing these tools as analogous to search engines, given their role in organizing and processing information within an information-rich digital environment.

Botero Arcila (2023) extends this analogy, arguing that generative AI applications perform fundamentally the same core functions as search engines, namely they support users in structuring and retrieving digital information. She suggests that the functional equivalence between generative AI applications and search engines justifies the application of DSA provisions related to search engines to generative AI. She acknowledges that generative AI applications display capabilities that go beyond those of traditional search engines. She insists, nonetheless, that the application of the DSA rules on search engines to generative AI applications is justified both *de lege lata* and *de lege ferenda*, while she posits that the advantages of her position are not merely academic but serve the resolution of very pressing practical problems of content moderation.

Finally, Lemoine and Vermeulen (2024) advocate for the most flexible approach, proposing that generative AI applications might be viewed as either hosting providers or search engines under the DSA, depending on the context. They argue that the input provided by users when prompting AI models could allow for a broader interpretation of the intermediary category of "hosting", thereby situating generative AI within its scope. Alternatively, they highlight the functional overlap between search engines and generative AI, suggesting that these tools could straddle both categories, necessitating a contextual and case-specific regulatory appraisal.

Finally, it is interesting to stress that there is no academic debate about the application of the content moderation rules of the DSA on traditional social media platforms, even when they host AI-generated content. All commentators agree that when AI-generated content appears in traditional social media platforms, the DSA provides an adequate framework for the moderation of such content. Nonetheless, the regulation of AI-generated content only during the dissemination phase over traditional social media applications is generally considered to be a partial and indirect solution to the problems raised by AI-generated content.

Assessment: AI Content Manipulation and the DSA

The divergence in scholarly opinion on whether the DSA can accommodate generative AI reflects a broader debate about the adaptability of legal frameworks to emergent technologies. On one end of the spectrum, the argument that the DSA was not designed with generative AI in mind highlights the challenges of applying its provisions to these tools. However, the recitals within the DSA itself suggest a more dynamic interpretative approach. Recital 28, for instance, acknowledges the evolving nature of digital services, while Recital 29 proposes that the classification of services should hinge on their technical functionalities, which may evolve over time. This invites a more fluid interpretation of the DSA, particularly regarding intermediary services that benefit from liability immunities while bearing responsibilities for content moderation.

The contention that generative AI should not be classified as hosting providers because they generate, rather than merely store, content is subject to further scrutiny. While generative AI undeniably plays a role in content creation, the extent of this role is mediated by the user's agency in directing the AI's outputs. The jurisprudence of the Court of Justice of the European Union (CJEU), particularly the judgement in Papasavvas (C-291/13, ECLI:EU:C:2014:2209), offers insight into how digital service providers might be situated within the intermediary taxonomy of the DSA. In this case, the Court's reasoning suggests that the decisive factor lies in who exercises control over content creation—a question that remains complex when considering the collaborative dynamic between user input and AI generation. In principle, ethical principles are to be applied by individuals and private entities in their own navigation of AI, as well as by authorities/governments.

Moreover, the argument that generative AI transcends the passive role envisaged by Article 6 of the DSA, particularly considering the CJEU's reasoning in L'Oreal (C-324/09, ECLI:EU:C:2011:474), must be revisited. The evolving case law, including the YouTube and Cyando cases (Joined Cases C-682/18 and C-683/18, ECLI:EU:C:2021:503), suggests a more nuanced approach, recognizing that automated processes, even when sophisticated, may still operate within the bounds of passive intermediary functions. This calls for a reassessment of the regulatory posture toward generative AI, acknowledging the intricate balance between automation and user agency.

Generative AI, with its capacity to produce content that blurs the line between reality and fabrication, presents formidable challenges for content moderation and regulatory frameworks. The DSA, as a legislative instrument, must be critically examined to determine whether it can adapt to the unique demands posed by these technologies. While some scholars remain skeptical of the DSA's suitability, others propose a more expansive interpretation that could encompass Generative AI under existing or reimagined intermediary categories.

In guiding the ethical development of AI, the **Ethics Guidelines for Trustworthy AI**, established by the High-Level Expert Group on AI Ethics for the European Commission[8], serve as a foundational framework. The guidelines are crafted in an effort to align AI development with democratic principles and the rule of law, forming the ethical backbone for EU legislative interventions in AI[9]. Importantly, the framework is intended to guide AI development in complex areas, such as deepfakes and content misuse, where the application of traditional content moderation regulatory tools remains open to debate, as previously detailed.

The Guidelines are grounded in the **EU Charter of Fundamental Rights** and **EU Treaties.** They aim to translate these fundamental rights into ethical principles that govern AI design, deployment, and oversight, with a central focus on maintaining democratic integrity and ensuring accountability within the rule of law. As AI-driven disinformation tactics, such as deepfakes, increasingly threaten democratic processes on social media, this framework aims to foster AI systems that protect democratic values while upholding transparency, fairness, and equality.

The Ethics Guidelines highlight democracy as a core principle, emphasizing that AI systems should actively support democratic values. In this context, several ethical tenets are modelled or derived from pre-existing fundamental rights. For example, the proposed ethical tenet of **respect for human autonomy**, is derived from the fundamental right to human dignity and requires that AI systems allow individuals to make independent, informed decisions, free from manipulation or coercion. In social media contexts, this principle addresses the risks posed by deepfakes and AI-generated disinformation, which can distort public perception and compromise democratic decision-making.

By prioritizing autonomy, trustworthy AI reinforces democratic freedoms, enabling citizens to engage in political discourse based on factual rather than manipulated information.

The principle of **prevention of harm** reflects the dedication of the Union to human security and well-being and is crucial for mitigating threats that AI-driven disinformation poses to the rule of law. Deepfakes, which can deceive the public and undermine trust in political and judicial institutions, are capable of destabilizing democratic systems. By embedding harm prevention within the ethical governance framework of AI applications, the EU aims to reduce the risks that AI systems could pose to democratic stability, political polarization, and institutional legitimacy.

Fairness and equality, as guiding principles are grounded in the Union's commitment to equality and non-discrimination. They demonstrate the need for AI systems that operate without bias, treating all individuals equitably. In the digital sphere, fairness in AI helps ensure that content moderation processes account for diverse voices, avoiding discriminatory practices that could marginalize specific groups or perspectives. Thus, fairness in AI platforms is instrumental in supporting democratic pluralism by fostering a diversity of views essential for democratic dialogue and preventing biases that could harm equitable representation.

Explicability, rooted in the rights to justice and accountability, is essential for building public trust and ensuring AI transparency within democratic frameworks. Explicability mandates that AI systems, especially those used in public information environments, provide clear, understandable explanations of their processes and decisions. This is particularly relevant for countering disinformation, where opaque algorithms could undermine trust in digital governance. By prioritizing explicability, AI systems support the rule of law by keeping decision-making processes open to public scrutiny, allowing both citizens and regulators to hold AI-driven platforms accountable for their impact on public discourse.

The work of the High Level Expert Group on Trustworthy AI demonstrates the strong commitment of the EU to embedding ethical oversight into AI. In that sense, respect for democratic principles and the rule of law should be guaranteed by an ethical design of generative AI applications. Next to traditional regulatory tools, therefore, the proliferation of a robust ethical arsenal of principles and governance mechanisms might prove a useful ally in the creation of safe and democratically sound AI spaces.

The Ethics of Algorithmic Power

This chapter has examined European perspectives on the development and regulation of rapidly evolving social platforms (and AI), framed by issues of corporate consolidation and an ongoing information crisis. Debates over the ethical implications of digital technologies vary considerably by geography (Hagerty & Rubinov, 2019). The European Union (EU) is in the distinctive position of leading the globe in developing regulatory standards, while hosting comparatively few major players in these two interconnected fields. A PwC (2017) study estimates that some 70% of the \$15.7 trillion predicted to be generated by applications of AI by 2030 will be concentrated in the United States and China, leaving the United Kingdom, the EU and other players far behind. The labor market dislocation and transformation engendered by AI will not be limited to the jurisdictions spearheading its development (Hagerty & Rubinov, 2019). As the International Labour Office (2019) warns, "Left to its current course, the digital economy is likely to widen both regional and gender divides." These trends highlight the importance of integrating ethical concerns for social justice into the decision-making processes of both private and public entities across a host of sectors. Investments in public social safety nets, education systems, reskilling and lifelong learning services will be particularly crucial given evidence that low-education workers and those relying on physical and manual skills will bear the brunt of job losses (Smit et al., 2023). For Europe, the challenges of ethically navigating fast-developing digital technologies will be compounded by a decline in the working-age population. The working-age population has decreased by 1.4% across EU member states since 2011 (Eurostat, 2024).

Thus far, EU action in the digital technology space, in the form of both regulation and investment, has been frequently framed in terms of "data protection" and an imperative to secure the "privacy and autonomy of the individual", exemplified by 2016's General Data Protection Regulation (Krasodomski et al., 2024). The European Union's Next Generation Internet initiative, which directs investment in decentralized and open-source solutions, is presented as a "human-centric, sustainable vision for digital society… to empower citizens and businesses" (European Commission, 2024).

This approach is not without critics. Brevini (2021) argues that the EU has consistently adopted a technologically determinist attitude to digital technologies, with AI the latest manifestation of a tendency to treat technology as engendering profound social change and simultaneously an opportunity, if harnessed correctly, to enhance individual freedoms. For Brevini (2021, p. 151), this embodies the "current neoliberal ideology of the current stage of capitalism" and echoes earlier optimism regarding the liberatory potential of the internet. If Brevini's fear is that treating the advance of AI as inexorable serves to legitimate commercial interests that view the "subordination of human labor to computers" as desirable, recent regulatory developments cast into doubt claims that the EU risks the "migration of strategic decisions and choices on the direction of AI development from government to corporate board rooms: the privatisation of public policy" (Brevini, 2021, p. 155).

While its legislation may influence policymaking in other regions, the ethical principles driving EU policy in this field—fundamental rights, human autonomy, prevention of harm, fairness and equality and explicability—will inevitably face contestation. Advocates of both decentralized solutions and regulatory approaches must recognize the diversity of ethical viewpoints, assessing where European best practices can address the risks and opportunities linked to AI, and where their applicability is limited. Public trust in state institutions shapes attitudes toward surveillance and personal data collection. Distinctions between "Western" concerns over privacy and "Eastern" confidence in state data use will be viewed critically (Hagerty & Rubinov, 2019).

Current and future European policy interventions in matters relating to AI, social media, and decentralization will inevitably have much to do with regional power dynamics and political action. These interventions must encourage critical scrutiny of the interests being served by current technological developments and how these may perpetuate or reshape inequalities (Daly et al., 2021). Both advocacy for decentralized platforms and support for AI regulation arise from dissatisfaction with the current trajectory of digital technology, particularly corporate dominance. Connolly (2024) argues that mainstream computer science has not sufficiently engaged with the ethical and societal impacts of AI advancements, especially in how commercial AI applications prioritize reducing labor costs over increasing productivity. He warns of "displacement with little to no reinstatement" (Connolly, 2024, p. 5). While a more politically and socially engaged computing community is necessary for steering AI development ethically, professional codes of ethics alone are not enough to substitute for regulation (Véliz, 2019).

BUILDING ETHICAL FRAMEWORKS FOR DIGITAL FUTURES

Various ethical frameworks have been proposed by policymakers, academics, and industry players to guide AI development. Data and Society's "AI on the Ground" initiative produces interdisciplinary research examining the impact of AI systems, targeting policymakers and regulators as well as developers, with a focus on the "social life of algorithmic harms". The "good data" vision by Daly et al. (2021) builds on the pillars of community, rights, usability, and politics, advocating for a shift beyond governments and corporations to include "initiatives and communities of change". Bremmer and Suleyman (2023) propose a "technoprudential" approach to AI regulation, aiming to pre-emptively address potential global stability disruptions. They suggest crafting international agreements that include private AI companies as signatories alongside states.

These proposals highlight the challenge of scale, emphasizing the need for global coalitions to credibly negotiate threats posed by digital technologies. Questions arise regarding the efficacy of regional-level regulations in keeping pace with AI development and achieving global consensus.

Scholars, practitioners, and citizens in the European Union must consider their positionality in this rapidly evolving landscape and explore ways to build effective coalitions with counterparts elsewhere. Meanwhile, developments in the United States, such as obstacles to passing a modest AI safety bill in California, suggest a reluctance to take proactive steps. Foley (2024) draws parallels to the trajectory of US banking regulation over recent decades, implying a reactive rather than preventative approach.

CONCLUSION

We are chasing a moving target with regulation. How do we govern a future shaped by rapidly evolving technologies, often regulated by institutions whose practices remain rooted in the past century? As AI becomes increasingly pervasive, a more nuanced understanding of the power dynamics embedded in digital technologies is critical. This requires ongoing examinations of the design, governance, and regulation of these systems, aligned with the goals of openness, transparency, and decentralization. Harari (2024) warns of the risks of technological dystopianism, cautioning against a future where unchecked AI erodes autonomy and societal cohesion.

The intersection of social media, AI, and decentralization presents both opportunities and challenges. Decentralized platforms promise alternatives to centralized control but also raise complex questions about governance, responsibility, and truth in digital spaces. Meanwhile, the European Union's regulatory approach, exemplified by the DSA and AI Act, attempts to balance innovation with human rights and democratic values. However, the effectiveness of regional regulation in a global digital ecosystem remains uncertain, as the "technipolar" world of dominant tech companies and competing national interests complicates the establishment of universal ethical frameworks for emerging technologies.

Addressing these challenges requires a multi-dimensional manner encompassing **laws, markets, norms, and tech architecture.** Four key challenges stand out as critical to navigating this landscape:

1. Developing adaptive regulatory frameworks that can keep pace with technological change.
2. Balancing decentralization with the need for accountability and oversight.
3. Integrating diverse cultural and philosophical perspectives on digital rights and responsibilities
4. Creating ethical frameworks to guide both technological development and governance.

Success will depend on unprecedented cooperation among technologists, policymakers, ethicists, and civil society. The path forward must embrace technological innovation while upholding human values, ensuring that digital spaces enhance rather than diminish human flourishing.

The long and winding road ahead demands careful navigation among competing interests and values. While perfect solutions may remain elusive, the frameworks and approaches discussed in this chapter offer important guideposts for this journey. By learning from past mistakes and remaining open to new possibilities, we can strive toward a digital future that prioritizes human flourishing and ethical technological advancement.

NOTES

1. In this chapter, the Internet is approached as both a landscape and a domain. As a landscape, the Internet is an interconnected space where information flows and interactions occur, encompassing social, cultural, and technological elements. As a domain, it refers to a specific sphere of activity or knowledge, often linked to particular topics, services, or websites (e.g., the domain of e-commerce or social media). Each perspective highlights different aspects of the internet's multifaceted nature.
2. See https://internethistory.org/
3. See https://www.eff.org/cyberspace-independence

4. See https://www.cluetrain.com/
5. The Cambridge Analytica scandal emerged in 2018 when it was revealed that the political consulting firm had harvested personal data from millions of Facebook users without their consent. This data was used to create psychological profiles for targeted political advertising, notably influencing the 2016 US presidential election and the Brexit referendum. The scandal raised significant concerns about data privacy, the ethics of micro-targeting in political campaigns, and the role of social media platforms in democracy. It led to widespread criticism of Facebook, regulatory investigations, and discussions on the need for stricter data protection laws.
6. Misinformation refers to false information that is not intended to cause harm. Disinformation refers to false information that is intended to manipulate, cause damage, or guide people, organizations, and countries in the wrong direction. Malinformation refers to information that stems from the truth but is often exaggerated in a way that misleads and causes potential harm.
7. In November 2023, the UK Government's AI Summit featured the then Prime Minister Rishi Sunak interviewing Elon Musk.
8. On the High Level Expert Group on AI see https://digital-strategy.ec.europa.eu/en/policies/expert-group-ai, while the text of the Guidelines can be accessed here https://digital-strategy.ec.europa.eu/en/library/ethics-guidelines-trustworthy-ai.
9. Including Article 50 of the EU AI, which puts emphasis on transparency for AI systems that can be deployed to produce content, including generative AI.

REFERENCES

Abbing, R., Diehm, C., & Warreth, S. (2023). Decentralised social media. *Internet Policy Review, 12*(1). https://doi.org/10.14763/2023.1.1681

Benkler, Y. (2006). *The wealth of networks.* Yale University Press.

Botero Arcila, B. (2023). Is it a platform? Is it a search engine? It's chat GPT! The European liability regime for large language models. *Journal of Free Speech Law, 3*(2), 455–488.

Bremmer, I., & Suleyman, M. (2023, August 16). The AI power paradox. *Foreign Affairs*, September/October 2023. https://www.foreignaffairs.com/world/artificial-intelligence-power-paradox

Brevini, B. (2021). Creating the technological saviour: Discourses on AI in Europe and the legitimation of super capitalism. In P. Verdegem (Ed.), *AI for everyone? Critical perspectives* (pp. 145–159). University of Westminster Press. https://doi.org/10.16997/book55.i

Carrigan, C., & Fatsis, N. (2021). *The public and their platforms: Public Sociology in an Era of Social Media.* Bristol University Press.

Castells, M. (2010). *The rise of the network society* (2nd ed.). Wiley-Blackwell.

Connolly, R. (2024). *Public computing intellectuals in the age of AI crisis.* https://arxiv.org/abs/2405.00860

Daly, A., Devitt, S. K., & Mann, M. (2021). AI ethics needs good data. In P. Verdegem (Ed.), *AI for everyone?: Critical perspectives* (pp. 103–122). University of Westminster Press. https://doi.org/10.16997/book55.g

Damro, C. (2015). Market power Europe: Exploring a dynamic conceptual framework. *Journal of European Public Policy, 22*(9), 1336–1354. https://doi.org/10.1080/13501763.2015.1046903

Edwards, L., Szpotakowski, I., Cifrodelli, G., Sangaré, J., & Stewart, J. (2024). *Private ordering and generative AI: What can we learn from model terms and conditions?* Create Working Paper. Retrieved September 30, 2024, from https://zenodo.org/records/11276105

Ellul, J. (2021). *Blockchain, decentralisation and the public interest: The need for a decentralisation conceptual framework for dApps.* https://www.um.edu.mt/library/oar/handle/123456789/110597

Ellul, J., Galea, J., Ganado, M., Mccarthy, S., & Pace, G. (2020). Regulating blockchain, DLT and smart contracts: A technology regulator's perspective. *Era Forum, 21*, 209–220. Springer Berlin Heidelberg, 2020.

Ellul, J., Grech, A., & Pace, G. (2021). Two sides to every story. The truth, post-truth, and the blockchain truth. In A. Grech (Ed.), *Media, Technology and Education in a Post-Truth Society* (pp. 243–253). Emerald Publishing.

Ellul, J., & Pace, G. (2019). *Blockchain and the common good reimagined.* arXiv preprint arXiv:1910.14415.

Ellul, J., Pace, G. J., Revolidis, I., & Schneider, G. (2023, February). When is good enough good enough? On software assurances. *ERA Forum, 23*(3), 337–360. Berlin/Heidelberg: Springer Berlin Heidelberg.

European Commission. (2018). *A definition of AI: Main capabilities and scientific disciplines.* The European Commission's High-Level Expert Group on Artificial Intelligence. https://ec.europa.eu/futurium/en/system/files/ged/ai_hleg_definition_of_ai_18_december_1.pdf

European Commission. (2024, September). *Next generation internet initiative*, https://digital-strategy.ec.europa.eu/en/policies/next-generation-internet-initiative

Eurostat. (2024). Eurostat regional yearbook. Publications Office of the European Union. https://op.europa.eu/en/publication-detail/-/publication/86bdd83d-7fa3-11ef-a67d-01aa75ed71a1/

Floridi, L. (2024). *Why the AI hype is another tech bubble*. Available at SSRN: https://ssrn.com/abstract=4960826 or http://dx.doi.org/10.2139/ssrn.4960826

Foley, J. (2024, October 2). AI may regret aping Wall Street's regulatory resistance. *Financial Times*. https://www.ft.com/content/cffd24e9-713a-424f-8fc0-0e421e78e5c9

Giansiracusa, N. (2021). *How algorithms create and prevent fake news: Exploring the impacts of social media, deepfakes, GPT-3, and more*. Apress.

Glomann, L., Schmid, M., Kitajewa, N. (2020). Improving the blockchain user experience – An approach to address blockchain mass adoption issues from a human-centred perspective. In: T. Ahram (Ed.), *Advances in artificial intelligence, software and systems engineering. AHFE 2019. Advances in intelligent systems and computing*, Vol. 965. Springer. https://doi.org/10.1007/978-3-030-20454-9_60

Grech, A., Sood, I., & Ariño, L. (2021). Blockchain, self-sovereign identity and digital credentials: Promise versus praxis in education. *Frontiers in Blockchain*, *4*, 616779. https://doi.org/10.3389/fbloc.2021.616779

Hacker, P., Engel, A., & List, T. (2023). *Understanding and regulating ChatGPT, and other large generative AI models*. Retrieved September 30, 2024, from https://verfassungsblog.de/chatgpt/

Hagerty, A., & Rubinov, I. (2019). *Global AI ethics: A review of the social impacts and ethical implications of artificial intelligence*. https://doi.org/10.48550/arxiv.1907.07892

Harari, Y. (2024). *Nexus: A brief history of information networks from the stone age to AI*. Random House.

International Labour Office (2019). *Work for a brighter future – Global Commission on the Future of Work*. Geneva: ILO. ISBN 978-92-2-132796-7 (web pdf)

Jenkins, H. (2006). *Convergence culture: Where old and new media collide*. New York University Press.

Krasodomski, A. et al. (2024). *Artificial intelligence and the challenge for global governance: Nine essays on achieving responsible AI*. Research Paper. Royal Institute of International Affairs. https://doi.org/10.55317/9781784136086

LaGrandeur, K. (2024). The consequences of AI hype. *AI Ethics*, *4*, 653–656. https://doi.org/10.1007/s43681-023-00352-y

Lemoine, L., & Vermeulen, M. (2024). *Assessing the extent to which generative artificial intelligence (AI) falls within the scope of the EU's digital services act: An initial analysis*. Retrieved September 30, 2024, from https://ssrn.com/abstract=4702422

Livitckaia, K., Iordanis, P., Votis, K., Revolidis, I., Ellul, J., Ferreira da Silva, C., Szegö, D., & Joshi, J. (2023). Decentralised social media. Available at SSRN 4636894.

Orr, D. (2024). *From naive optimism to nuanced realism: Some thoughts while reading Harari's new book*. https://digimusingsblog.wordpress.com/2024/09/25/from-naive-optimism-to-nuanced-realism-some-thoughts-while-reading-hararis-new-book/

Papacharissi, Z. (2010). *A private sphere: Democracy in a digital age*. Polity.

PwC Global Artificial Intelligence Study. (2017). *Sizing the Prize: What's the real value of AI for your business and how can you* capitalize? https://www.pwc.com/gx/en/issues/data-and-analytics/publications/artificial-intelligence-study.html

Rainie, L., & Wellman, B. (2012). *Networked: The new social operating system*. The MIT Press.

Raman, A., Joglekar, S., De Cristofaro, E., Sastry, N., & Tyson, G. (2019). Challenges in the decentralised web: The mastodon case. In *Proceedings of the internet measurement conference* (pp. 217–229).

Rheingold, H. (2000). *The virtual community. Homesteading on the electronic frontier* (2nd ed.). The MIT Press.

Rozenshtein, A. Z. (2023). Moderating the fediverse: Content moderation on distributed social media. *Journal of Free Speech Law*, *3*, 217.

Smit, S., Tacke, T., Lund, S., Manyika, J., & Thiel, L. (2023). *The future of work in Europe. Automation, workforce transitions, and the shifting geography of employment*. McKinsey Global Institute. https://www.mckinsey.com/~/media/mckinsey/featured%20insights/future%20of%20organizations/the%20future%20of%20work%20in%20europe/mgi-the-future-of-work-in-europe-discussion-paper.pdf

Stalla-Bourdillon, S. (2023). *What if ChatGPT was much more than a chatbox? What if LLM-as-a-service was a search engine?* Retrieved September 30, 2024, from https://peepbeep.blog/2023/04/03/what-if-chatgpt-was-much-more-than-a-chatbox-what-if-llm-as-a-service-was-a-search-engine/

Véliz, C. (2019). Three things digital ethics can learn from medical ethics. *Nature Electronics*, *2*(8), 316–318. https://doi.org/10.1038/s41928-019-0294-2

Verdegem, P. (Ed.) (2021). *AI for everyone? Critical perspectives.* University of Westminster Press. https://doi.org/10.16997/book55. License: CC-BY-NC-ND 4.0

Waldemarsson, C. (2020). *Disinformation, deepfakes & democracy.* Retrieved September 30, 2024, from https://www.allianceofdemocracies.org/wp-content/uploads/2020/04/Disinformation-Deepfakes-Democracy-Waldemarsson-2020.pdf

Wardle, C., & Derakshan, H. (2017, September 27). Information disorder: Toward an interdisciplinary framework for research and policy making. *Council of Europe.*

Yang, S. J. H., Ogata, H., Matsui, T., & Chen, N.-S. (2021). Human-centered artificial intelligence in education: Seeing the invisible through the visible. *Computers and Education: Artificial Intelligence, 2.* https://doi.org/10.1016/j.caeai.2021.100008

Zuboff, S. (2019). *The age of surveillance capitalism: The fight for a human future at the new frontier of power.* PublicAffairs.

19 AI Ethics and Societal Well-being

Perspectives from the Global North and South

Ram B. Ramachandran and Naresh Singh (as interlocutors)

INTRODUCTION

This chapter summarizes the key points arising from a colloquy comprising several global conversations conducted by the editors with selected thinkers and practitioners in the global AI ethics space. This was done in a rather unusual and innovative way to gather in a succinct manner ideas and issues that might go beyond the material covered in the individual expert chapters contained in this volume. The contributors and their affiliations are listed at the end of the chapter. The chapter is put together by the editors, who attempted to be as faithful as possible to the ideas discussed but must be responsible for any inaccuracies

These conversations aimed to explore the multifaceted implications of artificial intelligence (AI) on society, particularly through the lens of ethics and governance. It was organized as a colloquy that brought together diverse voices from the Global South and Global North, highlighting the need for a broader perspective beyond academic discourse. The discussion recognized that, while AI has the potential to drive significant advancements, it also raises critical ethical questions that require inputs from various stakeholders, including philosophers, technologists, and implementers. This collaborative approach was essential in addressing the challenges posed by AI, as its impact will not be uniform across different cultural and socio-economic contexts. To keep the sense and spirit of the conversations, the following reporting is discussed in the present tense.

Anjuli Bedi and Adah Parris provide insights from philosophical perspectives on AI guidelines and usage. They engage in a debate on how the intended application of AI—whether in social media, healthcare, or other sectors—shapes its structure and purpose. They emphasize the importance of considering not just practical use cases but also the broader societal implications of AI technologies. This includes addressing foundational issues such as data security, privacy concerns, and bias elimination. The speakers argue that while optimizing for user engagement is important, it should not come at the expense of promoting societal well-being.

Mammo Muchie focuses on the ethical implications of artificial intelligence through the lens of African philosophy, particularly Ubuntu. The discussion emphasizes the need for a collaborative approach to AI development that prioritizes humanity and ethical considerations over individualistic and competitive tendencies prevalent in Western paradigms. He argues that Ubuntu promotes values such as love, reconciliation, and forgiveness, which can guide AI to foster unity rather than division. Concerns are expressed about the current state of AI ethics, dominated by Western perspectives, and advocate for integrating diverse cultural philosophies to create a more inclusive ethical framework for AI applications across various sectors like education, health, and governance.

Thomas Koulopolous (TK) and John C. Havens engage in a critical dialogue on the technological implications of AI, emphasizing the need to reframe AI ethics beyond the prevailing Western-centric, profit-driven paradigms. John critiques existing AI systems for prioritizing national gross

DOI: 10.1201/9781003585527-24

domestic product (GDP) and shareholder value over societal well-being, advocating for a design approach that centers on humanity and sustainability. The conversation also highlights the importance of cultural pluralism in AI governance. This discussion underscores the necessity of preserving diverse ethical perspectives to foster resilience in AI systems. Ultimately, John and TK envision a future where AI enhances rather than undermines humanity's core values.

Kim Eunsong, Jibu Elias, and Mihir Kulkarni delve further in understanding the implementation and governance of AI, including the role of UNESCO and the unique challenges faced by the Global South. They highlight UNESCO's position as a key UN facilitator in AI ethics, bridging gaps between thought leaders, governments, and academic institutions. The speakers address critical concerns regarding data collection and consent in the Global South, noting limited legal protections and a lack of understanding about data implications. The conversation also explores bias in AI, presenting a dual perspective that acknowledges both harmful biases—such as facial recognition technologies failing to accurately identify darker skin tones—and potential positive applications, like prioritizing underserved communities during disasters. However, they caution against the environmental costs associated with addressing these biases. Finally, they argue for a shift away from profit-driven models toward visionary leadership that prioritizes long-term societal well-being.

RETHINKING AI: PHILOSOPHICAL DIALOGUES ON ETHICS AND HUMAN VALUES

(Perspectives of Anjuli Bedi and Adah Parris)

Dual Perspectives on AI Guidelines and Use

The conversation opens with a debate on the dual perspectives surrounding the implementation of AI guidelines. The first perspective focuses on practical use cases, emphasizing how the intended application—be it social media, advertising, or healthcare—determines the structure and purpose of AI systems. For instance, in healthcare, AI can enhance patient outcomes by providing personalized treatment recommendations based on data analysis. Conversely, in social media, algorithms may prioritize user engagement metrics, often leading to echo chambers and misinformation. The second perspective reflects on the broader societal impact of AI technologies, questioning whether they foster happiness, democracy, or social cohesion. These perspectives underscore the importance of addressing foundational issues like secure data curation, privacy concerns, bias elimination, and aligning business principles with ethical considerations. For example, while social media could optimize for user engagement, it could also be directed toward enhancing well-being or fostering constructive dialogues.

Ethical Framing Beyond Binary Constructs

The speakers Anjuli Bedi and Adah Parris critique the oversimplified "good versus bad" framing of ethics commonly observed in Western cultures. Instead of viewing ethical dilemmas through a binary lens, they propose evaluating AI through the concepts of responsible and irresponsible uses. This nuanced approach encourages deeper philosophical inquiries about AI's societal impact. Cultural differences significantly influence this framing; for instance, Western cultures often prefer binary solutions, while many indigenous cultures embrace multidimensional and cyclical perspectives rooted in intergenerational traditions. Anjuli draws parallels to Orwell's *1984*, highlighting how reducing language and ethical frameworks into simplistic binaries stifles critical thinking and intentional action. In this context, efficiency-driven decision-making often prioritizes shareholder value over social purpose, leading to technology that fails to serve broader human needs.

Societal Disconnect in AI Narratives

The dialogue highlights the exclusion of underrepresented communities from global conversations about technology. Communities such as Moroccan nomads are often labeled as "underdeveloped," despite their resilience and creativity; their contributions are frequently overlooked in the redesign of global systems. The speakers argue that profit-driven technology narratives erase the voices of those who physically create and sustain these systems. This disconnect is evident in how discussions around ethics fail to address broader consequences of technological advancements—such as geopolitical shifts and environmental degradation. To counteract this exclusion, there is a pressing need for inclusive narratives that prioritize human values and collective well-being over market-driven incentives.

Philosophical Foundations for Ethical Technology

The necessity of grounding technological progress in authentic philosophical principles is emphasized throughout the dialogue. Drawing from Rabindranath Tagore's 1917 essay on nationalism, Anjuli Bedi and Adah Parris argue that India could lead as a nation rooted in philosophical and spiritual values rather than materialistic ambitions. Indigenous cultures exemplify this principle by advocating for ethical pauses and reflective practices that encourage deeper connections with self and environment. Rituals such as vision quests promote nonlinear approaches to solving societal problems, contrasting sharply with Western methods that prioritize speed and efficiency—often at the expense of wisdom and introspection.

The Role of Bias and Systemic Inequalities

Bias is often viewed merely as a limitation within AI systems; however, it also reflects societal imperfections. For example, AI's struggle to accurately recognize darker skin tones exposes underlying racial inequities prevalent in society at large. Nevertheless, bias can also serve as a tool for positive change; designing algorithms that prioritize underserved communities during natural disasters illustrates how bias can be harnessed for good. The speakers critique profit-driven systems rooted in historic practices like British mercantilism and American capitalism that perpetuate inequities. They propose "flipping the script" to focus on societal well-being metrics—such as poverty reduction or environmental preservation—as desired outcomes for technological advancements.

Guardrails for Responsible AI

The establishment of universal ethical guardrails rooted in shared values across cultures and religions is suggested by the speakers to guide responsible AI development. Schwartz's theory of basic values serves as a potential framework to identify common principles—such as kindness, fairness, and respect—that could inform these guardrails (Schwartz, 2022). However, defining measurable parameters for responsible development remains a significant challenge, particularly in complex scenarios like human trafficking, where traditional metrics often fail to capture nuanced human-centric goals essential for building equitable AI systems.

Addressing Human Trafficking through AI

AI's potential to combat human trafficking exemplifies its capability for systemic solutions. Anjuli discusses how trafficking operates as a business model reliant on data flows from supply (victims) and demand (societal desires for cheap goods). AI tools can analyze patterns through linguistic or network analysis to detect trafficking activities effectively. However, cultural taboos complicate data collection efforts necessary for implementing preventative measures against trafficking practices. Ethical frameworks must integrate continuous feedback loops to ensure that AI systems learn from actions taken and adapt accordingly to avoid unintended harm.

Visionary Leadership for Transformative AI

The success of AI hinges on leaders willing to prioritize human-centric innovation over traditional profit-driven models. The speakers stress that systemic inertia among CEOs and policymakers stems from a lack of courage and creativity in rethinking priorities around technology's role in society. Embracing systems thinking—which incorporates relational and holistic approaches—could help reframe AI's objectives toward fostering societal harmony and sustainability. Achieving this requires moving beyond quantifiable metrics to cultivate genuine human connections and ethical progress within technological development.

In summary, these discussions highlight the multifaceted implications of artificial intelligence on society through various lenses: practical applications versus broader societal impacts; nuanced ethical framing beyond binary constructs; addressing societal disconnects; grounding technology in philosophical foundations; recognizing bias as both a limitation and an opportunity; establishing guardrails for responsible development; leveraging AI against pressing issues like human trafficking; and advocating for visionary leadership that prioritizes humanity over profit motives. Each point contributes significantly to understanding how we can navigate the complexities of AI while ensuring its alignment with ethical considerations essential for humanity's future.

HARNESSING UBUNTU: A PATHWAY TO ETHICAL AI DEVELOPMENT

(Perspectives of Mammo Muchie)

Relevance of Ubuntu Philosophy under AI Ethics

Emphasis has been laid on the relevance of Ubuntu, an African philosophy that advocates for community and interconnectedness, in shaping ethical frameworks for AI. Mammo Muchie articulates that Ubuntu promotes values such as love, reconciliation, and forgiveness, which can serve as guiding principles in AI development. Further, arguments have been made to for the Ubuntu values to replace hate and conflict with collaboration and mutual respect (Mhlambi, 2020). This perspective aligns with the idea that technology should enhance human relationships rather than exacerbate divisions, suggesting that AI can be a tool for unity if developed with an emphasis on communal well-being (Barroso, 2024).

AI Ethics and Its Challenges

A significant challenge identified is the dominance of Western ethical frameworks in AI governance, which often neglects non-Western philosophies. Today, approximately 63% of AI ethics guidelines are derived from Western thought, leading to a lack of representation for African principles like Ubuntu (Jobin et al., 2019). This imbalance results in ethical standards that fail to resonate with the realities faced by diverse cultures, perpetuating inequalities rather than addressing them. The absence of inclusive ethical guidelines poses a risk to the equitable application of AI technologies across different sectors (Barroso, 2024).

Need for Inclusive Dialogue

It is necessary to have open and inclusive dialogue among various cultures to develop ethical standards that reflect a broader range of values. Such inclusivity can be brought about by means of collaboration among policymakers, academics, and technologists to ensure that diverse philosophical perspectives are integrated into AI ethics. This approach is crucial for creating regulations that are culturally relevant and equitable. By fostering discussions that incorporate different cultural philosophies, stakeholders can work toward establishing ethical guidelines that genuinely represent global diversity (Gumede et al., 2021).

Real-life Applications of Ubuntu in AI

In today's world, the Ubuntu philosophy has been practically applied in sectors such as education and healthcare through AI technologies. For example, educational institutions could implement AI systems designed to promote collaborative learning environments that respect cultural differences while ensuring equitable access to resources (Mhlambi, 2020). Mhalambi further expounds the reasons for AI ethics to focus on relationality than rationality. He argues, "Any system, whether political, economic, or social, that is organized and structured around individualistic principles eventually produces harms". By embedding Ubuntu principles into AI applications, organizations can foster an atmosphere of mutual respect and cooperation, ultimately enhancing educational outcomes and addressing disparities within the system (Barroso, 2024).

Future Directions for Ethical AI Development

Mammo provides four specific recommendations for future AI design and development.

Policy Development: Advocate for global ethical charters integrating Ubuntu and other non-Western philosophies to balance AI's competitive nature with collaborative values.

Education Reform: Incorporate philosophical teachings like Ubuntu into curricula to promote ethical awareness in AI development.

Multi-Sector Collaboration: Encourage dialogue and partnerships across cultures to create AI systems that address societal needs without exacerbating conflicts or inequalities.

AI Design Principles: Develop AI applications that prioritize inclusivity, reduce harm, and foster cooperation among users and stakeholders.

AI'S JOURNEY: REDISCOVERING HUMANITY IN TECHNOLOGICAL ADVANCES

(Perspectives of John Havens and Thomas Koulopoulos)

Reframing AI Ethics

The discussion begins by critiquing the Western-centric, profit-driven approach that currently dominates AI ethics. John Havens emphasizes that existing systems prioritize GDP and shareholder value, often neglecting broader societal well-being. This narrow focus can lead to technological advancements that are misaligned with human values and needs. John advocates for a paradigm shift toward designing AI systems that center on humanity and sustainability. He argues for proactive ethical considerations during the design phase, rather than retrofitting them after deployment. By integrating concepts like gratitude, mindfulness, and positive psychology—many of which are rooted in non-Western traditions—John believes that AI development can better align with human-centric values. This approach not only enhances the ethical framework surrounding AI but also ensures that technology serves to uplift human experiences rather than diminish them, fostering a more compassionate and equitable technological landscape.

Cultural Pluralism vs. Centralized Frameworks

The discussants warn against the dangers of centralizing ethical AI standards, which can lead to the erasure of cultural diversity and local values. They highlight alternative philosophical frameworks, such as Ubuntu's relational ethos ("I am because we are"), which contrasts sharply with the individualistic perspectives prevalent in Western AI ethics. They also reference Donald Hoffman's work, which posits that human evolution prioritizes survival over perceiving reality accurately. This suggests that AI's mimicry of human behavior may obscure deeper truths about human existence. The speakers stress that preserving diverse cultural and ethical frameworks is essential for fostering resilience and inclusivity in AI governance. By embracing a pluralistic approach, AI can be designed to respect and reflect the values of various cultures, ultimately enriching global conversations about technology and ensuring that all voices are heard in shaping its future.

Existential and Philosophical Implications

The conversation delves into the philosophical implications of AI potentially surpassing human intelligence, raising existential questions about what it means to be human in an increasingly automated world. While some fear that this shift could diminish humanity by creating a divide between humans and machines, the speakers argue it presents an opportunity to rediscover deeper values such as love, community, and spirituality. Drawing parallels to Socrates' critique of the written word, TK cautions that while AI may further fragment human connection through digital interactions, this fragmentation could also prompt critical reflection on the human condition. Such reflection encourages individuals to cultivate meaningful relationships and emotional depth in their lives, ultimately leading to a richer understanding of what it means to be human amid technological advancements (Koulopoulos & Palmer, 2024).

Environmental and Resource Sustainability

John critiques the environmental costs associated with AI technologies, particularly highlighting the unchecked resource consumption of data centers as an example of irresponsible governance in technology development. He emphasizes the urgent need for energy-efficient AI designs that harmonize technological advancement with ecological preservation. For instance, leveraging renewable energy sources and reducing processing waste can significantly mitigate the environmental impact of AI systems, ensuring they coexist sustainably with our planet. This perspective calls for a paradigm shift in how technology is developed—one that recognizes environmental stewardship as integral to technological progress rather than as an afterthought. By prioritizing sustainable practices in AI development, we can create systems that not only drive innovation but also protect our natural resources for future generations.

Redefining Human Worth in the Age of AI

As AI continues to advance, it challenges traditional notions of human value tied primarily to intellect or productivity. The speakers argue that this shift could refocus humanity on intrinsic qualities such as relationality, moral agency, and spirituality—elements often overshadowed by metrics of efficiency and output in a technology-driven world. They highlight indigenous practices like vision quests as profound examples of self-discovery and connection with nature—experiences that AI cannot replicate or replace. These practices serve as reminders of the importance of preserving humanity's softer emotional dimensions amidst rapid technological change. By emphasizing these intrinsic qualities, society can foster a deeper appreciation for what it means to be human beyond mere productivity metrics.

Commodification of Data

John critiques the commodification of personal data, viewing it as a consequence of materialist systems that prioritize profit over people's rights and dignity. He calls for greater accountability in AI development by advocating for ethical frameworks prioritizing human dignity above profit motives (Havens, 2016). By fostering concepts like gratitude and mindfulness within AI systems, developers can align these technologies with collective well-being rather than exploitative tendencies inherent in centralized tech corporations. This shift emphasizes transparency regarding data usage and consent processes to ensure individuals maintain agency over their personal information. Ultimately, addressing data commodification is crucial for building trust between technology providers and users while safeguarding individual rights. TK refers to the explosive nature of tracking human behavior for monetization and not for furthering the basic human value systems (Koulopoulos & Achillias, 2018)

Ethical Relativity in AI Design

The speakers stress the importance of embedding pluralistic, multi-regional, and multidisciplinary perspectives into AI ethics to ensure inclusivity in design processes. This approach helps prevent homogenization of cultural identities or exacerbation of existing conflicts within societies influenced by technology. Instead of imposing a single narrative or standard across diverse populations, AI could be designed to celebrate cultural diversity by enriching global cultures and fostering mutual respect among different communities. By integrating various cultural narratives into the design process, developers can create systems that resonate more deeply with diverse populations while promoting understanding across different societal contexts.

Hope for Inclusive AI Frameworks

The discussion concludes with an optimistic vision for inclusive AI systems that prioritize relationality, environmental sustainability, and uniquely human qualities like love and consciousness. While acknowledging that AI has immense potential for growth and collaboration across sectors, the speakers emphasize the critical importance of embedding ethical frameworks that complement rather than undermine humanity's core values. This hopeful outlook envisions a future where technology enhances rather than replaces genuine human connections—where ethical considerations guide innovation toward outcomes beneficial for all members of society, ultimately leading to a more compassionate world shaped by shared values rather than profit alone.

These elaborations provide a deeper understanding of each point discussed regarding the implications of artificial intelligence on society through various lenses—ethical considerations, cultural diversity, existential reflections, environmental sustainability, redefined notions of worth, data commodification issues, inclusive design practices, and hopeful visions for future technologies aligned with humanity's best interests.

IMPLEMENTATION AND GOVERNANCE CHALLENGES: THE ROLE OF CORPORATIONS AND GOVERNMENTS IN AI REGULATION

(Perspectives of Eunsong Kim, Jibu Elias, and Mihir Kulkarni)

UNESCO's Role and Regional Contextual Challenges

UNESCO plays a pivotal role in the AI ethics landscape by acting as a convening body that bridges the gap between thought leaders, governments, and academic institutions. The organization emphasizes the importance of diverse perspectives in shaping AI policies and frameworks. However, there is a noticeable imbalance in global AI discussions, with disproportionate attention given to the Global North while the contributions and challenges faced by the Global South remain underrepresented. This imbalance becomes particularly pronounced in problem formulation, where issues relevant to the Global South may be overlooked or misrepresented. For instance, in India, there is a palpable sense of optimism regarding AI's potential; however, this enthusiasm is tempered by challenges such as "snake oiling," where AI's capabilities are overpromised. Additionally, low technological literacy among many individuals hinders effective engagement with AI technologies. Thus, it is crucial for AI frameworks in these regions to be grounded in scientific evidence demonstrating their use-case benefits to build trust and foster acceptance among local populations.

Data Collection and Consent in Global South Countries

The conversation highlights critical concerns regarding data collection practices in regions like the Global South. These nations often lack robust legal safeguards and have limited public understanding of data usage implications, leading to ethical dilemmas surrounding informed consent. The speakers argue that consented data collection may not always be genuinely informed due to these gaps in understanding. Furthermore, the labor force involved in creating and refining datasets for multinational corporations frequently faces exploitative practices, raising ethical questions about fairness and equity. For example, in India, tuberculosis data collection has provided valuable insights through behavioral analysis across gender and regional lines. However, this practice raises significant concerns about the fairness of labor conditions for those collecting this data and how contextual definitions of fairness differ across cultures and regions. Addressing these issues requires a nuanced understanding of local contexts and ethical standards that prioritize the rights and dignity of individuals involved in data collection.

Bias in AI: A Dual Perspective

The discussion on bias presents a nuanced view, arguing that biases within AI systems are not inherently negative but can be leveraged for positive outcomes. While harmful biases have been documented—such as facial recognition technologies struggling with darker skin tones—there are instances where biases can promote equity. For example, algorithms designed to prioritize assistance for underserved communities during natural disasters illustrate how bias can be harnessed to address systemic inequities effectively. However, addressing these biases also incurs significant environmental and resource costs, particularly given the immense energy consumption associated with training complex AI models. This aspect raises further ethical concerns about sustainability and resource allocation in AI development. Additionally, speculative fears surrounding "AI awakening" can divert attention from pressing safety issues that require immediate action. Therefore, it is essential to approach bias in AI with a balanced perspective that recognizes both its potential harms and benefits while ensuring that ethical considerations guide its application.

AI Exceptionalism and Its Societal Impacts

The concept of AI exceptionalism posits that artificial intelligence represents a transformative leap in technological evolution that compels society to confront its profound implications. Unlike previous technologies, AI has the potential to reshape fundamental societal constructs such as education, equality, and cultural identity on an unprecedented scale. However, the Global South faces significant obstacles in navigating these transformations due to limited access to concrete data points and resources necessary for informed policymaking. This lack of access makes it challenging for policymakers, researchers, and civil servants to craft meaningful solutions tailored to their specific contexts. The speakers also raise critical questions about how conflicting ethical principles should be addressed when they arise within AI systems and whether sufficient resources exist to resolve such dilemmas effectively. As societies grapple with these complexities, it becomes increasingly important to foster collaborative approaches that involve diverse stakeholders from both the Global North and South.

The Role of Governments and Corporations in AI Regulation

Governments and corporations play a crucial role in shaping the ethical landscape of AI through regulation and policymaking. The European Union has taken significant strides toward establishing high standards for AI regulation; however, many Southeast Asian countries remain hesitant to adopt stringent guidelines due to fears that such measures might hinder economic growth or technological innovation. Instead of developing their own regulatory frameworks, these nations often find themselves adopting regulations dictated by major global associations or corporations. The speakers critique this oligopolistic nature of AI development, where a small group of powerful corporations—often referred to as the "Magnificent Seven"—dominate the landscape. These corporations tend to prioritize shareholder profit over societal well-being, further entrenching systemic inequities within societies. This dynamic underscores the need for governments to assert their regulatory authority more robustly while ensuring that corporate practices align with ethical standards that promote social good.

Economic Systems and Profit-Driven Models

The speakers critique the deeply embedded profit-driven nature of current economic systems within which AI is developed. They highlight how prioritizing shareholder value often overrides considerations for equity and inclusivity in technological advancements. For instance, rather than focusing on affluent consumers with luxury products or services, AI systems could be harnessed to provide affordable healthcare or education solutions tailored for underserved populations. Such shifts require visionary leadership willing to prioritize long-term societal well-being over short-term economic gains driven by profit motives alone. By reimagining economic incentives around social impact rather than purely financial returns, stakeholders can create an environment where technology serves as a tool for positive change rather than exacerbating existing inequalities.

AI as a Reflection of Humanity

AI serves not only as a technological advancement but also acts as a mirror reflecting humanity's strengths and flaws back at us. It does not create societal problems but amplifies existing ones inherent within human behavior and societal structures. For instance, voice assistants' inability to recognize diverse accents reflects underlying language hierarchies and biases present in technological design processes—issues rooted deeply within societal norms rather than technology itself. Conversely, AI's potential role in combating human trafficking demonstrates its capacity for systemic solutions when applied thoughtfully; however, addressing complex issues like trafficking necessitates integrating diverse data sources while navigating cultural taboos and social stigmas that complicate data collection efforts. The speakers emphasize the importance of incorporating continuous feedback loops into AI systems so that actions taken can be refined over time based on real-world outcomes—ensuring that unintended harm is minimized while maximizing positive impact.

These elaborations provide an enriched understanding of each point discussed regarding UNESCO's role in AI ethics governance along with various challenges faced by both Global North and South countries concerning data practices, bias implications within artificial intelligence applications, regulatory frameworks established by governments/corporations alike—all contributing toward shaping more equitable futures through responsible use of technology across diverse contexts worldwide.

CONCLUSION

The discourse on ethical frameworks and responsibility in AI emphasizes the necessity for a dynamic and evolving approach that aligns with human needs rather than adhering to static rules. In healthcare, for instance, AI should prioritize compassion alongside efficiency, prompting questions about its impact on patients' lives, such as the emotional implications of a cancer diagnosis. The discussion also highlights the underrepresentation of Global South narratives in AI design, advocating for the inclusion of indigenous perspectives to create systems that truly reflect human diversity. Bias in AI is recognized as both a challenge and an opportunity; while it often mirrors societal inequities it can also be harnessed constructively to support underserved communities during crises.

Data usage and regulation are critical areas of concern, where ethical considerations must prioritize dignity and informed consent over exploitation. The need for sustainable practices is underscored, particularly regarding the environmental impact of data centers that consume vast resources. Philosophically, as AI advances, it challenges traditional notions of human worth tied to productivity, urging a focus on intrinsic qualities like relationality and moral agency. The conversation critiques profit-driven economic models that prioritize shareholder value over societal well-being, advocating for a shift toward equity-focused applications of AI.

AI is viewed as a reflection of humanity's strengths and flaws, amplifying existing societal issues rather than creating new ones. This perspective encourages a collaborative approach between humans and machines to tackle complex problems like human trafficking while respecting cultural sensitivities. Ultimately, the hope for transformation lies in reimagining AI's role in society—prioritizing human values, fostering inclusivity, and ensuring that technology serves as a tool for unity and growth rather than division or exploitation. By embedding ethical considerations into the fabric of AI development, leaders can pave the way for an equitable future where technology enhances rather than detracts from human dignity.

However, there are still some questions that we as readers are left to ponder on and address such as how does AI challenge our understanding? How can we ensure that AI ethics incorporates diverse cultural and philosophical perspectives? What are the implications of centralizing AI ethics in the hands of a few tech companies? These questions highlight the need for ongoing dialogue and critical reflection as we navigate the complexities of AI's role in society. Addressing them will be vital to fostering an ethical framework that not only guides technological advancement but also enhances human dignity and collective well-being.

REFERENCES

Barroso, L. R. (2024). *Artificial intelligence: Promises, risks, and regulation: Something new under the sun.* Carr Center for Human Rights Policy (Issues 2024–06) [Report]. https://www.hks.harvard.edu/sites/default/files/2024-12/24_Barroso_Digital_v3.pdf

Gumede, V., Muchie, M., & Shafi, A. (2024, February 1). *Indigenous systems and Africa's development.* https://repository.hsrc.ac.za/handle/20.500.11910/22841

Havens, J. C. (2016). *Heartificial intelligence: Embracing our humanity to maximize machines.* Penguin Random House.

Jobin, A., Ienca, M., & Vayena, E. (2019). The global landscape of AI ethics guidelines. *Nature Machine Intelligence*, *1*(9), 389–399. https://doi.org/10.1038/s42256-019-0088-2

Koulopoulos, T. M., & Achillias, G. (2018). *Revealing the invisible: How our hidden behaviors are becoming the most valuable commodity of the 21st century.* Post Hill Press.

Koulopoulos, T. M., & Palmer, N. (2024). *Gigatrends: Six forces that are changing the future for billions.* Post Hill Press.

Mhlambi, S. (2020). From rationality to relationality: Ubuntu as an ethical and human rights framework for artificial intelligence governance. *Carr Center for Human Rights Policy Discussion Paper Series, 9*, 31.

Schwartz, S. (2022, May 10). Understanding values: Schwartz theory of basic values. Integration and implementation insights. https://i2insights.org/2022/05/10/schwartz-theory-of-basic-values/

ADDITIONAL READINGS

Babajide, R. (2023, August 26). How the "Doorman Fallacy" has led to an overall decline in quality of living. *Medium.* https://medium.com/illumination-curated/how-the-doorman-fallacy-has-led-to-an-overall-decline-in-quality-of-living-2bedd5312cc0

Elias, J. (2024). *AI ethics: Global perspectives.* https://aiethicscourse.org/lectures/role-of-culture

McCandless, D. (2024, November 4). *Information is beautiful. Information is beautiful.* https://informationisbeautiful.net/visualizations/left-vs-right-world/

NYU Press. (2019, July 2). *Whose global village?* https://nyupress.org/9781479862962/whose-global-village/

Turkle, S. (2017). *Alone together: Why we expect more from technology and less from each other.* Hachette UK.

Verhulst, S. (2021, January 14). *From Rationality to relationality: Ubuntu as an ethical and human rights framework for artificial intelligence governance.* The Living Library. https://thelivinglib.org/from-rationality-to-relationality-ubuntu-as-an-ethical-and-human-rights-framework-for-artificial-intelligence-governance/

Index

Note: **Bold** and *Italic* page numbers refer to **tables** and *figures*. Page numbers followed by 'n' refer to notes.